Stats with Kittens

Growing up with Data, Charts, Surveys, Correlations, and Other Floofy Playthings.

Charles Kufs

Stats with Kittens. Growing up with Data, Charts, Surveys, Correlations, and Other Floofy Playthings.

ISBN: 979-8-218-89589-1

Library of Congress Control Number: 2025927596

First Edition

Independently published, Willow Grove, Pennsylvania, USA

Cover design by Miranda Kufs

Printed in the United States of America

Contents.

Wow. This is going to be interesting.

Contents.

OK, so turn the page.

Contents.

ACKNOWLEDGEMENT.

No one who achieves success does so without acknowledging the help of others. The wise and confident acknowledge this help with gratitude.

Alfred North Whitehead, 1861-1947,
English mathematician and philosopher.

We would like to thank the numerous family, friends, classmates, teachers, co-workers, bosses, clients, and professional buddies we've encountered over the years for the support and wisdom they have given to us. We would particularly like to acknowledge Maryanne Kufs, Bob and Dawn Drake, Jacob Walker, Ray Finkle, Rob Hopf, Al (George) Newnham, Don Messinger, Steve Jakatt, Levy Kroitoru, Rex Bryan, Ed Saltzberg, Geoff Smith, Bob Barcikowski, Sarah Funk Khalil, Stacy McBride, and Todd Chang for their support. I would also like to thank Chewy, Inc. for their indispensable support of my feline dependents.

We would also like to acknowledge all our furry friends who contributed to this book and to our lives just by being there: Betty, Brigitta, Buddha, Carter, Castiel, Cola, Critter, Dax, Dean, Flutterbye, Friedrich, Ghost, Gretl, Harry, Hermione, Jackson, Kerpow, Komorebi, Kurt, Liesl, Lou (Louisa), Lucifer, Magic, Maze (Mazikeen), Mazie, Marbles, Mojo, Martin (Marta), Mika, Morris, Moose, O'Neal, Obsidian (Sid), Odo, Onyx, Poofy, Ron, Sam, Sisko, Spice, Sugar, Teal'c, and Tiger.

Creating **Stats with Kittens** and **Stats with Cats** was truly a labor of love, and I'm grateful for the incredible team who helped bring my vision to life.

For my part, I came up with the initial concept of combining statistics with the playful antics of felines and wrote the text for **Stats with Kittens** and **Stats with Cats**. I created the technical graphics to present statistical insights and provided some of the kitten and cat images from my personal photo library.

Many paws make light work.

Krista Wolf – Krista's editing expertise was invaluable. From line editing to fact-checking, she refined every page with care. She also handled the layout, formatting, and managed all production details to ensure the final books were exactly what I envisioned.

Miranda Kufs – Miranda contributed her creativity through kitten and cat photography, infusing the book with character and charm. She also lent her talents to

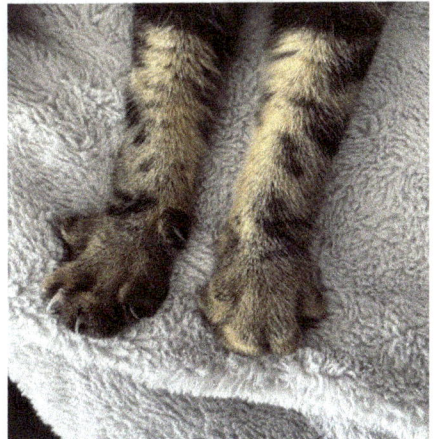

Acknowledgement.

cover design, coordinated with both the publisher and media, and led marketing efforts to help the books find their audiences.

Lisa Kufs – Lisa's video work provided most of the kitten images. Her ability to capture the essence of each moment made a lasting impression on the project. Beyond her creative contributions, Lisa also provided invaluable personal support throughout the process, managing the day-to-day tasks that allowed everything to stay on track behind the scenes.

Peter Kufs – Peter kept the project running smoothly with his behind-the-scenes IT and technology support, saving me countless hours in technical troubleshooting.

Nick Wolf – Nick's attention to detail in formatting, production, and cover design was the finishing touch these books needed, and his dedication to quality made sure every page looked its best.

This project was a team effort, and each person's unique talents helped make **Stats with Kittens** and **Stats with Cats** special. I couldn't be prouder of what we've created together.

Charlie Kufs
Willow Grove, PA

Harry and Ron playing with the big cats.

Acknowledgement.

FOREWORD.

You might have hated high school (HS) math. It was a challenge but not just for you. It was also a challenge for the teachers who had to teach you as well as the educators who had to create the HS curriculum for the teachers. Both the teachers and the curriculum developers had to decide what math (maths in Britain and some other countries) topics should be taught and how to teach them to every student—from those going to college to those entering the workforce after HS.

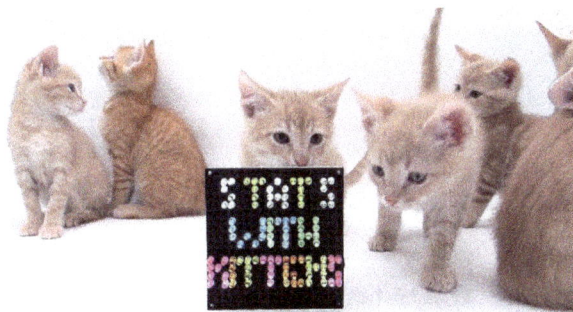

Don't panic. We're here to ease your statistical anxiety.

Even for those going on in education, the math curriculum had to prepare them for trades, business, liberal arts, and STEM (Science, Technology, Engineering, and Math). That's why you had to learn: areas and volumes of shapes; equations, inequalities, and functions; logarithms, roots, and exponents; geometric and trigonometric relationships; and many other challenging topics. How many of those topics do you think you use in your life after HS?

What would have been more useful to learn was statistics. You already had been exposed to some statistics in school. Your class grades were *averages* of scores you received for tests and other efforts. Your classes may have been graded on a curve, requiring the concepts of the *Normal distribution, standard deviations*, and *confidence limits*. Your scores on standardized tests, like the SAT, were presented in *percentiles*. You learned about *pie* and *bar charts*, *scatter plots*, and maybe other ways to display data.

> *The mathematics curriculum that we have is based on a foundation of arithmetic and algebra. And everything we learn after that is building up towards one subject. And at top of that pyramid, it's calculus. And I'm here to say that I think that that is the wrong summit of the pyramid ... that the correct summit—that all of our students, every high school graduate should know—should be statistics: probability and statistics.*
>
> Arthur Benjamin, "Teach statistics before calculus!," **TED Talk**, Jun 29, 2009.

You learned about equations for *lines* and some elementary *curves*. So, by the time you got to prom, you were expected to know enough statistics to gamble and read **USA Today**.

YOU NEED TO UNDERSTAND STATISTICS

Statistics is an integral part of everyday life in America. Without statistics, there would be no U.S. Census, Nielsen ratings of TV shows, political polls, and consumer preference surveys. Our society couldn't function without being able to figure out tax brackets, insurance rates, stock prices, and online matchmaking. We couldn't find DNA relatives or predict the outcome of elections before the polls closed. There would be no standardized tests. Online retail sites couldn't tell us what we want to buy. Baseball announcers would have nothing to talk about between pitches. It would be anarchy.

We're all in this together.

If you continue your education after HS, you might find that you are required to take an introductory course in statistics, call it *Stats 101*. Of the two million bachelor's degrees that are granted in the U.S. every year, probably two-thirds require completion of a statistics class. That's over a million and a half students taking Stats 101 every year. Statistics is certainly a requirement for degrees in STEM, economics, the social sciences, education, criminology, and even some degrees in history, culinary science, journalism, library science, and linguistics. It is also often a requirement for professional certifications in IT, data science, accounting, engineering, and health-related careers.

Statistics isn't just about mathematics. For consumers of statistical information, it doesn't matter if statistics is mathematics, or a subfield of mathematics, or just makes use of mathematics. Yes, the *theoretical statisticians* who develop statistical procedures and tests certainly have to know their mathematics. Even *applied statisticians*, the people who use those procedures and tests on real-world data, have to know a fair amount of math. But most consumers of statistics—which is EVERYONE—just need to know what the different types of statistics are, what they mean, and how they might be misleading. It isn't important to remember the actual formulas as long as you understand what they do. It's like driving a car, you don't have to know how the emission system works to stop at a red light.

Statistical skills are a fundamental component of what is called *data literacy* or *statistical literacy*—the ability to read, understand, argue with, and make decisions from numerical information. *Statistical literacy* is understanding, at least in general terms, the process of *statistical thinking*.

Statistical thinking is the process of how practitioners explore a phenomenon in a population. It's *inductive reasoning*, that is, creating new generalizations from individual bits of information. It's like assembling a picture puzzle from the individual puzzle pieces. *Statistical thinking* is inductive reasoning with numbers.

WHY READ STATS WITH KITTENS

The skills you need to read statistics presented in the media are different from the skills you need to conduct your own statistical analyses. This book is meant to help people who want to understand the science of statistics enough so they are not bamboozled by uninformed arguments and media presentations. Even that is a lofty goal, but you have to start somewhere … why not with kittens.

Stats with Kittens is not a textbook. There are few formulas or descriptions of statistical procedures (though there are several examples of calculations presented, especially in Chapter 2). There are only a few academic citations. There are no extended explanations of mathematical theorems and laws. There are no homework problems to test your understanding. You can find all those things in the thousands of traditional textbooks on introductory statistics.

Stats with Kittens is the prequel to those Stats 101 textbooks; **Stats with Cats** is the sequel.

Come for the cats; stay for the stats.

A word of caution—there is a HUGE amount of information in this book. None of it is particularly difficult conceptually, there's just a lot of it. Think of **Stats with Kittens** as a smorgasbord. Don't feel you have to partake in every detail or consume it all at once. Don't highlight every new morsel of information. It'll all be there whenever you want to come back and take in a little more.

This isn't a compendium of statistics, although it might seem like it. There's a lot more out there. Let the internet be your guide to even more statistical ideas, topics, equations, and applications whenever you need them.

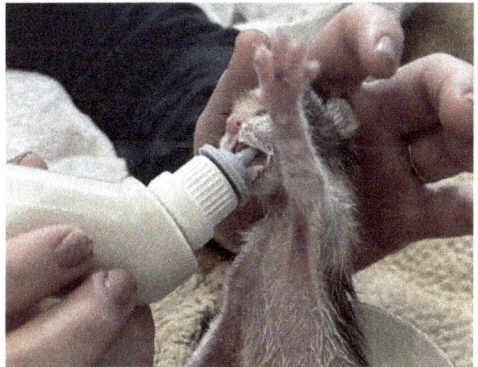

That's enough for now.
I'll come back for more later.

STATS WITH KITTENS USER GUIDE

Here are a few details about **Stats with Kittens** that will help you get more from your reading experience.

I put technical terms in *italics* the first time they appear in the text. This helped me to select those terms for inclusion in the Glossary. I also italicized some of the later appearances of certain terms that appear many times when they were important to subsequent discussions. I put non-technical words in **boldface** in situations when I felt they needed added emphasis, were needed for formatting, were key concepts, or were just pithy sayings. I also bolded characters that appear as mathematical symbols in the text.

Some statistical concepts—like data, samples, population, and variables—are reintroduced in almost every chapter. That's because I assumed that some readers would read the parts of the book that they thought might be interesting, rather than proceeding linearly through the entire text. Either approach is fine. You won't compromise the experience by skipping ahead to see how the story ends.

I sometimes go off on rants in **Stats with Kittens** and **Stats with Cats** about topics in which I don't share traditional approaches in statistics. One example involves how statistics, and mathematics in general, is taught in school. Take them for what they are worth to you.

Stats with Kittens explains a lot of the *jargon* you might hear about in statistics. Almost all of it involves words and phrases used primarily in statistics or math, even though some of it might look like common English words. Some jargon involves *repurposed words*, that is, English words with alternative meanings. These may be difficult to identify because you have to consider the context in which the words appear. The easiest to spot are *eponyms* (terms named after a person) and *special words* (words used exclusively in statistics). These really aren't all that common, at least until you get to more advanced levels of statistics. They're easier to deal with than you might think, though.

There's a *word cloud* (Chapter 4) at the beginning of the Glossary that highlights 780 key words used in **Stats with Kittens**. The terms appear 51,079 times, about 38% of the total text. The number one statistics-related word in **Stats with Kittens** is *data*, which appears 1,420 times. *Data* is plural and is pronounced day-ta, which rhymes with beta and zeta. The term *statistics* appears 1,058 times, *probability* appears 753 times, *population* appears 402 times, *sample* appears 681 times, *variable* appears 849 times, *scale* appears 310 times, *model* appears 646 times, and *relationship* appears 308 times.

Foreword.

There are over a thousand definitions included in the Glossary if you just want a quick reference. There's also a diagram in Chapter 9 showing how the concepts discussed in **Stats with Kittens** and **Stats with Cats** fit together. I have it on a t-shirt.

Are we similes, metaphors, analogies, or models?

If you enjoy learning from metaphors, similes, and analogies, you'll appreciate them in **Stats with Kittens**. There are comparisons about how aspects of statistics are like food, cooking and baking, tools, gardening, sports, superpowers, circus games, driving, high school, education, medicine, and of course, kittens. See if you can find why data are like potato chips.

There are also more examples than I could count. My software says I used the phrase "for example" 140 times. Many of the data I used in examples came from sports, mostly the National Football League (NFL), because those data are just so easy to find. Try looking, you'll see. Other data come from government, politics, business, medicine, psychology, geography, and environmental science. There are examples of over two dozen types of statistical graphics.

Most of the technical figures and tables were created in **Microsoft Excel**. Some were created in **Microsoft Word** or **Microsoft Visio**. A few were created in either specialty statistical analysis software, or from free-access templates, or on internet websites. These have been noted on the figures or in the text.

Wherever I thought it might be valuable to readers, I laid out steps for how to approach specific thought processes, including:

- What to look for in statistical surveys (6 steps in Chapter 5).
- How to decide if correlation implies causation (3 steps in Chapter 7).
- How to pick a model for a statistical analysis (5 steps in Chapter 8).
- How to pick a statistical method for an analysis (5 steps in Chapter 8).
- How to apply the scientific method to statistical studies (10 steps in Chapter 9).
- How to evaluate statistical presentations (4 steps in Chapter 9).
- How to become a critical thinker (5 steps in Chapter 9).

I included many hints for what to look for in evaluating data, statistical analyses, and presentations of analyses. For example, there are hints for determining which of Stevens' scales a variable uses, in Chapter 3. I hope the hints are helpful.

I tried to put things in every chapter that are unique and interesting. I know there are unique things because I created them as teaching devices when I tutored students. They are things like:

- 🐾 **5 Problem Solving Approaches in Statistics**—Describe, Classify, Test, Predict, and Explain (Chapter 1).
- 🐾 **4 Sources of Probabilities**—Logic, Data, Models, and Oracles (Chapter 2).
- 🐾 **The Data Dozen**—Twelve types of data used in statistical and other forms of analysis (Chapter 3).
- 🐾 **3 Elements of a Data Measurement**—Benchmark, Process, and Judgment (Chapter 3).
- 🐾 **3 Rs of Variance Control**—Reference, Replication, and Randomization (Chapter 3).
- 🐾 **3 Fs of Statistical Graphics**—Foundation, Framework, and Facade (Chapter 4).
- 🐾 **6 Reasons People Criticize Polls**—Too few participants, they didn't ask me, only landline users were interviewed, they asked the wrong questions, the results were predetermined, and the results were wrong (Chapter 5).
- 🐾 **9 Types of Data Relationships**—Direct Relationships, Feedback Relationships, Common-Cause Relationships, Mediated Relationships, Stimulated Relationships, Suppressed Relationships Inverse Relationships, Threshold Relationships, and Complex Relationships (Chapter 7).
- 🐾 **Six Ps of Establishing Causality**—Promote; Prevent; Prepare; Prosecute; Pontificate; and Probe.
- 🐾 **11 Reasons to Doubt a Regression Model**—Not enough samples, no intercept, stepwise regression, outliers, relationships, overfitting, misspecification, multicollinearity, unequal variances, autocorrelation, and weighting (Chapter 8).

You won't find these things in other books on statistics. I'm not sure about whether these items might be interesting to you, though. Maybe they're just interesting to me.

One point of confusion you should be aware of before reading further. Most people think of the word *statistics* as referring to the science of analyzing data to draw conclusions. However, the word *statistics* can also refer to a collection of individual numbers that have been calculated from a dataset, each calculated value being one *statistic*. In English, *statistics* the science is singular while *statistics* the collection of numbers is plural. This makes noun/verb agreement a challenge for grammar-checkers and readers alike. You have to determine which of the definitions apply by considering the context in which the term is used.

Stats with Kittens contains nine chapters that describe what you'll need to know to be an informed consumer of statistics.

Chapter 1. **Introduction**. Why you should learn about statistics and what you should know before starting.

Chapter 2. **Probability**. What probability and odds are, where they come from, and how they influence our lives.

Chapter 3. **Description**. How describing datasets is easier than describing people once you know what to look for.

Chapter 4. **Graphs**. What you need to know about statistical graphics to assess their validity. It's much more than you were taught in HS.

Wow. That's a LOT of information.

Chapter 5. **Surveys**. How to measure intangible, changing opinions. Everybody thinks surveys are easy to conduct but they are, in fact, the most difficult type of statistical analysis to get right.

Chapter 6. **Comparisons**. How to find significant and meaningful differences between populations represented by data. Statistical testing has been used for centuries but tests are complicated and easy to get wrong.

Chapter 7. **Relationships**. How data metrics can be related and why it's hard to tell when correlation implies causation.

Chapter 8. **Models**. How statistical models of a data relationship are created and how to spot common faults that others may overlook.

Chapter 9. **Literacy**. How to recognize possible issues in data and statistical analyses to assess the validity of information and arguments presented in technical reports and media stories.

And, it includes enough kittens so your experience will be tolerable if not enjoyable.

If you're interested in *readability*, **Stats with Kittens** has an average *Flesch-Kincaid Grade Level* of 11.48 (11th grade) which is considered fairly difficult but suitable for 16 to 17 year olds. The easiest chapters are Chapter 2 Probability (FKGL=9.64) and Chapter 4 Graphs (FKGL=10.32). The most difficult chapters are Chapter 7 Data Relationships (FKGL=12.39) and Chapter 8 Data Models (FKGL=12.55). You can calculate the readability of your own writing at readabilityformulas.com.

I hope this book will help you understand some of the often-bewildering terminology and concepts you will encounter as you embrace statistics in your life.

Good luck. Keep readings.

Foreword.

CHAPTER 1. FUNDAMENTALS.
Statistics In Your Everyday Life.

INTRODUCTION

You may have wondered what statistics is and why you should learn about it in school. Maybe more importantly, why do you have to take Stats 101 to complete a college degree.

Statistics is about numbers. It's about translating numbers into meaningfulness like translating a foreign language into your native tongue … only not so straightforward.

We're individual kittens but we're also a group of two kittens. We can be both.

Statistics is also about groups. Statisticians translate information about individuals, items, or entities into information about the groups they identify with. It's called *induction*.

The use of statistics is common to almost all fields of inquiry—social and natural sciences, sports, business, education, library and information sciences, and even music and art. Search the internet for "use of statistics in …" your specific career field or favorite pastime. You may be surprised with what you find.

The popularity of statistics is attributable at least in part to its applicability to any type of data. Statistical methods can be used for analyzing data based on natural laws, theories, or nothing in particular. If you can measure it, you can analyze it with statistics.

Most importantly, statistics is the one kind of higher math that you'll encounter every day in your life after you leave school. It's an important part of virtually everything in society from science and consumer marketing to social media and sports. It's a proficiency you simply must have.

That's what this chapter is about.

HISTORY OF STATISTICS

Statistics isn't a new thing. In fact, it dates back at least fifty centuries beginning as counts in the form of tally marks for keeping track of crops, animals, people, and

time. From there, it evolved with the demands of government and business, supported by academic inquiry and the growth of technology (Figure 1).

FIVE ERAS IN THE EVOLUTION OF PROBABILITY AND STATISTICS

PREHISTORIC STATISTICS, BEFORE 1300S
Starting with simple tallies, number systems evolved so that rulers could enumerate their citizenry, resources, and revenue.

EMERGING STATISTICS, 1300S-1700S
Continuing with governmental recordkeeping, the use of statistics spread to other uses in demographics, economics, insurance, and epidemiology. Probability matured with the introduction of the Normal distribution and Bayes' Theorem. The terms *statistic* and *statistics* were introduced during this era.

BLOOMING STATISTICS, 1800S-1940S
Development of statistics into a rigorous mathematical discipline including probability, random number generation, statistical sampling schemes, correlation and regression, hypothesis testing, confidence intervals, nonparametric statistics, and ANOVA.

MATURE STATISTICS, 1940S-1980S
Expansion of statistical concepts to many areas of science , technology, and business, especially in support of governance and war efforts. Introduction of many new statistical graphics and tests.

PERVASIVE STATISTICS, 1980S-PRESENT
Proliferation of computers and creation of the internet spread the use of probability and statistics everywhere.

Figure 1. Eras in the evolution of statistics.

As counting became more sophisticated in the ancient world, tally marks evolved into symbols. Those symbols evolved into numbers. Numbering systems were then created to allow for bigger numbers by using *place values* where the position of a digit determines its value.

Sumerians and later Babylonians had numbering systems around the 30th century BC. Roman numerals appeared in the 9th century BC while Egyptian and Greek numbering systems arose in the 4th and 5th centuries BC. Hindu-Arabic numerals originated in India in the 6th or 7th century and were introduced in Europe in the 12th century by Leonardo Pisano (aka Fibonacci), an Italian mathematician.

Ancient rulers used these numbering systems to perform censuses of their citizenry and inventories of their crops. Early scholars even used statistical principles to perform rudimentary calculations for solving real-world problems. No doubt, the governments also used numbers to administer taxes.

Statistics became established in society by the mid-1600s when governments began tabulating their resources and economies beyond just making a census of their citizenry. The 18th and 19th centuries were when the fundamental concepts of statistics and the mathematical underpinnings were developed.

Needless to say, the science of statistics has been around a long, long time.

Fundamentals.

Statistics has been around for a long time.

The 19th and 20th centuries brought statistics into a rigorous mathematical discipline, Even so, cynics railed against "lies, damn lies, and statistics," though nobody is quite sure who first said it (even Mark Twain says it wasn't him). The quote almost surely referred to simple statistics, like counts of church membership and agricultural inventories. A few decades earlier, English statistician Karl Pearson pointed out that "correlation does not imply causation." Although the era seemed to start with caution, even pessimism, the first few decades of the 1900s saw many advancements in methods of statistical analysis, not just in the sciences, but in industry and politics as well.

The modern era of statistics began in the 1940s when World War II brought an explosion of technology. Statisticians played their part in war efforts, benefiting greatly from the recognition, but the biggest boost for statistics came with the introduction of programmable computers. The government led the way in analyzing data, both census and business data, with their new computer resources. Universities, pollsters, and businesses followed. Suddenly, numbers were everywhere. Every aspect of American life was enhanced by statistics. Nine out of ten doctors said so.

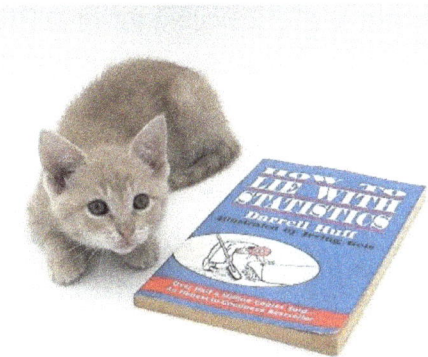

You need to understand statistics to decide what is a lie and what isn't.

Then in 1954, Darrell Huff published "How to Lie with Statistics," perhaps as an admonition against believing the results of statistical analyses without careful consideration. It is ironic that Huff, a journalist with no formal training in statistics, would later testify before Congress in the 1950s and the 1960s against the statistical relationship between cigarette smoking and lung disease.

Nevertheless, Huff's book became a best-selling book on statistics. It sold over a million copies in English by the 2000s and has been translated into over twenty languages. It has also spawned numerous books and articles echoing the same theme of caution in consuming information involving numbers. Unfortunately, many people take these books (or just their titles) to mean that all presentations that use statistics are dishonest. This unfounded belief has returned statistics to the false notoriety it suffered from over a century before.

Fundamentals.

But the popularity of statistics was poised to explode again, powered by the exponential proliferation of personal computers, statistical software and programming languages, and the rise of business use of statistics. Those advances led to the ascendency of *data science*, the melding of statistics with data engineering, programming, and domain expertise. It was thought by some to be a necessity in the age of computers, internet communications, *big data*, and high-stakes decision-making with profit motives. The creation of Amazon in 1994 ultimately led to an expansive use of statistics and *artificial intelligence* (AI) to increase sales, a trend that has spread to many other consumer businesses.

Academia also played a part in the rebirth of statistics after the 1950s. Statistics began to be a requirement for more and more degrees outside of STEM, including some degrees in history, archaeology, geography, agriculture, journalism, graphic communications, library science, culinary science, and linguistics.

One repercussion of academia's new focus on statistics has been that more people consider themselves to be competent in statistics after learning just a few concepts and formulas in HS. Some consider themselves to be experts after taking Stats 101. While not being expert enough to conduct an analysis of their own, they still consider themselves to be knowledgeable enough to argue using numbers on social media. That's still better than opinionated arm-waving or hearing *lies, damn lies, and statistics* all the time, though. Right?

At the same time that statistics was evolving into a bigger, more complex discipline, an even bigger revolution grew in mass communications.

Technology. The internet evolved from a secure communications system for military researchers in the 1960s to a consumer necessity in the 1990s. Satellites and wireless technologies are now spreading capabilities for communication everywhere.

Telephony. Wireless phones first appeared in the 1970s and evolved into smartphones by the 1990s. Mobile phones became as popular as landlines in the 2000s, creating issues with statistical surveys because the demographics of the owners were different. By the 2010s, more people owned wireless phones than landlines.

Not many people rely exclusively on landlines anymore.

Messaging. Instant messaging began in the 1970s and has evolved and expanded since. *Bulletin Boards Systems* (BBSs) were popular in the 1980s and 1990s. *Internet Relay Chat* (IRC) began in the 1990s and is still being used. Audio podcasts date back to the 1980s but didn't achieve much popularity until the 2000s. Text blogging began in the 1980s and expanded into video blogging in the 2000s. YouTube appeared in 2005.

Social Media. Social media began in the 1990s with GeoCities, Classmates, and SixDegrees. In the 2000s, Friendster and Myspace appeared only to be overtaken by Facebook in 2004 and Twitter in 2006. Two decades later, there are too many platforms to mention.

Books. Numbers differ, but there are from a half million to four million book titles published every year. eBooks represent only a fraction of that amount but while the growth of hardcopy book publishing is static, ebook publishing is increasing exponentially. eBook technologies have been around since the 1930s but didn't become popular until the 2000s. By 2010, Amazon was selling more ebooks than hardcopy books. Amazon currently holds over 60 thousand titles related to statistics, including over 2,000 introductory statistics titles, over 3,000 college statistics titles, and over 500 high school statistics titles.

Business. A critical change in how we receive news began in the 1970s when the major outlets decided to require their news divisions to be profitable. ABC led the way with non-traditional news programming, like 20/20 and Nightline, and other networks followed. The aim to be profitable pressured news outlets to sacrifice quality for quicker news releases. In the 1980s, all three major networks were bought by larger corporations, which increased demands for higher profits.

Government. Two major changes in Federal regulations transformed how information was made available to the public. From 1949 until 1987, the Fairness Doctrine required licensed radio and television broadcasters to devote some airtime to discussing controversial matters of public interest and to air contrasting views on those matters. The Telecommunications Act of 1996 increased the number of television stations that a single company could own, which led to a major consolidation of media outlets. Before the Act, fifty companies controlled the media in America; by 2011, only six did. These policies had the effect of limiting which news stories were presented and how they were framed. At

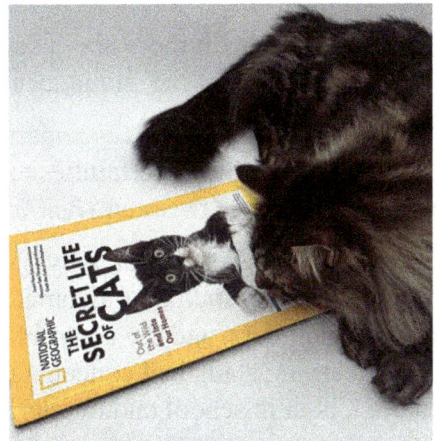

Who told them?

the same time, smaller information providers multiplied on the internet giving people greater latitude in picking their sources of information. Not all of those new information outlets provided the same level of veracity, however.

Fundamentals.

Society. There has always been gossip, rumors, unverified reports, propaganda, and other falsehoods passed from person-to-person in society. But today, mass communications has magnified the availability of those falsehoods leading to the proliferation of *fake news* in the media. The number of websites and podcasts spreading fake news increases every day with little regulation. There are even sites for generating fake news articles and memes, alternative facts, and spurious correlations. Sometimes it's difficult to distinguish between what is real information, what is satire and parody, and what is fake. This makes understanding and trusting statistics, and all of science in general, all the more difficult.

All of these changes happened in about forty years. Depending on when you were born, you may not even recognize how life worked before these advancements. Certainly, they have had an enormous impact on how the science of statistics is being presented in the media. This is why it is essential to become an informed consumer of statistical information.

We've only been around for a few weeks.

STATISTICAL THINKING VS. CONCEPTS VS. SKILLS

If you want to learn more about statistics, you should decide where your interests lie—statistical thinking, statistical concepts, or statistical skills.

Statistical thinking involves understanding **how** to define a problem in light of some objective, what the uncertainties and risks might be and how they can be controlled, and the difference between *significance* and *meaningfulness*.

Statistical concepts are the **why's** of statistics. They are the reasons why statistics work. Examples include fundamental concepts like populations, probability, the law of large numbers, and the central limit theorem … things you hear a lot about in Stats 101. Learning statistical concepts is beneficial to both statistics majors and non-majors, both in school and in later life.

Statistical skills are the **what's** of statistics. They involve calculations, like probabilities, descriptive statistics, and tests of hypotheses. Skills are learned by repetition. You learn them by doing the calculations in class and homework assignments. After Stats 101, skills like designing data matrices for a particular analysis are much more important than the calculations themselves, which are usually carried out by software.

Concepts and skills account for the majority of Stats 101 classes. This is perhaps unfortunate, for the greatest need in society is for people to understand statistical thinking.

Statistics is like sports. As a youngster, I would watch team sports with my father. I didn't understand all the rules but I liked the colorful uniforms and knew who won in the end. As a teenager, I understood many of the rules and how individuals with different body types would play different positions. As an adult, I understood how players might react differently based on subtle movements their opponents might make. But, there is a big difference between understanding and doing. I'm an avid fan of some sports but at no point in my life was I ever athletic enough to play a sport in real competition.

I did all the even-numbered homework problems.

Statistics is like sports in that there are great benefits to understanding what is going on even if you aren't actually doing it professionally. Analyzing data is a lot more involved than just understanding the results of a data analysis. Data have to be prepared so that they are appropriately formatted and error-free, processed in software using specialized formulas, and organized so the results are understandable to others. You don't have to be able to do all those things to benefit from statistical analyses just as you don't have to go through all the training of a professional athlete to enjoy a competitive sport.

Doing takes a lot more work than appreciating. Unfortunately, many instructors of Stats 101 don't see it that way. They'll want you to be able to analyze data far beyond what you might ever have to do in your life. You'll have to do dozens of calculations that you'll never have to do again. Do what you must to get a good grade, but if you're not planning a career in which you'll be conducting your own analyses, just be sure you can understand what an analysis means.

DESCRIPTIVE STATISTICS AND INFERENTIAL STATISTICS

Statistical analyses are often categorized as either descriptive or inferential (predictive).You will undoubtedly hear about this if you take Stats 101.

Some statisticians define this classification in terms of statistical techniques. Descriptive statistics (DS) involves characterizing data using graphs (Chapter 4) and statistical measures of frequency, central tendency, dispersion, and distribution form (Chapter 3). Inferential statistics (IS) involves exploring data using hypothesis testing (Chapter 6), regression analysis (Chapters 7 and 8), and more advanced techniques.

Some statisticians define the classification in terms of what the data represent. DS involves finite samples. IS involves populations.

Some statisticians define the classification in terms of objectives. DS paints a static picture of an existing reality, regardless of whether it is of an isolated set of values or a sample representing a population. IS reveals a general reality about a population that is not readily apparent.

I can be descriptive or inferential.

The DS-IS classification really embodies all of these ideas. However, the different perspectives lead to overlap and gaps between the definitions when taken in isolation. For instance, is it DS to estimate the central tendency and dispersion of a population? Is applying a regression equation to a sample without an underlying population an example of IS?

As it turns out, the distinction between DS and IS doesn't really matter much. It's just a convenient notion that has traditionally been used to facilitate teaching a very complex topic. By the time you understand concepts like central tendency and regression, the nuances between DS and IS are unimportant. But, that's just my opinion.

DON'T SWEAT THE JARGON

Like all professional disciplines, statistics has its share of esoteric jargon, well beyond the Greek letters and other symbols used to describe the mathematics of statistics.

The term **statistics** is defined as the science of analyzing *data* to infer knowledge about a *population* from a *sample*. What? In the definition, *data* refers to "numerical information." The *population* is a "group of entities having common characteristics." The *sample* is just a "representative part of the group." So, *statistics* refers to the science of analyzing numerical information to infer knowledge about a group of entities having common characteristics from a representative part of the group. Whew!

Jargon is just shorthand. If you know the meanings of the words, jargon can make reading and writing easier to understand than long explanations.

Fundamentals.

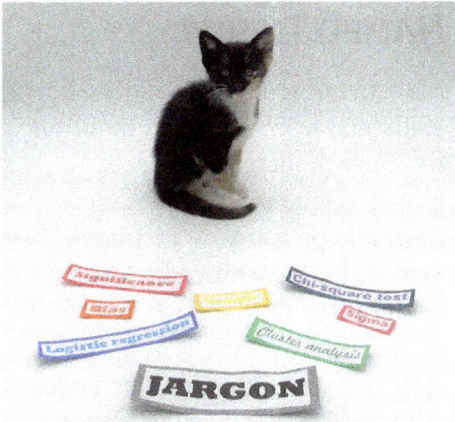

It's all Greek to me.

The word **statistic** (without the **s**) refers to a single number that has been calculated from a sample. If there is more than one *statistic*, they are called *statistics*. So, *statistics* may refer to the science of analyzing data OR to more than one number calculated from a dataset. This makes noun/verb agreement a challenge. You can tell which of the definitions apply by considering the context in which the term is used.

The word *data*, pieces of numerical information, is pronounced "day-ta," which rhymes with beta and zeta. Some people pronounce *data* as "daah-ta," which rhymes with strata. Fans of *Star Trek: The Next Generation* know that the first pronunciation is the only correct one. Data is the plural form of datum (that is "day-tum" not "daah-tum"). However, it is also used as a singular form, mainly by non-statisticians. Either is considered acceptable in common usage by linguists but professional statisticians prefer the plural form.

We are day-ta.

You'll see much more jargon as you read about statistical studies presented in the media, even more if you take Stats 101 in college. To simplify the statistical jargon you might hear, think of three distinctions:

- *Eponyms*, words named after people who discovered a statistical concept or developed a statistical procedure (e.g., *Bayesian statistics*, *Shapiro–Wilk test*)
- *Special words and phrases*, used primarily in statistics to convey a special meaning (e.g., *big data*, *infographics*, *data mining*). There are a few unique terms in statistics that are made-up words that don't appear anywhere else (*heteroscedasticity*, *variogramming*)
- *Repurposed words*, common words and phrases that are given alternative meanings (e.g., *mean*, *confidence*).

Fundamentals.

NAMED THINGS

Statistical procedures, especially statistical tests, are often modified to accommodate some special circumstance or to have some desirable property. When this occurs, the new procedure is sometimes eponymously named after the originators. There are statistical tests named after Tukey, Wilcoxon, Kolmogorov, Fisher, and many others. Other elements of statistics named after people include Bayesian statistics, kriging, and the Poisson and Bernoulli frequency distributions. If someone mentions a named distribution, test, or statistical procedure, don't panic. Nobody knows everything. Just ask what it is supposed to do. This type of statistical jargon could be much worse. When biologists name something after somebody, they do it in Latin.

Did I spell it right?

SPECIAL AND UNIQUE WORDS

Some statistical jargon might just as well be a foreign language because the words or phrases have different common meanings in the English language outside of statistics. Some appear to be innocent but convey a special meaning in statistics, like *bell curve* and *big data*. A few words are truly unique, like *kurtosis, skewness, covariance, autoregressive,* and my personal favorites, the multi-syllable *multicollinearity* and *homoscedasticity* (pronounced ho-mo-ski-das-tis-e-ti). It takes practice to even pronounce some of these terms correctly.

Statisticians who use these words with innocent civilians either don't understand their audiences or are sadists. Dealing with such special statistical terms is straightforward; just ask the statistician using them what they mean. Ask in a foreign language just to prove the point.

ALTERNATIVE MEANINGS

The most confusing statistical jargon just might be words in most people's everyday vocabulary that have a very different statistical meaning. For example, when you hear the word *mean*, your mind has to sort out the word's connotation. It can signify "to intend" as in *say what you mean*. It can be used "to associate" as in *spring means flowers*. It can refer to "resources or methods" as in *by any means*. It can indicate "character" as in *she has a mean streak*. It can imply "exceptional skill" as in *he has a mean fastball*. And of course in statistics, *mean* means "average."

If you don't realize that some words in English have different meanings in statistics, you can get confused very quickly. I've had well-meaning report editors change *Normal* to "typical," *median* to "medium," *gradient* to "grade," *histogram to "histograph,"* and *nonsignificant* to "insignificant." The term *significant*, in particular, has a very strict definition in statistics involving probability. It doesn't just mean "important."

Table 1 provides a few examples of words and phrases that have different meanings in statistics and English.

Table 1. Words that have a different meaning in statistics.

Word	Meaning to a Statistician	Meaning to a Non-statistician
Bagging	A method for combining predictions from many data mining models	What the cashier does with your groceries when you're done paying
Blocking	A technique for controlling variation in ANOVA	What the offensive line does during football season
Brushing	Interactively selecting data points on an on-screen graph to access other information associated with the point	What you do with your toothpaste and toothbrush
Breakdown	Splitting data into groups to calculate descriptive statistics and correlations	What happens to your car when you're in a hurry to get somewhere
Censoring	Data with a real but undetermined value, usually less than or greater than all other values in a dataset.	Restricting free speech; removing material considered to be offensive from books or other media
Confidence	Absence of type I errors	Ego stability
Discriminate	Classify observations by a statistical model; a good thing.	To make distinctions based on race, creed, ethnicity, age, or other category without regard to individual merit; a bad thing.
Errors	Differences between observed values and values predicted from a statistical model; residuals	Mistakes
Mode	The most frequently appearing number in a set of numbers	A manner of acting, such as being in "relaxation mode."
Monte Carlo	A simulation procedure for evaluating the properties or performance of a statistic	The quarter of Monaco known for its resorts and casinos; a hotel in Las Vegas
Normal	Follows a Gaussian (bell-shaped) distribution	Typical, routine, sane
Residual	Differences between observed values and values predicted from a statistical model; errors	Money made by musicians and actors when their works are replayed.
Sample	An individual observation or multiple observations that are part of a population	A piece, a bit, a taste.

You're not alone in the quagmire of statistical jargon; even statisticians can be baffled by some technical vocabulary. Like dialects of the English language, different statistical specialties have their own jargon and ways of expressing ideas. *Data mining, time-series forecasting, quality control, nonlinear modeling, biometrics, econometrics,* and *geostatistics* are all examples of statistical specialties that use terms not used in other specialties. Some terms are relatively recent so even statisticians have to learn new vocabulary.

Fundamentals.

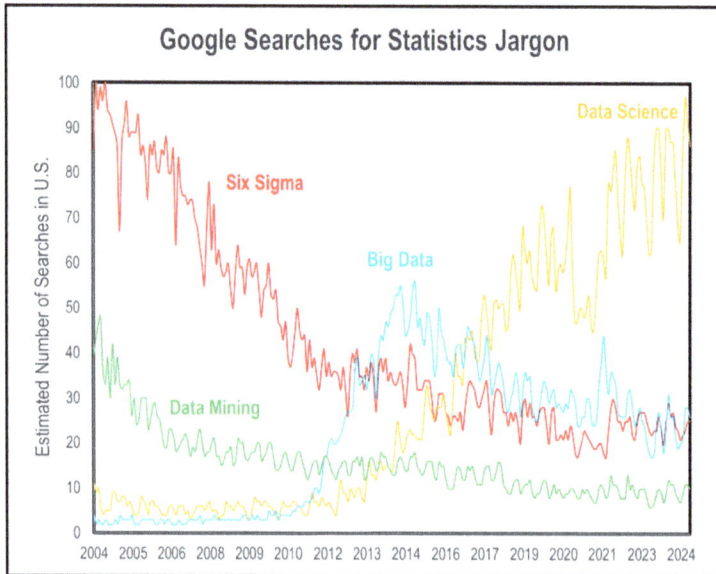

**Figure 2. Google searches for four popular
statistical terms.**

There are also new ideas that arise from time to time, like *six sigma*, *data science*, *data mining*, and *big data*, that lead to evolving capabilities and changing points-of-view in both the professional and user communities. Figure 2 shows how many times these terms have been searched for over the last two decades. *Six sigma* originated in the 1980s and *data mining* originated in the 1990s, so searches for those terms were declining by 2000. *Big data* had a clear though evolving definition so searches for the term peaked after just a few years. In contrast, *data science* came into prominence in the 2010s and searches for the term are still increasing, perhaps because there are different opinions about what the field encompasses.

If you need a quick refresher on a word used in **Stats with Kittens**, visit the Glossary at the end of the book. Of the thousand definitions, a bit less than two-thirds are used primarily in statistics or math, even though some of them might look like common English words. Only about 5% of those words are unique, made-up statistical terms, like *platykurtic*. About 17% of the words in the Glossary are words that have distinctly different meanings in statistics and English. Only 5% are eponyms. The remaining 14% are non-statistical, such as kinds of fallacies and bad science.

So if you are a newcomer to statistics, your biggest challenge won't be statistical tests named after 19th century statisticians or bizarre made-up words like *heteroscedasticity.* Those terms are a minority of what you will run into. Beware of terms that look like common English but aren't. If you go on to advanced levels of statistics, the proportions of eponyms, and special and unique words and phrases increase a lot.

Fundamentals.

If you don't understand a word or phrase used in a statistical presentation, either ask someone who knows more statistics than you do, search for it on the internet, or just ignore it. It probably won't matter.

FIVE FUNDAMENTAL THINGS YOU SHOULD KNOW ABOUT STATISTICS

Whether you're taking an introductory course in statistics or just trying to understand the *sabermetrics* (baseball analytics) of your favorite team, there are a few things that you should understand.

FUNDAMENTAL 1. EVERYTHING IS UNCERTAIN.

The fundamental difference between statistics and most other types of data analysis is that in statistics everything is uncertain. Input data and output results have variabilities associated with them. If they don't, they are of no interest to a statistician. As a consequence, statistical results are always expressed in terms of probabilities.

In statistics, everything is uncertain.

Every data measurement is a composite, consisting of:

- 🐾 **Characteristic of Population.** This is the part of a data value that you would measure if there were no variability. It's the "true" population value.
- 🐾 **Natural Variability.** This part of a data value is the inherent uncertainty or variability within the population. In a completely deterministic (predictable) world, there would be no natural variability.
- 🐾 **Sampling Variability.** This part of a data value is the difference between a sample and the population that is attributable to how uncharacteristic (non-representative) the sample is of the population. It reflects bias in sample selection.
- 🐾 **Measurement Variability.** This part of a data value reflects differences that are attributable to how data were measured or otherwise generated.
- 🐾 **Environmental Variability.** This is the difference between a sample and the population that is attributable to extraneous, often unexpected or even unknown, factors.

The goal of most statistical procedures is to estimate the characteristic of the population, characterize the natural variability, control and minimize the sampling and measurement variabilities, and hope the environmental variability is small. Minimizing variance can be difficult because there are so many causes and because the causes are often impossible to anticipate or control.

In statistics, accuracy comes easy, the procedures are designed to minimize bias. Precision, on the other hand, is dependent on the data. Consequently, ***precision***

Fundamentals.

trumps accuracy in an analysis. You can't understand the data without controlling variance. You can't control variance without understanding the data. Variance doesn't go away just because you ignore it (Chapter 3).

The bottom line is, if you're going to understand a statistical analysis, you'll need to appreciate the importance of variability.

Accuracy comes easy; precision is a challenge.

FUNDAMENTAL 2. STATISTICS ♥ MODELS.

Models are representations of something, usually an ideal or a standard. They can be physical (like a mannequin), written (like a business model), drawn (like a geologic cross-section), or consist of mathematical equations (like a radioactive decay model), or computer programming (like a weather model). They may be based on data having either uncertainty or deterministic laws having none (Chapter 8).

Statistics and models are closely intertwined. Models serve as both inputs and outputs of statistical analyses. Statistical analyses begin and end with models.

Statistics uses *distribution models* (equations) to describe how frequently data would occur if they were a perfect representation of a *population*. If data follow a particular distribution model, like the Normal distribution, the model can be used as a template for the data to calculate data frequencies and error rates. This is the basis of *parametric statistics*; you evaluate your data as if they came from a population described by a theoretical model of the frequency distribution.

In contrast, *nonparametric statistics*, also called *distribution-free statistics*, don't rely on any assumptions concerning the frequency distribution of the data. They use ranks or other imposed orderings of the data in the analysis.

This is my favorite model.

Statistical techniques are also used to build models from data. Statistical analyses estimate the mathematical coefficients (*parameters*) for the terms (*variables*) in the model, and include an *error term* to incorporate the effects of variation. The resulting statistical model then provides an estimate of the metric being modeled along with the probability that the model might have occurred by chance, based on the distribution model.

Statistics uses models to create other models.

Fundamentals.

FUNDAMENTAL 3. MEASUREMENT SCALES SHAPE ANALYSES.

You may not know anything about measurement scales (yet) but they are fundamental to statistics. You should at least become aware of the difference between nominal scales, ordinal scales, and continuous scales.

Nominal scales, also called grouping or categorical scales, are like stepping stones. Each value of a nominal scale, called a level, is different from and unrelated to the other scale levels. Ordinal scales are like steps in that each level of the scale has a distinct break from the next level, which is either higher or lower. Continuous scales are like ramps; each value of the scale is just a fraction higher or lower than the next value.

There are many more types of scales, especially for time scales. You can learn more about them on the internet if you ever need to.

Discrete scales are like steps.

The reason measurement scales are so important is that they help guide which graph or statistical procedure is most appropriate for an analysis. In some situations, you can't even conduct a particular statistical procedure if the data scales are not appropriate (Chapter 3).

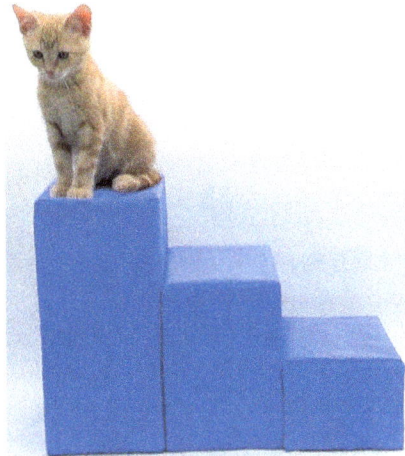

FUNDAMENTAL 4. EVERYTHING STARTS WITH A MATRIX.

All statistical procedures involve a matrix. Matrices, pronounced "may-triss-eez," are convenient ways to assemble data so that computers can perform mathematical calculations. If you study more advanced college statistics, you'll learn a lot about matrix algebra, but for Stats 101, all you have to know is that a matrix is very much like a spreadsheet. Specifically, in a spreadsheet there are rows and columns that define rectangular areas, called cells.

In statistics, the rows of the spreadsheet usually represent individual samples, cases, records, observations, entities, sample collection points, survey respondents, organisms, or any other location or object on which information is collected. The columns usually represent variables, the metrics, measurements, or types of information being recorded. The columns can correspond to instrument readings, survey responses, biological parameters, meteorological data, economic or business

Fundamentals.

measures, or any other types of information. Usually, there are several sets of variables for a given set of samples.

Together, the rows and the columns of the spreadsheet define the cells where the data are stored. Samples (rows), variables (columns), and data (cells) are the matrix that goes into a statistical analysis.

If you understand data matrices, you'll be able to conduct your own statistical analyses if you ever want to.

It's a matrix, three-dimensional, no less.

FUNDAMENTAL 5. THERE'S MORE TO STATISTICS THAN YOU THINK.

Statistics is much more than mind-numbing number-crunching that just creates yet more numbers. Believing that is like touring a cabinetmaker's shop and seeing the sawdust but ignoring the exquisite beauty and function of the products.

Your whole perception of statistics may be based on just a few applications that you're familiar with. You know about averages from your grades in school. You might know about probability from games, especially games of chance. You've definitely heard about statistical surveys if you've ever followed politics. But no matter how you may be familiar with statistics, there are many, many more applications.

With statistics you can describe datasets, compare and test differences between groups of numbers, identify and classify individuals and groups, make predictions from statistical models, and explain complex phenomena from data relationships. The range of questions that can be answered using statistics is astonishing. You'll see all these applications reported every day in news stories although the connection to statistical analyses is not always elucidated.

SIX MISCONCEPTIONS YOU MAY HAVE ABOUT STATISTICS

You undoubtedly have heard stories about statistics, otherwise the word wouldn't invoke such dread. Many of those stories are misleading. Here are six misconceptions you may have gotten.

Fundamentals.

MISCONCEPTION 1. "STATISTICS IS ALL MATH."

How could you not believe that statistics is all about math? In school, you were told that you could take a statistics class to fulfill a math requirement. It was taught by a math teacher and the homework assignments all involved numbers and calculations. If an introductory course in statistics was all math, statistics must be all math too.

I should be doing my statistics homework instead of watching this.

Reality – Statistics uses numbers but numbers are not the primary focus of statistics, at least to practitioners. The point of using statistics is to discover new knowledge and solve problems through the use of inductive reasoning (reasoning that starts with individual data to discover general concepts) involving numbers. It uses math as a language and one of its tools but it also uses sorting for ranks, filtering for classification, and all kinds of graphics.

That's why statistics is required for careers in business, social sciences, and many other disciplines. That's why it's taught by professors in all of those disciplines, too. Yes, it's required for math degrees and is taught by math professors at many schools. That's so there will be mathematical statisticians who will invent statistical tools for the applied statisticians to use. You can love statistics and be good at statistical thinking even if you think you hate math.

MISCONCEPTION 2. "STATISTICS REQUIRES A LOT OF DATA."

Statistics doesn't analyze individual pieces of information, like a solitary measurement, or a picture, or eyewitness testimony. Statistics uses data, lots of data, the more data the better. The number of data points (called the *sample size*) is part of almost every equation in statistics. And anyway, that's what the *Law of Large Numbers* says, the more data the better the results.

Reality – The number of samples you really **need** for a statistical analysis is contingent on how much **resolution** you want. Think of the resolving power of a telescope or a microscope, or the number of pixels in a computer image (Figure 3). The greater the resolution, the more detail you'll see. But, too much resolution also has drawbacks. An extremely high-resolution computer image can be too large to be manageable. It's the same way with statistics.

Fundamentals.

81 pixels

9 pixels per inch

324 pixels

18 pixels per inch

1,296 pixels

36 pixels per inch

5,184 pixels

72 pixels per inch

20,736 pixels

144 pixels per inch

Figure 3. Resolution differences for Kerpow.

What's more important than the number of data points is the quality of the data points. In statistics, the quality of a set of data points is how accurately and precisely the data points represent the population from which they are drawn. But, representative data can be incredibly difficult to generate. How do you decide which registered voters are actually likely to vote in the next election? How do you decide who might use a product you want to sell?

The number of items in a sample is easy to determine exactly, just count them. The representativeness of the items is virtually impossible to determine exactly. Nevertheless, what you should remember is that more data may be better but better data are always best.

More data are better but better data are best.

MISCONCEPTION 3. "DATA ARE DEPENDABLE."

Even in Stats 101, you'll do a lot of number crunching. You'll use small datasets and big datasets, real data and manufactured data. But, you'll never be told to delete data, so you figured that data are like facts. You don't delete them for any reason or you will bias your results.

Reality – Datasets are messy. Most newly generated datasets have errors, missing observations, and unrepresentative samples. Some population properties may be under-represented or over-represented. Some data measurements may involve inconsistent judgments by the individuals who record the data. There may be samples that should not be included in the analysis, like quality-control samples and metadata. All these problems with data require a lot of processing before an analysis can begin. In fact, data scrubbing often consumes the majority of time in an analysis. Nonetheless, you have to do it anyway.

Get the right data and get the data right before you start the analysis. Conducting a sophisticated analysis of poor data is like painting rotted wood. It won't hold up to even a cursory inspection.

Fundamentals.

MISCONCEPTION 4. "STATISTICS PROVIDES UNIQUE SOLUTIONS."

In all the problems teachers solve in class, all the homework assignments you did, and all the exams you took, there was only one right answer to a question. So, any statistical analysis should provide the same results no matter who does it.

Reality – Even if two statisticians start with identical datasets, they may not come to identical results, and sometimes, even identical conclusions. This is because they have to make many decisions about how to handle mathematical *assumptions* and *scrub* the data. Furthermore, there may be more than one way, even many ways, to approach a problem. There may also be different statistical analysis techniques that can be used, or even different options within the same technique.

It is actually more surprising for two statisticians to calculate exactly the same results from a dataset than for them to have some differences. It

Data scrubbing is always the most work in a data analysis.

would be like two artists representing a landscape differently on their canvases. Just like many problems in the real world, there is usually more than one *right* (acceptable) answer from a statistical analysis.

MISCONCEPTION 5. "STATISTICS PROVIDES UNAMBIGUOUS RESULTS."

Results are either *significant* or they're not. That's pretty unambiguous.

Reality – Statistical results are based on data and assumptions about the data. Change the number of samples and you change the resolution of the statistical procedure. Change the data or the assumptions and you change the estimates of variability. Change the resolution or the estimates of variability and you have different results.

There is uncertainty in uncertainty and that uncertainty brings ambiguity.

Is there really a meaningful difference between estimates of 49 and 51 errors out of a thousand possibilities? Many decision makers who never studied statistics think so. Even trained statisticians may not agree. One statistician might take a firm stance and say a result is *significant* and another might say, maybe not. Results have

Fundamentals.

uncertainty, interpretations have ambiguity, and decisions have risks. That's statistics.

MISCONCEPTION 6. *"IT'S EASY TO LIE WITH STATISTICS."*

Darrell Huff wrote "How to Lie with Statistics" in 1954. Michael Wheeler wrote "Lies, Damn Lies, and Statistics: The Manipulation of Public Opinion in America" in 1976. John Allen Paulos wrote "Innumeracy: Mathematical Illiteracy and Its Consequences" in 1988. Joel Best wrote "Damned Lies and Statistics: Untangling Numbers from the Media, Politicians, and Activists" in 2001. So, lying with statistics must be easy and commonplace since everybody is talking about it.

It's easy to get fooled if you don't understand statistics.

Reality – It's hard to do statistics right but it's even harder to do them wrong on purpose, in a particular way, without being obvious about it. You have to collect data, crunch the numbers, and cook up your story, or perhaps cook up your story, make up the data, and schedule a press conference. If you just want to mislead an audience, it's much easier to use made-up facts, phony anecdotes, and illogical conjectures.

> *In listening to stories, we tend to suspend disbelief in order to be entertained, whereas in evaluating statistics we generally have an opposite inclination to suspend belief in order not to be beguiled.*
>
> John Allen Paulos, American mathematician and writer.

So why do so many people, particularly politicians and biased media sites, even bother lying with statistics? It's because numbers provide credibility. If you have little credibility yourself, using numbers can confer the illusion of veracity and expertise. And that is why some people use statistics in the first place.

Those authors aren't saying that it's easy to lie with statistics, they're saying that it's easy for you to get fooled if you don't understand statistics. Big difference.

It's not that it's easy to lie with statistics; it's that it's easy for you to get fooled if you don't understand statistics.

Don't be paranoid that every poll or report on a scientific discovery is meant to mislead. Some writers may do that but real journalists don't. They'll at least try to get a story right.

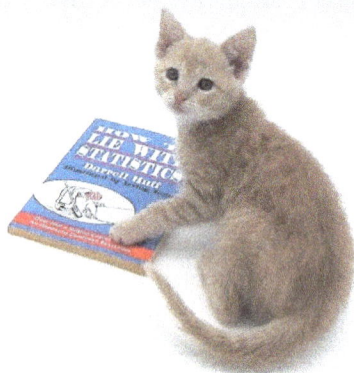

Fundamentals.

Scientists and statisticians typically don't hype their work to the public. They'll make optimistic claims to their peers, but not the public. They're more likely to hedge on the side of caution. Scientists and statisticians who do promote their work directly to the public too often have an ulterior motive.

Remember also that all experts don't have the same level of statistical expertise. Even professional statisticians don't all have the same knowledge base in statistics. Like doctors, they specialize in some topics and know only the basics about others.

SIX WAYS STATISTICS CAN BENEFIT YOU

So why do colleges want you to take statistics, even if you want to be a nurse, teacher, journalist, or some other career in which you'll never have to conduct your own analyses? It's because they want you to understand why people use statistics, why they don't just reason by text and pictures. They want you to understand that it means something when people use statistics, and it means something else when they don't.

If you understand a little about what statistics can do, you'll be better equipped to appreciate the big picture even if the details might escape you. Here are a few of the reasons why people who do use statistics use them whenever they can.

There are a lot of ways that you can use statistics.

BENEFIT 1. STATISTICS PROVIDES A STARTING POINT AND A COURSE OF ACTION.

If you pursue a career in the natural sciences or engineering, you'll have some basic principles, laws, or at least theories to start with in analyzing data. Even some of those, though, were discovered or verified by statistical observation. If you pursue a career in the social sciences, business, economics, or most other fields, though, you've got little to go on besides statistics. Anecdotes aren't worth much.

Statistics are antidotes for anecdotes.

Statistics gives you a place to start by having you focus on the *population*, so you think about what to *sample*, and the *phenomenon*, so you think about what to *measure* and how to measure it. Once you have laid this groundwork, statistics helps you define alternative hypotheses to weigh and provide a variety of methods to analyze the data.

Fundamentals.

BENEFIT 2. STATISTICS GIVES YOU MORE WAYS TO ANALYZE DATA.

Statistics isn't just a toolbox, it's a whole workshop filled with toolboxes holding more tools than you could ever use in a career. Statistics allows you to describe, correlate, detect differences, group, separate, reorganize, identify, predict, smooth, and model.

And, it's not just the variety of tools for doing *different* things, there are also many tools for doing the *same* thing in different ways. Want to find the center of a data distribution? You can use the *arithmetic mean*, the *geometric mean*, the *harmonic mean*, *trimmed* and *winsorized means*, *weighted means*, the *median*, and the *trimean*. Each has its own special use, like the variety of types of wrenches used by a mechanic.

With a statistician's toolbox, you can gain far more insight from your data than you might from any other type of analysis.

There's natural variability even in littermates.

BENEFIT 3. STATISTICS EXAMINES BOTH ACCURACY AND PRECISION.

Any marksman will tell you that it's not enough to be able to hit a target. You have to be able to hit it where you aim and do it consistently. That's *accuracy* and *precision*. Many analytical techniques focus on accuracy and forget all about precision. Variability, uncertainty, and risk don't go away by ignoring them.

Statistics is all about understanding variability.

BENEFIT 4. STATISTICS EXAMINES BOTH TRENDS AND ANOMALIES.

Most forms of analysis focus on finding similarities and patterns in data. Statistics, in particular, can be used to find linear and nonlinear *trends*, *cycles*, *steps*, *shocks*, *clusters*, and many other types of groupings. What's more, statistics can be used to identify and explore divergent or anomalous cases that don't fit general patterns. Sometimes it is these *outliers* rather than the patterns that reveal the information most crucial in an analysis.

Fundamentals.

BENEFIT 5. STATISTICS TELLS YOU HOW MUCH INFORMATION YOU NEED.

In data analysis, more is not always better. It's not too unusual to have too much data to make sense of using only graphs and tables. Statistics provides a variety of ways to help you decide about how many samples you need to achieve a certain objective. Statistics provides ways to judge the quality of the data and compensate for misleading variability. Statistics can also tell you if your data are redundant, and if so, provide ways to reassemble the data more efficiently.

BENEFIT 6. STATISTICS PROVIDES STANDARDIZATION.

Statisticians can usually convince people who are reviewing their work that their data analysis is legitimate because it uses well-known, professionally-accepted procedures. Likewise, it's easier to require analysts to use some established statistical procedure rather than assuming whatever they might do is legitimate. For example, Government regulations frequently require the use of some particular statistics to report and analyze datasets, such as crime rates, pharmaceutical effectiveness, environmental impact, occupational safety, public health, and educational testing.

> *A judicious man looks at Statistics, not to get knowledge, but to save himself from having ignorance foisted 'on him.*
>
> Thomas Carlyle, 1795-1881,
> Scottish satirist and historian,
> from **Chartism, Chapter II** (1839).

So you see, statistics has a lot to offer you, whether there is a strong theoretical basis to your field of practice or not. That's why your college advisers want you to learn about it.

FIVE PROBLEM-SOLVING APPROACHES USING STATISTICS

Now that you know a bit about the statistics, what it is and what it is not, and how you might benefit from understanding why statistics are useful, here are the types of tasks that statistics can accomplish.

Statistical analyses usually aim at achieving one of five objectives:

- Describe
- Identify or Classify
- Compare or Test
- Predict
- Explain.

Fundamentals.

These five categories encompass just about every problem-solving approach you could want.

APPROACH 1. DESCRIPTION.

Describing a dataset is relatively straightforward, it's even easier than describing a person. It involves characterizing populations and samples using descriptive statistics and statistical graphics, requiring an understanding of fundamental concepts of measurement scales, measures of *central tendency* and *dispersion*, and data *frequency distributions*. You can even do all the calculations on spreadsheet software. Eventually, description can become more sophisticated as you learn about *sampling schemes*, controlling *uncertainty*, and methods for dealing with *outliers* and *missing data*.

APPROACH 2. IDENTIFICATION AND CLASSIFICATION.

Can you tell us apart?

Identification and classification involves finding a sought-after data point or group of data points. This process may involve using *descriptive statistics*, *statistical graphics*, *statistical comparisons* (i.e., *hypothesis tests* and *intervals*), and for more sophisticated applications, multi-variable techniques such as *cluster analysis*.

Analyses range from simple visual recognition from graphs to the exploration of arcane mathematical dimensions where only bold number-crunchers venture. It's like finding Waldo. At a convention of funeral directors, one look would be all you needed. If he were making American flags in a candy cane forest, you might need some non-visual clues.

As another example, you could determine a person's height by looking at them but not from a table of eye and hair colors. On the other hand, you couldn't tell who the best players were on a sports team from their pictures but you could from a table of their performance statistics.

However you do it, identification is the gateway to classification. If you can do one, you can probably do the other.

APPROACH 3. COMPARISON.

Comparison involves detecting statistically-significant differences between statistical populations or reference values. This objective involves more than just a visual comparison of values, it involves consideration of both *central tendency* and *dispersion*. Statistical methods might involve using simple *hypothesis tests*, *statistical intervals* (i.e., average plus or minus some number of standard deviations), or more specialized techniques like the *analysis-of-variance* (ANOVA).

This objective is tougher than description or classification even though there is ample software available for most analyses. You need to know what test to run or what ANOVA design to use as well as understand *probability*, *effect size*, and *violations of assumptions*. There's a much greater chance of something going wrong because there are so many ways that a comparison analysis can be designed and more assumptions that have to be met.

APPROACH 4. PREDICTION.

Prediction involves estimating new values by creating statistical models from data, characterized by *attributes* (i.e., *variables*) of *samples* (i.e., *observations*). using sophisticated techniques such as statistical *regression* and *neural networks*. Forecasting future values using *time-series modeling* techniques, and interpolating spatial data using *geostatistics* or classical methods are also aims of prediction.

Compared to my paw, your paw is HUGE

In addition to all the description and comparison techniques, you'll need to know how to use a variety of model building and assessments methods and understand the morass of *prediction error*.

It's easy to make a prediction.
It's hard to make an accurate prediction.
It's damn near impossible to make
an accurate prediction that is also precise.

Even if you did nothing wrong statistically, it's easy to produce a poor prediction, and a poor prediction will eventually be noticed. One really good prediction and a psychic is famous; one really bad prediction and a statistician is relegated to selling insurance.

APPROACH 5. EXPLANATION.

Explaining latent aspects of phenomena is the toughest of all statistical objectives to attempt because it is based on data (involving *induction*) rather than known scientific laws (involving *deduction*). First, you need to understand some very sophisticated statistical methods, like *regression*, *cluster analysis*, *discriminant analysis*, *factor*

Fundamentals.

analysis, *canonical correlation*, and *data mining* techniques (see glossary). Second, you have to understand the conceptual framework of the systems the data come from. Third, you have to have the talent to apply the knowledge creatively. Explanation is where statistical science meets statistical art. It is why explanation is the most difficult of all the objectives of statistics to pursue.

For example, to explain a statistical model of stream contamination, you must know something about fluvial hydraulics, hydrogeology, meteorology, and environmental geochemistry. To explain customer satisfaction, you must know something about demographics, marketing, business, and consumer psychology.

The field of *data science* is an attempt to integrate statistics, domain expertise (knowledge in a pertinent field), and programming skills. Just what is needed for *explanation*.

> *The combination of some data and an aching desire for an answer does not ensure that a reasonable answer can be extracted from a given body of data.*
>
> John Tukey, 1915-2000, American mathematician and statistician, from **Sunset salvo**, The American Statistician 40 (1), (1986), page 74.

Data explanation is also a challenge because there's no guidance for where to start or where to finish. That leads to the real danger of biasing the analysis. This biasing may be manifested by finding obscure relationships (*beating the data until it confesses*) that don't appear in independent samples of the population (*capitalizing on chance*) and then manipulating the samples in the analysis by adding, deleting, or editing them to produce significant results (*p-hacking*). That is a real challenge even professionals wrestle with.

Sophisticated explanation models require data analysts to integrate diverse information and think of it in ways that may have never been thought of before. Explanation can create fundamental wisdom, although often, results will be humdrum. If something of true consequence is discovered, though, it is often difficult to explain in a manner that most people can understand.

WHAT'S NEXT

Don't be discouraged if all this information seems obscure or different from what you knew … or thought you knew. It will all make more sense as you read further in **Stats with Kittens.** But at least now you should understand better why you have to take Stats 101 to complete a college degree. It's essential for most careers and almost anything else you do in life.

The rest of **Stats with Kittens** consists of eight chapters that aim to help you understand the kinds of statistics you see every day. If you think you might want to go on and take a Stats 101 course, the chapters will give you some hints about what you'll be learning in the course.

Fundamentals.

- **Chapter 2. Probability.** Where probabilities and odds come from, probability models, and tips on what to look for in probabilities.
- **Chapter 3. Description.** The fundamentals of data description, terminology, variance, scales, and tips on what to look for in data descriptions.
- **Chapter 4. Graphics.** Fundamentals of graphing, graphs you ought to know, tips on what to look for in statistical graphics.
- **Chapter 5. Surveys.** Elements of statistical surveys and tips on what to look for in surveys.
- **Chapter 6. Comparisons.** The fundamentals of statistical testing, terminology, assumptions, procedures, and tips on what to look for in statistical tests.
- **Chapter 7. Relationships.** Types and patterns of relationships, correlation and causation, tips on what to look for in data relationships.
- **Chapter 8. Models.** Types of models, model components, diagnostics, and tips on what to look for in models.
- **Chapter 9. Literacy.** How to evaluate statistical presentations.

After you complete Stats 101, you might enjoy **Stats with Cats**, which provides tips for using statistics in your career and in your life.

Good luck.

How hard could it be? I don't even need the ball.

Fundamentals.

Why do we have to start with probability?

Fundamentals.

CHAPTER 2. PROBABILITY.
Probability Is Simple … Kinda.

INTRODUCTION

The science of probability began thousands of years ago, perhaps over games of chance. The first book about gambling probabilities was written in the 16th century. And, although we've learned a lot about probability since then, we still buy lottery tickets.

One of these tickets ought to be a winner.

The one concept that sets statistics apart from the rest of mathematics is that everything is uncertain. Data are variable and calculations have ranges of possible results. Statistics deals with these uncertainties quantitatively rather than just saying *likely, probably, possibly, maybe,* or *who knows*. That's where **probability** comes in.

For the most part, probability is a simple concept with straightforward formulas. The applications, however, can be mind-bending. There are aspects of probability associated with every type of statistical analysis. It is complex and inescapable. That's why there are thousands of books and websites devoted to it. You simply must have some understanding of probability to function in modern times.

That's what this chapter is about.

TWO SCHOOLS OF PROBABILITY

There are two types of probability used in statistics, *frequentist probability* (also called *classical probability*) and *Bayesian probability*. Both date back to the 1700s.

Frequentist probability is based on the objective, long-run frequency of an event occurring in repeated experiments. It relies solely on the numbers of occurrences of events, hence the reference to frequency. Frequentist probability does not allow for knowledge other than frequencies to be considered. Frequentist probabilities are also fixed once calculated.

Bayesian probability is based on beliefs about an event in the form of probabilities. While frequentist probabilities are calculated from fixed frequencies, Bayesian probabilities are calculated from probabilities that reflect distributions of possible values. Bayesian probability allows calculations to be updated when more

Probability.

information becomes available. It is often used in situations where frequency data are limited or when important non-frequency information is available.

This chapter focuses on the frequentist interpretation of probability as it is simpler and more commonly used. If you take Stats 101, you'll be taught probability from the frequentist perspective. There is an example of the use of Bayesian probability later in this Chapter in the section on medical testing.

WHERE DO PROBABILITIES COME FROM?

I thought probabilities came from the grocery store.

One of the first questions you may have about probabilities is "where do they come from?" They sure don't come from the arm-waving know-it-alls at work or anonymous memes on the internet. Real probabilities come from four sources—logic, data, models, and oracles.

LOGIC

Some probabilities are calculated from the number of logical possibilities for a situation. For example, there are two sides to a coin so there are only two possibilities, each having a probability of occurrence of 50%. Standard dice have six sides so there are six possibilities, each having a probability of occurrence of about 17%. A standard deck of playing cards has 52 cards so there are 52 possibilities, each having a probability of occurrence of about 2%. The formulas for calculating probabilities aren't that difficult. By the time you finish reading this chapter, you'll be able to calculate all kinds of probabilities based on logic.

DATA

Some probabilities are based on surveys of large populations or experiments repeated a large number of times. For example, the probability of a random American having a blood type of A-positive is 0.34 because approximately 34% of the people in the U.S. have been determined to have that blood type. Likewise, there are probabilities that a person is a male (49.2%), has brown eyes (34%), and has brown hair (58%). The internet has more data than you can imagine for calculating probabilities.

MODELS

A great many probabilities are derived from mathematical models, built by experts from data and scientific principles. You hear meteorological probabilities developed from models reported every night on the news, for instance. These models help you plan your daily life, so their impact is great. Perhaps even more importantly, though,

Probability.

are the probabilistic models that serve as the foundation of statistics itself, like the *Normal distribution*.

ORACLES

Some probabilities come from the heads of experts, not the arm-wavers on the internet, but real professionals educated in a data-driven specialty where there may not be any underlying science. Experts abound. Sports gurus live in Las Vegas, survey builders reside largely in business and academia, and political prognosticators dwell everywhere.

Oracles make predictions based on their knowledge, often using probability models as a starting point. The difference between models and oracles is that the key elements of probabilities based on models all have a theoretical foundation whereas one or more of the key elements of probabilities from oracles are based on conjectures of the oracle because there is no theoretical foundation. Predictions

What will my crystal ball tell me to do?

involving stocks and sports are good examples. Models from oracles are often based on *Bayesian statistics*.

TURNING PROBABLY INTO PROBABILITY

Probabilities are discussed in terms of events or alternatives or outcomes. They all refer to the same thing, something that can happen. The probability of any outcome or event can range only from 0 (no chance) to 1 (certainty).

The probability of an event is calculated as follows.

- The probability of a single event occurring is equal to 1 divided by the number of possible events.

- The probability of any one of multiple, independent events occurring is equal to the number of favorable events divided by the number of possible events.

These rules are referred to as simple probability because they apply to the probability of a single *independent, disjoint* event from a single *trial*. (Independent and disjoint events are described in the following paragraphs.) A trial is the activity you perform to determine the probability of an event. Trials are also referred to as tests or experiments.

If the probability involves more than one trial, it is called a *joint probability*. Joint probabilities are calculated by multiplying together the relevant simple probabilities. Joint probabilities also range from 0 to 1.

Probability.

For example, if the probability of event A is 0.3 (30%) and the probability of event B is 0.6 (60%), the joint probability of the two events occurring is 0.3 times 0.6, or 0.18 (18%).

Using data from the internet, the probability of a brown-eyed (0.34), brown-haired (0.58), non-Hispanic-white (0.63) male (0.49) having A-positive blood (0.34) is about 0.02 (2%) in the U.S., or about one in 48.

Events can be independent of each other or dependent on other events. For example, if you roll a dice or flip a coin, there is no connection between what happens in each roll or flip. They are *independent events*.

On the other hand, if you draw a card from a standard deck of playing cards, your next draw will be different from the first because there are now fewer cards in the deck. Those two draws are called *dependent events*. Calculating the probability of dependent events has to account for changes in the number of total possible outcomes. Other than that, the formulas for probability calculations are the same.

This one time at band camp, I tried to explain that I'm not a "Jellicle Cat."

Some outcomes don't overlap. They are one-or-the-other. They both can't occur at the same time. These outcomes are said to be *mutually-exclusive* or *disjoint*. Examples of disjoint outcomes might involve coin flips, dice rolls, card draws, or any event that can be described as either-or. For a collection of disjoint events, the sum of the probabilities is equal to 1. This is called the *Rule of Complementary Events* or the *Rule of Special Addition*.

Some outcomes do overlap. They can both occur at the same time. These outcomes are called *non-disjoint*. Examples of non-disjoint outcomes include a student getting a grade of B in two different courses, a used car having heated seats and a manual transmission, and a playing card being a queen and in a red suit. For a collection of non-disjoint events, the sum of the simple probabilities minus the probability in common for the events is equal to 1. This is called the *Rule of General Addition*. The joint probability of non-disjoint events is called a *Conditional Probability*.

There is a LOT more to probability than that, but that's enough for you to understand news stories in the media. Read through these examples to see how probability calculations work.

Probability.

These used cars have manual transmissions but no heated seats.

FIVE PROBABILITIES YOU MIGHT HAVE THOUGHT ABOUT

Probability does not indicate what *will* happen, it only suggests how *likely*, on a fixed scale of 0 to 1, something is to happen. If it were definitive, the likelihoods would be called certainties not probabilities. Here are some examples.

1. COINS.

Find a coin with two different sides, call one side A and the other side B.

What is the probability that B will land facing upward if you flip the coin and let it land on the ground?

    Probability = number of favorable events / total number of events
    Probability = 1 / 2
    Probability = 50% or 0.5 or 1 out of 2.

This is a probability calculation for two independent, disjoint outcomes.

To make a coin flip legitimate, you have to toss the coin so it spins from heads to tails in the air many times. This is a randomization process that ensures that both sides have an equal chance to be on top. If you just tossed the coin up so it remained flat, there wouldn't be an equal chance for both sides to be on top. Then the two events wouldn't be independent and the probability calculation wouldn't be valid.

Coin edges aren't included in the total-number-of-events because the probability of flipping a coin so it lands on an edge is much smaller than the probability of flipping a coin so it lands on a side. Alternative events have to have an observable (non-zero) probability of occurring for a calculation to be valid.

Probability can be expressed in several ways—as a percentage, as a decimal, as a fraction, or as the relative frequency of occurrence. Odds are not the same thing as probabilities, as discussed in a later section.

Using the same coin, record the results of 100 coin-flips. Count the number of times the results were the A-side and how many times the results were the B-side. Then flip the coin one more time.

What is the probability that side B will land facing upward?

    Probability = number of favorable events / total number of events
    Probability = 1 / 2
    Probability = 50% or 0.5 or ½ or 1 out of 2.

Probability.

Each coin flip is independent of the results of every other coin flip. So, whether you flip the coin 100 times or a million times, the probability of the next flip will always be 1 out of 2, based on logic. When you flipped the coin 100 times, for instance, you might have recorded 53 B-sides and 47 A-sides. The probability of the B-side facing upward after a flip would NOT be 53/100 because the flips are *independent* of each other. What happens on one flip has no bearing on any other flip. If you recorded results appreciably different from a 50%/50% split, say 70 B-sides and 30 A-sides, you might consider whether your coin or the way you flipped it are biased in some way.

I'd bet all my money on this.

2. TOAST.

Make two pieces of toast and spread butter on one side of each. Eat one and toss the other into the air.

What is the probability that the buttered side will land facing upward?

> Probability = number of favorable events / total number of events
> Probability = 1 / 2
> Probability = 50% or 0.5 or 1 out of 2.

WRONG … maybe. You might think this was wrong because in your experience the buttered side of a piece of toast usually lands facing down. Some people think that is the result of the buttered side being heavier than the unbuttered side so that the two sides aren't the same in terms of characteristics that will dictate how they will fall. Other people believe it is because the toast hasn't flipped enough in the air so that each side doesn't have a random and equal chance of being on top.

Watch out for falling toast.

For probability calculations to be valid, each event has to have a known, constant chance of occurring. Unlike a typical coin flip, the two possible outcomes of a toast-flip wouldn't have a known, constant chance of occurring either because the two sides aren't identical or because they aren't randomly on top in the air the same amount of time.

This is one reason why some probability calculations don't match the results that are observed. Sometimes, it is difficult to

Probability.

ensure that the underlying conditions are exactly comparable for all alternatives. The calculations are easy; meeting the assumptions of the calculations may not be. If you see a news story that discusses the likelihood of competing events, think about whether the two events have equivalent characteristics.

Probability calculations are easy;
Meeting the assumptions of the calculations isn't.

Say you knew the buttered side of your toast landed downward 85% of the time (based on data rather than logic). In other words, the two alternatives are NOT equally likely for some reason.

What is the probability that the buttered side of your next toast flip will land facing upward?

Probability = number of favorable events / total number of events
Probability = (100-85) / 100 = 15 / 100
Probability = 15% or 0.15 or 1 out of 6⅔.

To make this calculation valid, you would have to establish that, when tossed into the air, buttered toast will land with its unbuttered-side upward a **constant** percentage of the time. So, start by making 100 pieces of toast and butter one side of each …. Let me know how this turns out.

3. DICE.

STANDARD DICE

Standard dice are six-sided with a number (from 1 to 6) or a set of 1 to 6 small dots (called *pips*) on each side. The numbers (or number of pips) on opposite sides sum to 7.

What is the probability that a 6 (or 6 pips) will land facing upward when you toss the dice?

Probability = number of favorable events / total number of events
Probability = 1 / 6
Probability = 17% or 0.167 or 1 of 6.

More falling toast ...

This is a probability calculation for 6 independent, disjoint outcomes.

What is the probability that an even number (2, 4, or 6) will land facing upward when you toss the dice?

Probability = number of favorable events / total number of events
Probability = 3 / 6
Probability = 50% or 0.5 or 1 out of 2.

This calculation considers 3 sides of the dice to be favorable outcomes.

Probability.

What is the probability that a 1 will land face upward on two consecutive tosses of the dice?

> Probability = (Probability of Event A) times (Probability of Event B)
> Probability = (1 / 6) times (1 / 6)
> Probability = 3% or 0.28 or 1 out of 36.

This calculation estimates the *joint probability* of 2 independent, disjoint outcomes occurring based on 2 rolls of 1 dice.

Dice? OHHH, we thought you said mice.

What is the probability that, using 2 dice, you will roll "snake eyes" (only 1 pip on each dice)?

> Probability = (Probability of Event A) times (Probability of Event B)
> Probability = (1 / 6) times (1 / 6)
> Probability = 3% or 0.28 or 1 out of 36.

This calculation also estimates the joint probability of 2 independent, disjoint outcomes occurring based on 1 roll of 2 dice.

What is the probability that a 6 will land face upward on 3 consecutive rolls of the dice?

> Probability = (Probability of Event A) times (Probability of Event B) times (Probability of Event C)
> Probability = (1 / 6) times (1 / 6) times (1 / 6)
> Probability = 0.5% or 0.0046 or 1 out of 216.

It doesn't matter if you roll 1 dice 3 times or 3 dice 1 time, the joint probability is the same. If you roll three 6s consecutively, either the dice is loaded or the Devil is messing with you.

DnD Dice

Dice used to play Dungeons and Dragons (DnD) have different numbers of sides, usually 4, 6, 8, 10, 12, and 20. The formula to calculate the probabilities remains the same, only the number of events is different for each type of dice.

What is the probability that 6 will land facing upward when you throw a 20-sided (icosahedron) dice?

> Probability = number of favorable events / total number of events
> Probability = 1 / 20
> Probability = 5% or 0.05 or 1 out of 20.

This is a probability calculation for 20 independent, disjoint outcomes.

Probability.

What is the probability that a 6 will land face upward on 3 consecutive tosses of the 20-sided (icosahedron) dice?

Probability = (Probability of Event A) times (Probability of Event B) times (Probability of Event C)
Probability = (1 / 20) times (1 / 20) times (1 / 20)
Probability = 0.01% or 0.000125 or 1 out of 8,000.

So, you have a smaller chance of summoning the Devil by rolling 6-6-6 if you use a 20-sided DnD dice instead of a standard 6-sided dice. Good to know.

What is the probability that you will roll a 6-6-6 using an 8-sided, a 12-sided, and a 20-sided dice together?

Probability = (Probability of Event A) times (Probability of Event B) times (Probability of Event C)
Probability = (1 / 8) times (1 / 12) times (1 / 20)
Probability = (0.125) times (0.083) times (0.05)
Probability = 0.052% or 0.00052 or 1 out of 1,920.

It's my turn to roll the dice ...
I just can't pick them up.

Obviously, probability is what makes playing DnD so entertaining.

What is the probability that 6 will land facing upward when you throw a 4-sided (tetrahedron, Caltrop) dice?

Probability = number of favorable events / total number of events
Probability = 0 / 4
Probability = 0% or 0.0 or 0 out of 4.

There is no 6 on the 4-sided dice. Not everything in life is possible.

4. CANDY BARS.

Your son has just returned from trick-or-treating. He inventories his stash and has: 5 Snickers; 6 Hershey's bars; 4 Pay Days; 5 Kit Kats; 3 Butterfingers; 2 Charleston Chews; 5 Tootsie Roll bars; a box of raisins; and an apple. You throw away the apple because it's probably full of razor blades. He throws away the raisins because … they're raisins.

After the boy is asleep, you sneak into his room and, without turning on the light, find his stash. Putting your hand quietly into the bag, you realize that it's

Mine ... all mine.

too dark to see and all the bars feel alike.

What's the probability that you'll pull a Snickers out of the bag?

> Probability = number of favorable events / total number of events
> Probability = 5 / 30
> Probability = 17% or 0.167 or 1 out of 6.

This is a probability calculation for 30 independent disjoint outcomes. The draw is independent because only 1 bar is being drawn.

What's the probability that you'll pull out a Snickers on your next attempt if you put back any bar you pull out that isn't a Snickers?

> Probability = number of favorable events / total number of events
> Probability = 5 / 30
> Probability = 17% or 0.167 or 1 out of 6.

This is called *probability with replacement* because by returning the non-Snickers bars to the bag, you are restoring the original total number of bars. The outcomes are independent of each other.

How many bars do you have to pull out before you have at least a 50% probability of getting a Snickers if you put the bars you pull out that aren't Snickers into a separate pile (not back into the bag)?

> Probability = number of favorable events / total number of events
> 1st bar pulled Snickers probability = 5 / 30 = 17%
> 2nd bar pulled Snickers probability = 5 / 29 = 17%
> 3rd bar pulled Snickers probability = 5 / 28 = 18%
> 4th bar pulled Snickers probability = 5 / 27 = 19%
> 5th bar pulled Snickers probability = 5 / 26 = 19%
> 6th bar pulled Snickers probability = 5 / 25 = 20%
> 7th bar pulled Snickers probability = 5 / 24 = 21%
> 8th bar pulled Snickers probability = 5 / 23 = 22%
> 9th bar pulled Snickers probability = 5 / 22 = 23%
> 10th bar pulled Snickers probability = 5 / 21 = 24%
> 11th bar pulled Snickers probability = 5 / 20 = 25%
> 12th bar pulled Snickers probability = 5 / 19 = 26%
> 13th bar pulled Snickers probability = 5 / 18 = 28%
> 14th bar pulled Snickers probability = 5 / 17 = 29%
> 15th bar pulled Snickers probability = 5 / 16 = 31%
> 16th bar pulled Snickers probability = 5 / 15 = 33%
> 17th bar pulled Snickers probability = 5 / 14 = 36%
> 18th bar pulled Snickers probability = 5 / 13 = 38%
> 19th bar pulled Snickers probability = 5 / 12 = 42%
> 20th bar pulled Snickers probability = 5 / 11 = 45%
> 21th bar pulled Snickers probability = 5 / 10 = 50%

Stay away from my candy.

This is called *probability without replacement.* The outcomes are dependent on how many bars have already been taken out of the bag. You would have to try 21 times until you get to 50% probability. 11 Tries will

Probability.

get you to 25% probability. Still, if you returned the bars to the bag, you would never have better than a 17% chance of grabbing a Snickers.

Say you picked a Snickers on your first grab.

What's the probability that you'll pull out a Snickers on subsequent grabs if you don't replace the bars you took out of the bag?

> Probability = number of favorable events / total number of events
> 1st bar pulled is a Snickers
> 2nd bar pulled Snickers probability = 4 / 29 = 14%
> 3rd bar pulled Snickers probability = 4 / 28 = 14%
> 4th bar pulled Snickers probability = 4 / 27 = 15%
> 5th bar pulled Snickers probability = 4 / 26 = 15%
> 6th bar pulled Snickers probability = 4 / 25 = 16%
> 7th bar pulled Snickers probability = 4 / 24 = 17%
> 8th bar pulled Snickers probability = 4 / 23 = 17%
> 9th bar pulled Snickers probability = 4 / 22 = 18%
> 10th bar pulled Snickers probability = 4 / 21 = 19%
> 11th bar pulled Snickers probability = 4 / 20 = 20%

Once you do grab a Snickers, the probability that you'll get another goes down from 17% to 14% because there are fewer Snickers in the bag. However, as you remove more bars from the bag, the probability will go up because there are fewer total bars. So, the lesson is: "being greedy is a lot of work!"

You are allergic to peanuts.

What's the probability that you'll pull out a peanut-free bar (i.e., Charleston Chews, Tootsie Rolls, Kit Kats, or Hershey's bars, see Figure 42)?

> Probability = number of favorable events /
> total number of events
> Probability = 18 / 30
> Probability = 60% or 0.6 or 1 out of 1.67.

Watch out for the bars that may have been produced in facilities that also process peanuts.

What's the probability that your son will notice you raided his stash?

> Probability = 1.0 or 100%.

Are you kidding? You wouldn't even need Bayesian statistics to figure this out.

5. CARDS.

I finally got a Snickers and I'm not putting it back.

STANDARD PLAYING CARDS

A standard deck of playing cards consists of 52 cards in 13 ranks (an Ace, the numbers from 2 to 10, plus a Jack, a Queen, and a King) in each of 4 suits—black

Probability.

clubs (♣), red diamonds (♦), red hearts (♥) and black spades (♠). The Ace and number cards are called pip cards. The Jack, Queen, and King are called court or face cards. So, there are 26 **Black** cards, 26 **Red** cards, 13 Club (♣) cards, 13 Diamond (♦) cards, 13 Heart (♥) cards, 13 Spade (♠) cards, 40 pip cards, 12 face cards, and 4 each of the number cards and Aces. Playing cards are *non-disjoint alternatives* because each card has a color, suit, and rank that overlap. However, they can be treated as *disjoint alternatives* if the probability calculation is framed as just 52 different cards.

The next six examples show probability calculations for disjoint events involving cards.

What is the probability that you will draw a **red card** from a complete deck?

> Probability = number of favorable events / total number of events
> Probability = 26 / 52
> Probability = 50% or 0.50 or 1 out of 2.

This is a probability calculation for 26 favorable outcomes out of 52 because there are 26 **red** cards in the deck.

What is the probability that you will draw a club (♣), from a complete deck?

> Probability = number of favorable events / total number of events
> Probability = 13 / 52
> Probability = 25% or 0.25 or 1 out of 4.

Pick a card ... any card.

This is a probability calculation for 13 favorable outcomes out of 52 because there are 13 club cards in the deck.

What is the probability that you will draw a Queen from a complete deck?

> Probability = number of favorable events / total number of events
> Probability = 4 / 52
> Probability = 7.7% or 0.077 or 1 out of 13.

This is a probability calculation for 4 favorable outcomes out of 52 because there are 4 Queen cards in the deck.

What is the probability that you will draw a **red** Queen of hearts (♥), from a complete deck?

> Probability = number of favorable events / total number of events
> Probability = 1 / 52
> Probability = 1.9% or 0.019 or 1 out of 52.

Probability.

This is a probability calculation for 1 favorable disjoint outcome out of 52 because there is only one **red** Queen of hearts (♥).

What is the probability that you will draw a **red** Queen of spades (♠), from a complete deck?

    Probability = number of favorable
    events / total number of events
    Probability = 0 / 52
    Probability = 0% or 0.0 or 0 out of 52.

There are no red spade cards; they are all black. Learning probability is more challenging if you don't play cards. In fact, the same is true of all probability calculations. You have to know something

I'm not counting cards. I'm just a kitten.

about the events in order to know how to formulate the calculations.

What is the probability that you will draw any particular card from a partial deck?

    Probability = number of favorable events / total number of events
    Probability = ???

You can't calculate that probability without knowing what cards are in the partial deck. That's why card games are so challenging unless you are good at card counting.

These examples all involve calculations involving disjoint alternatives, the probabilities of drawing particular cards. That's not how card games are played, though. Take poker, for instance, which involves each player having five cards. There are 2,598,960 possible combinations of five cards in a standard 52-card deck (i.e., no Joker). There are nine patterns of cards, called hands. Each hand has a rank that determines where it stands compared to other hands. For example, a Full-House beats Two-Pair.

You can calculate the probability of getting a particular hand, but it involves more sophisticated math. All those probabilities, however, have already been calculated so that's not a big deal. What's more problematic is that, in a game, there are other players who are also getting cards, so not all cards would necessarily be available to be selected. That's what makes card games interesting. You can consider probabilities but there will always be other factors that prevent the calculations from being absolute. The same is true in media stories you may read that involve calculated probabilities.

Always question how a probability was formulated.

Probability.

TAROT CARDS

There are different types of cards but frequentist probabilities are calculated in the same way. I suppose it's possible that Tarot card readers use Bayesian statistics.

A deck of Tarot cards consists of 78 cards, 22 in the Major Arcana and 56 in the Minor Arcana. The cards of the Minor Arcana are like the cards of a standard deck except that the Jack is also called a Knight, there are 4 additional cards called Pages, 1 in each suit, and the suits are Wands (Clubs), Pentacles (Diamonds), Cups (Hearts), and Swords (Spades).

What is the probability that you will draw a Major Arcana card from a complete deck?

> Probability = number of favorable events / total number of events
> Probability = 22 / 78
> Probability = 28% or 0.282 or about
> 1 out of 4.

This is a probability calculation for 22 disjoint outcomes. The draw is independent because only 1 card is being drawn.

What is the probability that you will draw a Knight from a complete deck?

> Probability = number of favorable events / total number of events
> Probability = 4 / 78
> Probability = 5% or 0.051 or about 1 out of 20.

This is probably a good card.

This is a probability calculation for 4 disjoint outcomes. The draw is independent because only 1 card is being drawn.

What is the probability that you will draw Death (a Major Arcana card) from a complete deck?

> Probability = number of favorable events / total number of events
> Probability = 1 / 78
> Probability = 1% or 0.0128 or 1 out of 78.

If the Death card turns up a lot more than 1% of the time, maybe seek professional help.

Assuming you have already drawn Death …

What is the probability that you will draw either The Tower, Judgment, or The Devil (other Major Arcana cards) from the same deck?

> Probability = number of favorable events / total number of events
> Probability = 3 / 77 (NOTE: there are 77 cards instead of 78 because you have
>   already drawn Death)

Probability.

Probability = 4% or 0.039 or about 1 out of 26.

This is a probability calculation for 77 disjoint outcomes. The draw is dependent because one card (Death) has already been drawn, leaving the deck with 77 cards. If a The Tower, Judgment, or The Devil card does turn up after you have already drawn Death, definitely get professional help. DO NOT use a Ouija Board to summon help.

What is the probability that you will draw Death followed by Judgment from a complete deck on sequential draws?

> Probability = (Probability of Event A) times (Probability of Event B)
> Probability = (1 / 78) times (1 / 77)
> Probability = 0.02% or 0.0002 or 1 out of 6,006.

What are the chances I'll get one of these?

If you draw Death and Judgment consecutively, you are toast. Refer to the second example.

REMEMBER THESE THINGS

So, those are a few examples of how to calculate probabilities based on logical alternatives. Alternatives can be disjoint or non-disjoint, independent or dependent. You just have to identify the correct combination of characteristics to identify the appropriate formula (Figure 4).

The way you calculate a probability will also depend on how you formulate the trial. You have to specify what you want to know very precisely. "Will I win the lottery?" can be formulated as YES-NO, 1 in 2, or as one divided by the number of people expected to buy lottery tickets. The first answer is not very helpful; the second one is.

	Disjoint Events	*Non-Disjoint Events*
Independent Events	**Simple Probability** Number of favorable events *divided by* Total number of events	**General Addition Probability** Sum of the simple probabilities *minus* Probability of events in common
Dependent Events	**Joint Probability** Probability of first event *times* Probability of second event	**Conditional Probability** Joint probability of first and second events *divided by* Probability of first event

Figure 4. Calculating the probability of events.

PROBABILITIES IN LIFE

The last section provided a few examples of how probabilities can be calculated using *logic*. This section provides a few examples of how data and information from

Probability.

the internet can be used to find probabilities. These are all probabilities you might encounter in your life.

PEOPLE LIKE YOU

The section on probabilities coming from data mentioned that there is a 2% chance of there being a brown-eyed (0.34), brown-haired (0.58), non-Hispanic-white (0.63) male (0.49) having A-positive blood (0.34) in the U.S.. You can do this too. It's easy.

Define the characteristics of the person you're interested in, then for each characteristic, search the internet for the percentages of the people in the country that have that characteristic. Multiply all the percentages together and you'll have your probability.

Do this for your own characteristics to see how unique you are. Do it for a friend or someone else you know. Find out your chances of finding Mr. Right.

What's the probability that there is another kitten just like me?

FUN PROBABILITY MODELS

Using the approach to calculating non-disjoint probabilities, you can create your own probability models for fun and profit. The approach is easy. Come up with a decision or event that has several contingent events, decisions, or inputs. These events would be things that you can assign a probability of happening. They just have to happen before the event you're interested in. Then assign a probability to each of those events. Don't worry about finding data on the internet, just make an educated guess. Then multiply all the probabilities together and you'll have the probability of the event you're interested in. This is a bit like how oracles do it.

Don't get hung up on the exact probabilities of the events, their independence, or any other important mathematical consideration. This is for fun.

For example, say you are trying to decide if you should go to the prom. You might consider the following events and their probabilities:

1. You'll have the money to get the clothes and transportation. The probability is 99%. Dad is good for a loan if you need it.

Will I find love at Prom?

Probability.

2. You'll be able to get the night off from work. The probability is 90%. You'll be able to schedule this week well in advance.
3. It will matter that you don't know how to dance. Probability is 40%. You don't know. Maybe yes; maybe no.
4. You'll be able to get a date. The probability is 60%. You have no one in mind. The person you ask might not be your first choice. But in a pinch, you have a cousin who owes you a favor.

Multiply all the probabilities together and you have a 21% chance that you should go to the prom. Consider this your baseline. Look at the events that are least likely to go in your favor.

We have to learn how to dance before Prom.

If you can't dance, ask a family member or friend to teach you. There are people who like to dance everywhere who would love to have you as a partner. Even if you only learn the "jock shuffle" and the "hold-and-sway," it'll be something. You could raise this probability to 90% without too much effort.

If you don't have a date, start asking around early. Most people will go to the prom if somebody would just ask them. Maybe the person who taught you to dance would go. You could raise this probability to 75% just by asking. That makes your new chance that you should go to the prom 60%. You're ready to go.

You get the idea? Now, create a probability model for something you're interested in, like going on a vacation, attending a family event, or getting a job.

SHARED BIRTHDAYS

Have you ever been to a party on your birthday and wondered if anyone else there was also celebrating their birthday? You might have wondered how big the party would have to be for another partygoer to share your birthday.

We all have the same birthday.

As it turns out, the answer isn't in the hundreds, it's 23 people. In a room of just 23 people there's a 50-50 chance that at least two people have the same birthday. In a room of 75 people there's almost a 100% chance that at least two people share a birthday. The math is a bit

 Probability.

challenging even assuming there are only 365 possible birthdays (i.e., ignoring leap years, twins, and a few other things) because the calculations require a *permutation*. If you are interested in seeing the formula, search the internet for the *Birthday Problem*.

CAREERS

Rather than calculating probabilities, sometimes you can find that the probabilities have already been determined and posted on the internet. Here are a few examples.

COLLEGE GRADUATE

The probability that someone will graduate from college varies with the number of years attending, the type of school, and a few other things. The official four-year graduation rate for students attending public colleges and universities is 33.3%. One in three freshmen will graduate. The six-year rate is 57.6%. Keep trying and the probability that you will graduate will go up. At private colleges and universities, the four-year graduation rate is 52.8%, and 65.4% earn a degree in six years.

NFL PLAYER

As you might expect, the percentage of college players that make it into the NFL is incredibly low.

There are over a million high school football players, of which over 300,000 are Seniors. About 70,000 make it into college football, 6.5%. There are about 15,000 college Seniors, of which 6,500 are scouted by the NFL and 256 are eventually drafted, 1.6%.

The NFL consists of 32 teams and each team has anywhere between 65 and 70 players at once (depending on how many players are injured at that time). On game days, there are 1,696 active players in the NFL—1,504 of which are on an active roster—and 384 players on practice squads.

I'm going to be the first feline in the NFL. I want to play for the Panthers, Jaguars, Lions, or Bengals.

With 6.5% of high school Seniors making it into college football and 1.6% of college Seniors making it into the NFL, the probability of a high school player making it to the NFL is about 0.1%, one tenth of one percent. Half of those players last as long as four years in the league.

ASTRONAUT

NASA receives quite a few applications from people who want to be astronauts. Between the 6,300 in 2012 and the 18,300 in 2017, only 8 to 14 applicants are

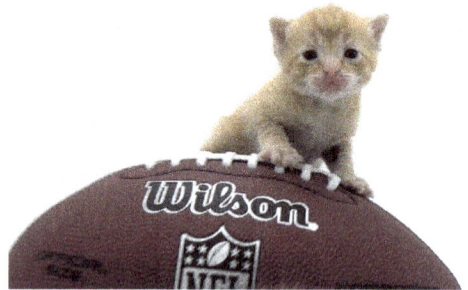

selected for the program. There is between a 0.04% and 0.08% chance of getting selected to go to the next round.

You would have a better chance of becoming an NFL player.

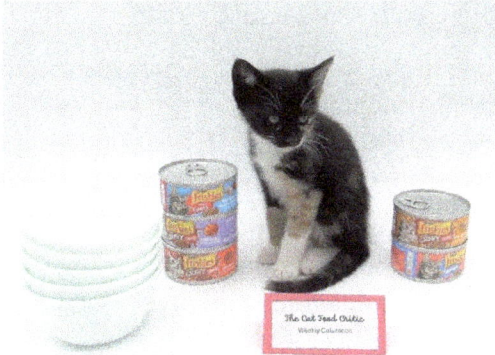

You shouldn't criticize something you don't understand.

AUTHOR

Between 500,000 and 1 million books are published each year, excluding self-published books. Within the book publishing industry, it is agreed that the odds of an author getting their work published stands between 1% and 2%. It beats playing football or going into space.

CHOICES

If you ever watched the game show Let's Make a Deal, you'll remember that contestants would be given the choice of three doors to receive a prize. Their probability of getting the prize was 1 in 3. The host, Monty Hall (1963-1986), would then reveal that one of the doors not selected did not contain the prize. Contestants were then asked to keep the door they had previously chosen or pick the other door.

As it turns out, the probability that either of the two doors contained the prize would not be 1 in 2. Monty Hall's intervention changed the situation from a simple-probability problem into a *conditional-probability* problem because the choices are no longer *random* and *independent*.

One way to think about the problem is this. The probability that the originally-selected door contained the prize would have been 1 in 3 as would the probabilities that each of the other two doors containing the prize (i.e., equal probabilities for all three doors). Once Monty Hall eliminated a door, the probability that the originally-selected door containing the prize still had a probability of 1 in 3 but the probability of the final door containing the prize would become 2 in 3. The probability associated with the third door changed because it is now contingent on the second door being eliminated. Thus, contestants should change their choice and select the third door.

I'm learning stuff from television ... like how to nap.

Although the Monty Hall problem was made famous in 1990 by Marilyn Vos Savant in her Ask Marilyn

Probability.

column in Parade Magazine, it was originally posed in 1975 by Steve Selvin in the American Statistician as a Letter to the Editor called "A Problem in Probability." There are many more detailed explanations of the problem available on the internet than the one presented here. Don't feel intimidated if you don't follow the explanation of the problem. In 1990, even some professional mathematicians failed to understand Vos Savant's explanation and criticized it as incorrect.

Calculating probabilities is easy. Visualizing how probabilities work can be a challenge. Who said you can't learn anything from watching television?

MEDICAL TESTING

Beware, this section gets complicated, both conceptually and computationally. The example involves medical tests that you might get periodically if not routinely, so there is some medical jargon to hurdle. Moreover, the calculations involve probabilities using Bayesian rather than frequentist statistics. if you are just skimming through **Stats with Kittens** for tips and pithy sayings, this might be a section you could skip.

There are thousands of medical tests that patients undergo to reveal the presence of a disease or condition, whether it be for pregnancy, diabetes, intoxication, celiac disease, or Covid. Tests are not 100% correct. Different kinds of tests for the same condition aren't even equally "good" otherwise there would be no need for different tests. Every test has performance characteristics based, of course, on probabilities.

There are two concepts that characterize a test's performance ability—*sensitivity* and *specificity*.

The ability of a test to detect a condition is called the test's *sensitivity* or its *True Positive Rate* (TPR) or its *probability of detection*. It is

That's a lot of tests for one little kitten.

the tests' ability to avoid false positive results (see Chapter 6). The *false positive rate* (FPR) is the probability that the test will produce an incorrect positive result. Sensitivity equals 100% minus the false positive rate. TPRs tend to be above 90%; FPRs tend to be below 10%.

The ability of a test to detect the absence of a condition is called the test's *specificity* or its *True Negative Rate* (TNR) or its *probability of nondetection*. It is the tests' ability to avoid false negative results. The *false negative rate* (FNR) is the probability that the test will produce an incorrect negative result. Specificity equals 100%-false negative rate. TNRs tend to be above 80%; FNRs tend to be below 20%.

While sensitivity and specificity are statistics that characterize diagnostic tests, *Predictive Value* is a statistic that characterizes test results. It is based on test sensitivity and specificity, and incorporates the rate of affliction in the population being tested. It is calculated using *Bayesian statistics* (eponym honoring 18th-century statistician Reverend Thomas Bayes). Positive predictive value (PPV) is the probability that a positive test result really is positive. Negative predictive value (NPV) is the probability that a negative test result really is negative.

Medical statistics (and other disciplines) often employ Bayesian probability. In *frequentist probability,* the probability of an event is the number of times that that event occurs over many trials. Probabilities are based on frequencies (counts) having no uncertainty. Bayesian statistics, on the other hand, are based on probabilities that include uncertainty. Bayesian statistics don't just allow but even invite the use of relevant non-frequency information to estimate probabilities.

To interpret medical test results from a Bayesian perspective, you must know three things:

- **Sensitivity.** The false positive rate for the test.
- **Specificity.** The false negative rate for the test.
- **Prevalence.** The percentage of people in the population who have the disease or condition.

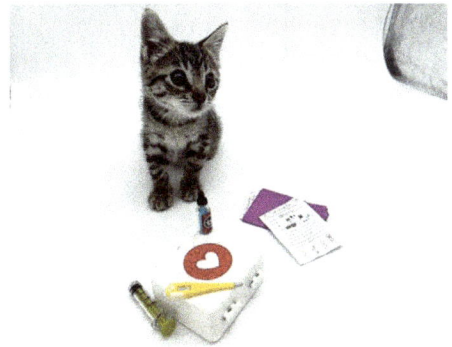

I hope I don't test positive.

The *Bayesian probability* that a positive test is correct that you have the disease or condition, called the *Positive Predictive Value,* is equal to:

$$\frac{\text{Sensitivity} * \text{Prevalence}}{(\text{Sensitivity} * \text{Prevalence}) + \{(1\text{-Specificity}) * (1\text{-Prevalence})\}}$$

The *Bayesian probability* that a negative test is correct that you do not have the disease or condition, called the *Negative Predictive Value,* is equal to:

$$\frac{\{\text{Specificity} * (1\text{-Prevalence})\}}{\{\text{Prevalence} * (1\text{-Sensitivity})\} + \{\text{Specificity} * (1\text{-Prevalence})\}}$$

All of the terms in the formula for the Bayesian probability represent probabilities themselves rather than concrete counts that classical probability relies on.

Figure 5 shows the Bayesian probability that a positive medical test truly indicates the presence of a disease or condition, the Positive Predictive Value, for five combinations of sensitivity and specificity, over a range of prevalence rates in the population. In the chart, the vertical axis represents the calculated Bayesian probability that a positive test is correct, that is, the PPV. The value of 100% on the

 Probability.

vertical axis means that all the tests are likely to be correct. The horizontal axis represents the prevalence of the disease in the population (whatever the population might be). The prevalence axis ranges from 0% to 25%.

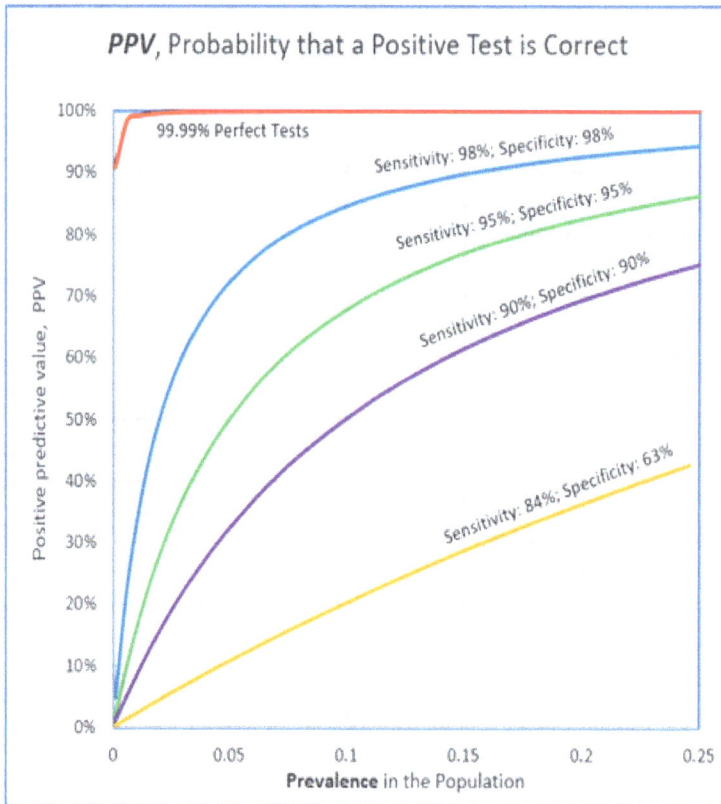

Figure 5. The Bayesian Probability that a Medical Test is Correct.

The five curves represent tests with different sensitivities and specificities. The red line at the top of the plot represents a near-perfect test having a sensitivity and a specificity of 99.99%. The probability of such a test providing a correct positive result would be 100% unless the prevalence was below about 2%. The yellow line at the bottom of the plot represents a test with a sensitivity of 84% and a specificity of 64%, which was from a test for Covid. The probability that a positive result on this test is correct is only 10% if the prevalence is as high as 5%. The remaining three lines represent tests having sensitivities and specificities of 90%, 95%, and 98%.

What the plot shows is that even the best tests available, with sensitivity and selectivity over 98%, will provide incorrect positive results in almost 20% of cases if the prevalence of the infection in the population is less than 5%. Half of the positive test results will probably be false if the prevalence in the population is 1%.

Probability.

In general, the rarer the disease is in the population, the lower the probability that a positive result indicates a real instance of a disease despite a test's high sensitivity. This means that many people who test positive for a disease may not have it at all. Stories of spontaneous recoveries after a positive test result may actually be a reflection of imperfect tests.

And it gets even more convoluted.

People who appear to have a disease and people who don't are not equally likely to be tested because testing resources are limited. This means the population being tested is not the same as the general population. Data on the rate of infection in a population doesn't apply to all the people in the population, but rather, applies only to the people who took a test.

Whaddya you mean I have to take another test?

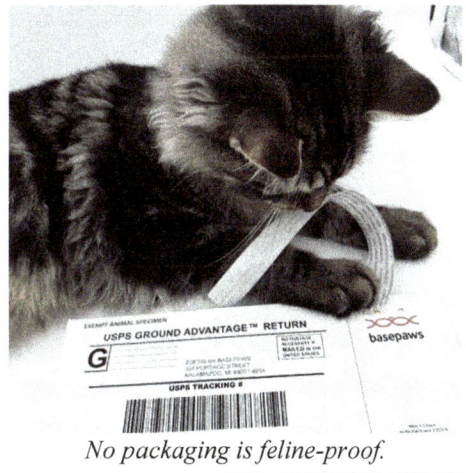

 Furthermore, another assumption is that the tests of the individuals in the population are independent of each other. This is certainly not true. Some individuals are tested repeatedly, like medical workers who are subject to recurring exposures. The prevalence in the population may actually be much different than is estimated because of replicate testing.

Even under the most carefully controlled conditions, individuals present unique challenges to testing. Their peculiarities also introduce uncontrolled variation into results. This is a reason why doctors often want to repeat medical tests they prescribe.

Don't sweat the details, just remember that sometimes probabilities are really, really tough to calculate and even experts are not always correct with their *estimates*.

DNA ANCESTRY

You've probably heard about those home test kits that analyze your DNA and judge what your ancestry might be. How do they do that?

Consumer DNA testing began in the 2000s and millions of people have taken a test. It involves spitting into a test tube or swabbing the inside of your cheek and sending the sample to one of the dozens of commercial testing laboratories.

No packaging is feline-proof.

Probability.

So, will this tell me where I came from?

In the lab, technicians extract DNA from the sample to reveal the individual's *genome*. They measure specific positions on the DNA strands looking for small differences, called *variants*, between individuals. The company that provides the DNA service then compares information from individuals' DNA to information from their *reference panel* on the variants associated with different ancestries. Each company has its own reference panel, developed from public databases and the company's own customer base using, you guessed it, statistical methods.

For humans, reference panels contain information linking DNA variants to individuals having ancestors who lived in specific geographic locations. The assumption is that if an individual carries a certain variant in their DNA, they probably had an ancestor from the geographic areas associated with the variant.

Other species also have DNA that can be tested, including felines. With felines, as with humans, it cannot be assumed that a cat was descended from a mixture of purebred lines. Instead, the parts of a cat's genome that are similar to the genomes of modern day purebred cats are identified. This information is analyzed statistically and compiled into a test provider's reference panel.

One testing company, **Basepaws**, used its genomic reference panel of purebred cats to classify feline breeds into four groups—Western, Eastern, Exotic, and Persian. We used **Basepaws** to test our kitten, Moose.

> *I found myself scrolling through local animal rescues looking for kittens to adopt, when I stumbled upon a litter named "Melody's Kittens." My five year old daughter, who is also named Melody, wanted a kitten so I had no choice but to apply. Melody's kittens' names were all music-themed: Octave, Tempo, Vibrato, Harmony, and Lyric. There was quite a variety in appearance: three were grayish/brown tabbies, one was gray with white socks, and one was all black. Of those colors, three were short haired, and two were medium-length haired. I've never owned a fluffy cat before, so we decided to adopt the more social of the two, Vibrato. The trickiest part of adopting him was coming up with his family name, as all the names in our family, humans included, begin with M. My five year old decided his name should be Moose because it rhymes with FLOOF.*

Basepaws extracted Moose's DNA from a swab sample and performed low-coverage whole genome sequencing (WGS). Then they compared the results to the genomes of

21 popular purebred feline breeds, categorized into the four main breed groups represented in their reference panel.

Moose was assigned the following ancestry:

- 🐾 92.92% Western, including 35.65% Broadly Western, 25.84% Maine Coon, 15.27% American Shorthair, 8.15% Siberian, 5.21% Norwegian Forest Cat, and 2.8% Russian Blue.

- 🐾 5.81% Polycat, representing many generations of mixed breeding between different types of cats.

- 🐾 1.27% Persian.

Human ancestry and pet breed breakdown both rely on reference panels developed using statistical methods. It's an example of the statistical objective of *classification*, in which known DNA information from thousands of subjects is correlated to ancestry or breed. DNA information from new subjects are then classified by comparing the new information to the established information in the panel. The percentage to which DNA variants match allows the probabilities of ancestry or breed to be calculated. It is an example of a probability estimated from *data*.

Understanding heritage has many benefits, from answering questions about origins to anticipating how the subject might mature in time. It is both informative and fun. All brought to you by science.

Give me a minute. I can get this open.

Breeds Group

Western 92.92%

Broadly Western	35.65%
Maine Coon	25.84%
American Shorthair	15.27%
Siberian	8.15%
Norwegian Forest Cat	5.21%
Russian Blue	2.8%
Ragdoll	0%
Abyssinian	0%
Turkish Van	0%
Turkish Angora	0%

Eastern 0%

Broadly Eastern	0%
Oriental	0%
Peterbald	0%
Burmese	0%
Birman	0%
Thai	0%

Persian 1.27%

Broadly Persian	1.27%
Persian	0%
Exotic Shorthair	0%
British Shorthair	0%
Himalayan	0%

Exotic 0%

Broadly Exotic	0%
Bengal	0%
Savannah	0%
Egyptian Mau	0%

Polycat 5.81%

A domestic polycat is a remarkable result of many generations of mixed breeding between different types of cats, which is why the ancestry and origin of these kitties can be very difficult to determine.

✕✕✕ basepaws | Genetic Report

**Figure 6.
Basepaws report
for Moose.**

*It's been so interesting to see Moose grow so FAST and how his medium-length hair has continued to grow long, thick, and silky. His paws feel so beefy yet his body remains rolly-polly. He seeks out water to splash in, from open toilets, to soaking dishwater, to the tub after someone has showered. By 6 months old he appeared the same size as my 3 and 4 year old cats. Despite his mom appearing to be the typical American shorthair, I had to wonder what mysterious genes his dad must have brought to his DNA. The results from **Basepaws** Genetic Report were so exciting, and confirmed my suspicion that Moose had to have some kind of "spicy mix" I was unaware of. My family and I are so excited to see how big he'll grow, what tricks we can teach him, and if he'll ever lose his fascination with water.*

HOW YOU MIGHT DIE

Some people fear death and some people are fascinated by it. If you've ever wondered how likely you are to die from a specific cause, visit the U.S. Centers for Disease Control and Prevention (CDC) website at wonder.cdc.gov. The data are based on the International Statistical Classification of Diseases and Related Health Problems (ICD), which specified codes for 6,131 causes of death.

The CDC database has information on 53,422,612 deaths that occurred in the U.S. from 1999 to 2020 (before Covid). The database is searchable, so you can explore the nine ways you can die from hemorrhoids.

Don't worry, we're just dressed up for Halloween.

Causes of death are determined by medical personnel. There is always a primary cause determined, although sometimes a cause may be listed as "unknown etiology." Sometimes, there are also secondary or contributing causes. This makes it even more complicated for statisticians trying to write about medical topics they know little about.

The most common causes of death were circulatory and respiratory diseases, and cancer, which accounted for two-thirds of all U.S. deaths over the twenty-year period. Table 2 shows the ten most common general causes of death.

Within those general categories, are specific causes of death categorized in the CDC database. Table 3 are eleven of the most common specific causes of death.

Some deaths are not considered to be "natural." Deaths attributable to external causes—like overdoses, traffic accidents, assaults, and suicides—account for 7.5% of the recorded deaths. External causes of death are shown in Table 4.

Probability.

Table 2. Probabilities of general causes of death.

General Causes of Death	Deaths		Probability
Circulatory system	17,855,050	33.4%	1 in 3
Cancer	12,349,887	23.1%	1 in 4
Respiratory system	5,187,487	9.7%	1 in 10
External causes	4,027,488	7.5%	1 in 13
Nervous system	3,054,820	5.7%	1 in 17
Metabolic diseases	2,254,532	4.2%	1 in 24
Mental disorders	2,167,257	4.1%	1 in 25
Digestive system	1,979,096	3.7%	1 in 27
Infectious diseases	1,400,955	2.6%	1 in 38
Genitourinary system	1,337,306	2.5%	1 in 40
Unknown etiology	711,692	1.3%	1 in 75

Table 3. Probabilities of specific causes of death.

11 Specific Causes of Death	Deaths		Probability
Atherosclerotic heart disease	5,207,240	9.75%	1 in 10
Lung cancer	3,217,804	6.02%	1 in 17
Acute myocardial infarction	2,874,877	5.38%	1 in 19
COPD	2,198,462	4.12%	1 in 24
Alzheimer disease	1,683,409	3.15%	1 in 32
Dementia	1,681,537	3.15%	1 in 32
Stroke	1,494,522	2.80%	1 in 36
Congestive heart failure	1,181,942	2.21%	1 in 45
Pneumonia	999,306	1.87%	1 in 53
Colon cancer	878,608	1.64%	1 in 61
Breast cancer	875,417	1.64%	1 in 61

Some external causes may appear to be questionable. For example, the low number of deaths by earthquake may be attributable in part to more seismically-resistant structures in the U.S. or it may be that the cause of death for someone who dies in an earthquake is coded as a different cause, like a fall or a crush injury. In other words, the cause of death may be different because of a *judgment* (see Chapter 3) made by the medical personnel involved. A specific cause might be used if 10 people die in an

Probability.

Table 4. Probabilities of external causes of death.

11 External Cause of Deaths	Deaths	Probability	
Hornets, wasps and bees	1,299	0.002432%	1 in 41,126
Lightning	782	0.001464%	1 in 68,315
Cardiac catheterization	205	0.000384%	1 in 260,598
Spiders	138	0.000258%	1 in 387,120
Use of cannabinoids	132	0.000247%	1 in 404,717
Snakes and lizards	124	0.000232%	1 in 430,828
Human stampede	23	0.000043%	1 in 2,322,722
Crocodiles and alligators	13	0.000024%	1 in 4,109,432
Earthquake	8	0.000015%	1 in 6,677,827
Scorpions	7	0.000013%	1 in 7,631,802
Volcanic eruption	3	0.000006%	1 in 17,807,537

Table 5. Probabilities of rare causes of death.

11 Rare Causes of Death	Deaths	Probability	
Hemorrhoids	352	0.0006589%	1 in 151,769
Migraines	86	0.0001610%	1 in 621,193
Syphilis	5	0.0000094%	1 in 10,684,522
Hepatitis B	3	0.0000056%	1 in 17,807,537
Acne	2	0.0000037%	1 in 26,711,306
Ebola	1	0.0000019%	1 in 53,422,612
Anal spasm	1	0.0000019%	1 in 53,422,612
Second degree sunburn	1	0.0000019%	1 in 53,422,612
Pathological gambling	1	0.0000019%	1 in 53,422,612
Mouth breathing	1	0.0000019%	1 in 53,422,612
Sneezing	1	0.0000019%	1 in 53,422,612

earthquake in the U.S. but it wouldn't be possible if 10,000 people were to die in an earthquake in another part of the world. That situation might require the more general cause of "earthquake" to be used.

The frequency of some causes may be somewhat surprising. For example, deaths attributable to "Mental and behavioral disorders due to use of cannabinoids," which is further classified as acute intoxication (1 death), harmful use (26 deaths), dependence syndrome (18 deaths), psychotic disorder (1 deaths), and unspecified (86 deaths), for a total of 132 deaths. Still, it is more likely for you to die from an insect bite, a lightning strike, hemorrhoids, or a cardiac test than from smoking marijuana. You have to consider the data in context.

Probability.

You have to consider data in context.

Rare causes of death can be somewhat surprising. There were 720 of the 6,131 causes of death that accounted for only one death in twenty years. There were 2,669 causes that accounted for only one death per year, on average.

Table 5 is a list of eleven causes of death that are extremely rare, including migraines, STDs, and acne. Who knew you could die from mouth breathing?

The important thing to remember is that there is a wealth of data available on the internet, covering almost every topic imaginable, that you can use to develop your own probability scenarios.

WINNING

What would a discussion about probability be without some mention of winning events. What is the probability of winning an election, or better, winning the lottery?

ELECTIONS

The probability that a certain candidate will win an election would be based on, but not determined by, polls of voters' preferences for the candidates. There are quite a few factors that affect the legitimacy and the validity of such polls (see Chapter 5).

Polls don't predict outcomes, they only characterize current opinions of people who may or may not actually vote. Political pundits take these poll results and augment them with other information they feel is important, like projections of voter turnout and even the weather.

Pundit forecasts are examples of probabilities produced by an *oracle* because both the data from the polls and judgments concerning non-poll information are combined in some organized manner unique to each pundit.

To estimate the probability of a candidate winning an election, pundits start by identifying one or more legitimate polls from reputable sources. Some pundits use

weighted averages (Chapter 3) of several polls, called *aggregated poll results*. Notable independent poll aggregators include: Real Clear Politics; 270 To Win; FiveThirtyEight; and Ballotpedia.

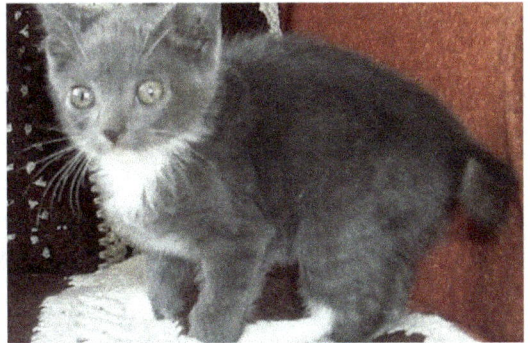

Then the pundit incorporates the poll results into their proprietary prediction models. It's much more complex than just looking at the results of a single poll, which is why they make the big bucks.

Polls don't predict, but they are a good place to start.

Table 6 summarizes examples from the 2016 and 2020 U.S. Presidential

Probability.

elections. A few things to note about these examples. First and foremost, **political preference polls estimate the popular vote not the electoral vote**. This is important because as of 2020, there have been five times in U.S. history that a candidate has won the popular vote but was awarded fewer than 270 electoral votes and lost the election—in 1824, 1876, 1888, 2000, and 2016. To assess electoral votes, polls for each State have to be reviewed. That's a lot of work.

Another pattern to be aware of is that many voters do not decide on who they will vote for until very close to the election (see Chapter 5). Sometimes voters change their plans based on late poll results. In the examples, the proportions of voters who were undecided or changed their voting plans, especially to vote for third-party candidates, accounted for shifts of several percentage points.

To be effective for forecasting election results, polls have to be conducted shortly before an election, no more than a few weeks if not just days. Polls conducted months before an election are useful for candidates in deciding how to allocate their campaign resources but not of much value in forecasting election results.

Finally, forecast election results are examples of what MIGHT happen in the eyes of an *oracle*. The probability forecasts don't carry the same weight as probabilities calculated from *logic*.

Table 6. Aggregator and actual results from the 2016 and 2020 U.S. Presidential elections.

2016 Election	Clinton	Trump	Other
Aggregators	45.2% to 46.8%	41.8% to 44.3%	9.6% to 12.5%
Results	48.2%	46.1%	5.7%
2020 Election	**Biden**	**Trump**	**Other**
Aggregators	50.6% to 51.8%	43.1% to 43.4%	4.8% to 6.2%
Results	51.30%	46.80%	1.90%

LOTTERY

It wouldn't be a Chapter on probability if there were no mention of lotteries. In fact, the probability of winning a lottery game may be the easiest of all probabilities to determine—they're all posted on the official website for the game.

In a *drawing*, all the tickets sold or chances taken are unique; there are no duplicates. A drawing has only one winner for the game. Thus, the probability of winning a drawing is equal to one divided by the number of chances available. That isn't difficult to calculate, you just have to know the number of tickets sold or chances taken.

In a *lottery*, players pick their own chances; they are not unique. A lottery can have more than one winner of a prize. Thus, the probability of winning a lottery is equal to one divided by the number of possible chances.

Probability.

The math for calculating the probability for winning a lottery is a bit more of a challenge. Calculating the number of possible chances involves *factorials*, denoted by an exclamation point (!), which most people don't use after high school. I won't resurrect old memories here; you can look it up if you're interested.

I've got scratchers for these scratchers.

The formula for the number of possible *combinations* in a lottery requiring players to pick **r** numbers from a set of **n** possible numbers is equal to the factorial-of-the-number-of-items (**n!**) divided by the factorial of the quantity (the number-of-items minus how-many-items-are-taken-at-a-time (**r**), all then multiplied by the factorial of how-many-items-are-taken-at-a-time. This is a good example of how mathematical notation can be less confusing than words.

In mathematical notation, **n** is the number of items (the numbers in the set you can pick from), and **r** is the actual number of items that you select from the set. The number of *combinations* (**C**) of **r** items selected from a set of **n** items (i.e., **nCr**) is:

$$nCr = n! / (r! * (n - r)!)$$

Most spreadsheet software provides built-in functions for factorials and even combinations, so the calculations aren't that difficult if you know **n** and **r**.

So for example, in a lottery requiring players to select 5 numbers from a set of 70 numbers, there will be 12,103,014 possible combinations.

$$70C5 = 70! / (5! * (70-5)!)$$
$$= 12,103,014$$

The probability of winning that lottery prize would be 1 in 12,103,014 or 0.00000826%, which is about the same probability as dying from carpal-tunnel syndrome.

Big lotteries have many ways to win besides just picking the correct numbers, though. There are many levels of prizes, each of which have their own probabilities of winning. That's what makes the games engaging. And although you now know a bit about how the probabilities of winning are calculated, it's still just easier to get the probabilities from the website for the games.

EVENTS IN NATURE

Some events are predictable and some are not. Some events are predictable only if the conditions under which a prediction applies are narrowly and precisely defined. That's the case with many events in nature.

Probability.

As with the probabilities involving medicine, probabilities involving events in nature are conceptually complex. Calculations require professional training. Still, it's good to know something about some of the most important probabilities you'll encounter every day of your life.

PRECIPITATION

Think of a weather forecast you might have seen recently on TV or the internet. The statistic everyone wants to hear about is the *probability of precipitation* (PoP). PoP will always be presented as a probability, expressed as a percentage, linked to a specific location for a specific time period.

I better bring my umbrella. I hate getting wet.

For example, "Philadelphia will have a 60% probability of precipitation tomorrow from 1PM to 7PM." This does not mean that it will rain in 60% of Philadelphia or that it will rain 60% of that six-hour time period. What it does mean is that there is a 60% probability that some part of Philadelphia will receive at least 0.01-inch of precipitation (usually enough rain to form puddles) in that six-hour time period. It's a bit more involved than many people understand, which is probably why there are so many complaints about bad weather forecasts.

PoP is determined by multiplying two factors—the forecaster's confidence that rain will fall in the area and the percentage of the area where 0.01-inches of rain is expected to occur. These estimates are derived from complex mathematical models, usually the Global Forecast System (GFS), which is maintained by the United States' National Weather Service (NWS).

Weather forecasting is technically complicated, data intensive, and expensive, which is why the Federal government takes the lead in maintaining the system as a service for itself and for private-sector businesses that use weather information.

Weather models are for the most part mixed deterministic (based on scientific laws), numerical (based on increments in a range of likely values), and stochastic (based on randomly generated data) models (see Chapter 8). They are based on theoretical equations of atmospheric and environmental processes that are solved iteratively for times and locations using current data on a variety of environmental metrics (e.g., temperatures, winds, precipitation, soil moisture, and atmospheric ozone concentration).

These models provide weather-persons with rainfall-probability information related to a specified area. The weather-persons enhance the information to take local conditions into account and reformat the results to make it understandable for consumers. That's why weather forecasts for the same area and time may not be identical for all weather-persons.

Weather prediction is a probability from an *oracle* even though most of the inputs come from *models*. Did you expect anything else? It's a big effort but well worth it.

FLOODING

If you receive too much precipitation, too fast, for too long, you might end up being in a flood. You might hear the weather-person describe it as a 20-year flood, or a 100-year flood, or … gasp … a 500-year flood.

Does that mean it's the largest flood in the past 20, 100, or 500 years? Does it mean if you have a flood that big this year, you won't see another one like it for a long time? No, not at all.

C'mon Magic 8-Ball, give me my expert opinion.

The meaning of an *m-year flood* is that the probabilities of rivers or streams having discharges that large occurring in any year are **1/m**. For example, a 20-year flood will have a probability of 1/20 occurring in any year, a 100-year flood will have a probability of 1/100 occurring in any year, and a 500-year flood will have a probability of 1/500 occurring in any year. This is essential information to have if you live anywhere near a body of water.

So how do hydrologists calculate the probabilities that a flood may occur in a given location? It's a complicated, time-consuming, and expensive process.

First, stream gauges, which measure the height of the water in a river (called *stage*) and the quantity of flow (called *discharge*), are installed and calibrated. The U.S. Geological Survey (USGS) operates more than 10,000 stream gauges nationwide for this purpose. At least 30 years of data on a river's stage and discharge over time are required before they can be used to calculate flood probabilities. Some rivers have been monitored for a century.

Once enough data are available for a river's stage and discharge, *flood frequency curves* are created that depict the relationship between the two metrics for the location of the stream gauge. Then, the average time interval between the occurrence of two discharge events of a given or greater size, called the *recurrence interval,* **RI**, or *return period,* are calculated using the formula

$$RI=(n+1)/m$$

where **m** is the size of the discharge of interest and **n** is the number of years in the data record.

Probability.

The probability of a flood of a given discharge is the reciprocal of the RI (i.e., **1/RI**).

Flood probability = 1/RI

The cumulative probability of an m-year flood occurring over a span of years is equal to:

1-(1-Flood Probability)^{Years})

So, the probability that a 100-year flood (probability of 1/100) will reoccur at least once in a decade is:

$$1-(1-0.01)^{10})$$
$$1-(0.99^{10})$$
$$10\%.$$

Complicated but useful. Still, it's even a lot more involved than that. Flood frequency analysis also uses statistical frequency distributions to fill-in for sparse data records, for instance.

Flood probabilities are essential to know for anyone living near a stream (or insuring someone who does). Flood probabilities are critical to society for designing structures to withstand flood events and for minimizing risks to humans. Nevertheless, they are often misunderstood, and even ignored, by homeowners despite their importance.

Things to remember about flood probabilities are that they are based on historical river discharge data collected over long periods in many places across the country. Thus, the probabilities are based on *models* that use *data*.

You'll never have to calculate the probability of a flood occurring in a stream near you. The USGS and other government agencies do that for you. Even so, you should at least appreciate the work that goes into the determination. And if you think calculating the probability of a flood is incomprehensible, imagine what scientists go through to estimate the probability of a volcano erupting, an earthquake occurring, or a celestial body striking the Earth.

PROBABILITIES ARE EVERYWHERE

Probabilities come from four seeds—logic, data, models, and oracles—sometimes alone and sometimes in combinations. "Shared Birthdays" and "Choices" (Monte Hall Problems) are based on logic. "How You Might Die" and "Careers" (College Graduate, NFL Player, Astronaut, Author) are based on historical data. "Winning" might be based on data (elections) or logic (lottery). Finding "People Like You" is based on data and logic. "Medical Testing" and "Events in Nature" are based on data and models. With "Fun Probability Models," you too can be an oracle.

Probabilities are everywhere in life; you just have to be aware of what to look for.

Probability.

WHAT ARE THE ODDS

Probability is the core of statistics. You'll definitely hear about it in any course in statistics. You might also hear the phrase "what are the odds," although perhaps not in Stats 101.

Odds are the preferred way of expressing uncertainty in gambling. What are the odds of winning the lottery? What are the odds my team will win the game? It sounds just like probability, but it's not. Probability and odds are related but different.

While probability is defined as the number of favorable events divided by the total number of events, odds are defined as:

What are the odds of rolling a strike?

Number of favorable events / Number of unfavorable events

This is the same as saying that odds are defined as the probability that the event will occur divided by the probability that the event will not occur (Figure 7).

Probabilities range from 0 (0%) to 1 (100%). Odds range from 0 to infinity.

To convert from a probability to odds, divide the probability by one minus that probability. To convert from odds to a probability, divide the odds by one plus the odds. Whether probabilities or odds are used depends in part on what is traditionally used—in gambling, it's odds; in statistics, it's probability.

A probability of 0 is the same as odds of 0. Probabilities between 0 and 0.5 equal odds less than 1.0. A probability of 0.5 is the same as odds of 1.0. As the probability goes up from 0.5 to 1.0, the odds increase from 1.0 to infinity.

Probability and odds are expressed in several ways: as a fraction, as a decimal, as a percentage, or in words. Words used to express probability are "out of" and "in." For example, a probability could be expressed as

Crystal ball, tell me who's going to win the big game.

1/5, 0.2, 20%, 1 out of 5, or 1 in 5. The words used to express odds are "-" and "to." For example, odds could be expressed as 1/4, 0.25, 25%, 1-4, or 1 to 4. Figure 7 shows the differences between probability and odds.

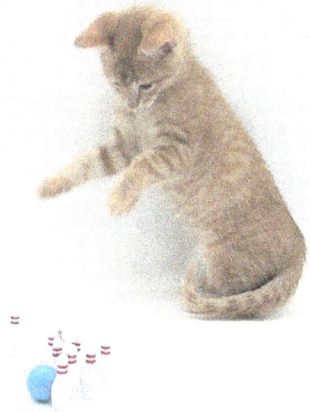

Probability.

There are even more ways to express odds for gambling, but you probably won't even hear about expressing uncertainty using odds in an introductory statistics course. Odds become important in more advanced statistics classes in which *odds ratio* and *relative risk* are discussed.

Odds, like probabilities, come from four sources—logic, data, models, and oracles—but oracles use them the most. Perhaps the best-known examples of odds creation come from the oracles in Las Vegas. Those oracles, called bookmakers, decide what the odds should be for a particular bet.

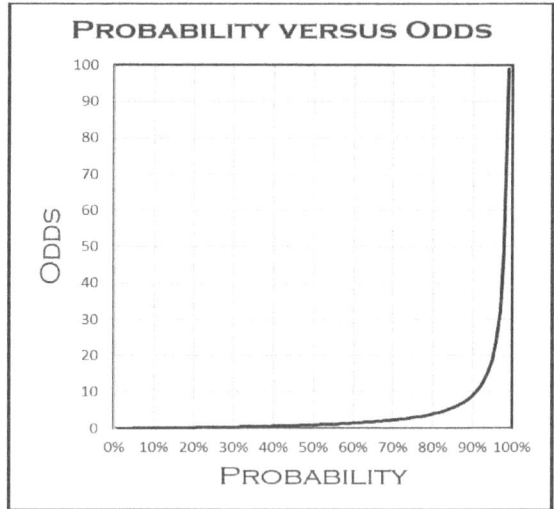

Figure 7. Comparison of odds to probabilities.

Historically, before computers, bookmakers would compile as much information as they could about prior occurrences of an event and the conditions related to the bet, then make educated guesses about what might happen based on their experience, intuition, and judgment. With computers, this job has become bigger, faster, more complicated, and more reliant on statistics and technology.

The field of sports statistics, in particular, has exploded since the 1970s when baseball statistics (*sabermetrics*) emerged. Now, virtually every sport is awash in statistics and probabilities. That knowledge is often incorporated into gambling odds.

Modern gambling odds even fluctuate to take account of wagers placed before the actual event. To keep up with the demand, bookmakers now employ teams of *Traders* who compile the data and use statistical models to estimate the odds. But, you'll never have to know that for Stats 101.

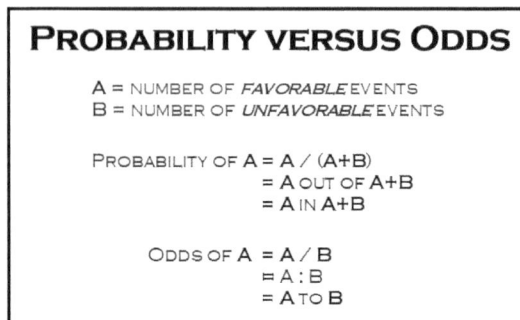

Figure 8. The relationship between probability and odds.

Probability.

Of course, that leaves out a lot of other odds you may have heard about.

> **SPOCK**: *Captain, there are approximately one hundred of us engaged in this search, against one creature* [the Horta]. *The odds against you and I both being killed are 2,228.7 to 1.*

> **KIRK**: *2,228.7 to 1? Those are pretty good odds, Mister Spock.*

The Devil In The Dark,
Star Trek, Season 1, Episode 25, 1967.

PROBABILITY AND MODELS

If calculating the probability or odds of an event were the only ways the concept of chance was used in statistics, you might think that it was a great tool with a well-defined but limited role. The truth is that these uses are barely the tip of a very large iceberg of uncertainty. Probability is actually most often used in statistics in relation to models.

There are two ways models are used with probabilities. The first is as a way to describe the likelihood of obtaining possible values for an event from a mathematical equation (*model*) of data characteristics (*parameters*). You'll hear a lot about a few of the many different *probability distributions* if you take Stats 101, but you'll already have heard about one in high school, the *Normal Distribution*, which is used to grade on-a-curve among other things.

The second way models are used with probabilities is as a way to predict the likelihood of an event from the probabilities of related events (see **Fun Probability Models**). This is just essentially an application of calculating joint probabilities from simple probabilities. This type of probability model isn't used often and you won't see it at all in Stats 101.

WHAT PROBABILITY DISTRIBUTIONS ARE

Probability distributions are fundamental to most statistical analyses. There are well over a hundred different probability distributions with new ones being defined all the time. Most are specialized, applicable to only certain cases, slight variations of more general distributions, or just plain esoteric.

For example, the *Normal distribution* is used to estimate the probability of a *mean* occurring in a certain interval. It's used commonly for a variety of purposes in both descriptive and inferential statistics. A related distribution, the *t-distribution*, is used in place of the Normal distribution when sample sizes are small.

The *Chi-squared distribution* is used to estimate the probability of a population *standard deviation* (rather than a mean) occurring in a certain interval and for conducting *chi-squared tests* of *contingency tables*. Its usage is less prevalent than the Normal distribution but it is not uncommon.

The *Weibull distribution* is used to estimate the time it takes for something to fail. The *Gamma distribution* is used to predict the wait time until a future event occurs. The *Poisson distribution* is used to predict how many times an event is likely to occur. These distributions and many others are used in special situations.

I'm a unimodal distribution with a head on the left tail and a tail on the right tail.

Probability distributions are either *discrete* or *continuous*.

Discrete distributions assume data can only take on certain values, usually integers, like *ordinal scales*. Discrete distributions you may hear about in an introductory course in statistics include the *Binomial* and the *Poisson distributions.*

Continuous distributions assume data can take on any value within a specified range. Continuous distributions you will hear about in an introductory course include the *Normal distribution*, the *Chi-square distribution*, the *t distribution*, and possibly the *Lognormal distribution* and the *F distribution*. There's also the *Uniform distribution,* but that's a bit trivial in that all values have an equal probability of occurring.

Probability distributions are models of ideal data frequencies, defined by equations. Each probability distribution has its own equation, which includes terms called *parameters* that define the distribution.

For example, the parameters for the Normal and Lognormal distributions are the *mean* and the *standard deviation*. The parameters for the Chi-square, F, and t distributions are *degrees-of-freedom*, which is related to the sample size. The parameters for the Binomial distributions are the *number of independent tests* and the *probability of success*. The parameter for the Poisson distribution is the *mean number of events*.

The Normal distribution is by far the most useful probability distribution because of the *Central Limit Theorem*. The Central Limit Theorem says that estimates of the mean of any sample of measurements from a population will be Normally-distributed regardless of the underlying frequency distribution of the measurement

Whaddya mean I have to count them before I eat them?

 Probability.

themselves, so long as enough samples are used. This means that the Normal distribution can be used as a basis for statistical comparisons for virtually any continuous-scale data (see Chapter 6).

Table 7 lists some of the more commonly used probability distributions.

The important thing to understand is that probability distributions are idealized standards. They are used to characterize data distributions so that probability statements can be made about the data. They are like templates or molds. In a statistical analysis, they are surrogates for the distribution of the data.

Probability distributions are idealized standards.
They are only useful if the data fits the distribution.

WHAT PROBABILITY DISTRIBUTIONS ARE NOT

If you take an introductory course in statistics, you'll spend a considerable amount of time calculating the probability that certain values will occur assuming a specific probability distribution (i.e., a model), usually the Normal distribution. It's different from calculating probabilities based on the number of alternatives (i.e., logic) in that the frequencies come from the equations of the theoretical probability distributions. It sounds formidable but it's not that difficult once you understand the procedure.

Table 7. Commonly used probability distributions.

Distribution	Type	Description and Usage
Common Probability Distributions		
Benford	Continuous	Describes the frequency of the first digit of many naturally occurring data. Used in fraud investigations involving data.
Bernoulli	Discrete	Describes the probability distribution of a binary random variable which takes the value 1 with probability p and the value 0 with probability 1-p. It is a special case of the binomial distribution in which a single trial is conducted.
Binomial	Discrete	Describes the number of successes in a series of independent Yes/No experiments all with the same probability of success.
Chi-squared	Continuous	Describes the sum of the squares of $\underline{n}$ independent ,Normally -distributed, random variables, used in goodness-of-fit tests.
Exponential	Continuous	Describes the distance between events in a Poisson process.
F	Continuous	Describes the ratio of two normalized chi-squared-distributed random variables, used in the analysis of variance.
Gamma	Continuous	Describes the time until n consecutive rare random events occur in a process with no memory
Geometric	Discrete	Describes the number of attempts needed to get the first success in a series of independent Bernoulli trials.
Normal or Gaussian	Continuous	Described by the mean and the standard deviation. Also called the bell curve because of its symmetrical unimodal asymptotic shape. It is common because of its association with the central limit theorem in which it is used as a model for the sampling distribution of many statistics.
Poisson	Discrete	Describes the number of successes in a series of independent Yes/No experiments with different success probabilities. Also used to describe a very large number of individually unlikely events that happen in a certain time interval.
t	Continuous	Used to estimating unknown means of Normal populations from small sample sizes.
Uniform or Rectangular	Discrete	Describes data in which all values are equally likely.
Weibull	Continuous	Describes time to failure, time between events, particle size distribution and other processes and events.

Probability.

Calculating the probability that certain values will occur given a certain theoretical distribution only involves reading values from a table and doing some arithmetic. Your Stats 101 textbook and your instructor will explain the process in detail. You'll also do dozens of homework problems to ensure that you understand the process. At least, that's the way probability distributions have been taught for the past fifty years.

There's one issue, though. Nobody except Stats 101 students calculates probabilities from distributions that way anymore. It's all done using software that works with the equations of the distributions rather than using tabulated values based on the equations. Even professional statisticians rely on the calculations provided by software.

So, what's the point of having students do all those calculations? It's so students will understand how theoretical models are used to calculate probabilities. It seems to be a lot of effort when even

I'll probably be able to do this.

spreadsheet software has functions that can do those calculations for you if you know how to set them up.

The reason probability distributions are taught to the extent that they are is that every result of a statistical analysis, every hypothesis test, every regression model, every time a statistician refers to *significance*, a calculation involving a probability distribution was done first. Probability distributions are the whole reason that statistics can make the news with some eye-catching revelation. You won't need to remember how to do the calculation after you complete Stats 101, you'll only need to know how important it is.

What you may not understand from Stats 101 is that calculations involving probability distributions are idealized. If the data distribution doesn't **exactly** match the assumed theoretical probability distribution, which they never do, the calculation will be misleading, maybe by a little and maybe by a lot. It's like if you tried to fill a gelatin mold with something that isn't a liquid. Instead of filling the mold completely, there would be gaps and overages. The results wouldn't look **exactly** like the mold. That's why you have to be cautious about probabilities calculated from a theoretical distribution.

WHAT TO LOOK FOR IN PROBABILITIES

It's one thing when Jim at work spouts off about the probability that the Mayor will get reelected but it's quite another thing when a media pundit makes the same claim. You can laugh at Jim, but there are a lot of people who will believe the pundit

 Probability.

because they don't understand probability. For probabilities and odds related to events like the Mayoral election, consider these questions:

- 🐾 **Is the claim expressed as probability or odds?** That should be easy to tell from the context of the claim.
- 🐾 **How is the probability being used?** If it's for entertainment only, don't worry about it. If it is a critical part of an important argument, ask for information about how it was created. If no information is provided, consider the claim to be unsupported.
- 🐾 **How was the probability created?** If the probability is based on *logic*, consider how comprehensive and convincing the alternatives are. If the probability is based on *data*, find the source of the data and consider the hints about sources provided in Chapter 3. If the probability is based on a *model*, consider the expertise of the source of the model. Organizations with established reputations to protect tend to be more reliable than isolated claimants. If the probability is based on an *oracle*, find out who determined the number. If you can't assess their

Know where the probabilities you see come from.

methods, look at their *track-record* of success. If the probability is based on a *theoretical distribution*, be sure the correct distribution is being used as a reference and, if you can, check to be sure the data fit the distribution.

- 🐾 **What is the claim?** Is the claim a simple expectation in which the events are independent of each other, *disjoint*, and have equal likelihoods of occurring? More likely, the claim will involve events that are at least non-independent, non-disjoint, or don't have equal likelihoods of occurring. That would make the calculation more difficult. Look for some description of how it was done.
- 🐾 **Search the internet**. Look for "What is the probability of _____." You might be surprised at what you find.

More challenging probabilities that you might encounter are the probabilities reported as the results of a statistical analysis. The tipoff is usually when the report talks about *significance*.

A *significant* result of a statistical analysis means that the results are not likely to have occurred by chance. These reports are based on statistical comparisons of samples from a population (more information on this topic is provided in Chapter 6). Determining significance requires the use of a probabilistic distribution model.

For probabilities related to distribution models, consider these questions:

🐈‍⬛ Probability.

- ❧ **What distribution model was used to calculate the test probability?** Usually, the model is the *Normal Distribution*. The reason for this is the *Central Limit Theorem*, which you will learn about in Stats 101 if you take the course. Other common distributions that might be mentioned are the *t-distribution*, the *F-distribution*, and the *Chi-Squared distribution*.
- ❧ **How good is the fit of the data to the model?** This is the key thing you'll need to know and the thing that almost certainly will not be available to you. You would need access to the report on the statistical analysis and look at a *Q-Q plot of residuals* (see Chapter 4). Don't worry about it. Assume the fit is OK but far from perfect, which is usually the case.
- ❧ **What is the calculated probability versus the expectation?** If you are looking for a *false-positive error rate* of no more than 0.05 and the calculated rate is 0.04, feel free to be skeptical. If the calculated rate is closer to 0.001, you can be more assured that the result is truly *significant*.

Evaluating some probability distributions is beyond even Stats 101. It's not something that could easily be put into a succinct and understandable checklist for a non-statistician. Don't worry about it. Some things just require more knowledge than you're likely to have as a statistical kitten.

Don't be discouraged if this sounds incredibly complex … it is. Probability is a topic that takes most people a while to warm up to. Just be aware that the concept of chance pervades everything in life in one way or another. You'll see it everywhere once you learn more about it.

I can tell kittens and toys apart but data all look the same to me.

Probability.

CHAPTER 3. DESCRIPTION.
Either Not-As-Bad-As Or Much-Worse-Than You Imagined.

INTRODUCTION

If you're ever asked to describe a person that you've seen, you wouldn't think twice about mentioning a dozen or more individual characteristics including some that are relatively hard to assess. There's height, weight, sex, age, complexion, hair, eye color, tattoos, and birthmarks. There's clothing, behavior, speech, and odors. But all of these characteristics are second nature; we all have them ourselves. We use them all the time to describe other people. But if asked to describe a dataset, you might panic. Where would you begin?

As it turns out, describing a dataset is easy, much easier than describing a human. But, there's a catch. You have to understand some basic statistical concepts and jargon, things like *population, sample, phenomenon, variable, measurement scales, bias* and *variability, central tendency, dispersion,* and *frequency distributions.*

Once you learn the four characteristics you need to describe a dataset, though, it's hardly a challenge to get it right.

How would you describe me?

That's what this chapter is about.

Note: In many introductory textbooks on statistics, the chapter on data description precedes the chapter on probability. The reason for this may be that some authors believe teaching data description is easier than teaching probability. I can't disagree. In **Stats with Kittens,** though, I put the chapter on probability first because it is a foundational concept of all of statistics. Descriptive statistics depends on probability but probability doesn't depend on descriptive statistics. Read the two chapters in whatever order you prefer.

Description.

PHENOMENA

To describe a dataset, you have to understand its origin as a set of measurements about a *phenomenon* taken on samples from a *population*.

A *phenomenon* is any idea, topic, event, process, entity, condition, or other aspect of reality that motivates a researcher to conduct research. A phenomenon is sometimes called the *subject* of the research except that the term *subject* is also used to refer to an individual in the statistical population being studied. It is sometimes called the *target* of the research except that the term *target* is also used to refer to the statistical population that a researcher studies. Some textbooks don't even use a specific term to refer to phenomena. This approach is confusing to some readers. Consequently, in **Stats with Kittens** and **Stats with Cats**, the term *phenomenon* refers to the reason why research is conducted. It is the reason behind why the targeted statistical population is being studied.

Information about a *phenomenon* is collected from each entity in a *sample* of a targeted statistical *population*. This information is contained in separate groupings of the same measurements. These information groupings are referred to as *metrics*, *measures*, *measurements*, *attributes*, and most often, *variables*. Information for each variable is collected in the same way using specific *scales-of-measurement* and having the same *units-of-measurement*. Variables are what are used in statistical analyses to characterize a phenomenon in a population.

**Variables characterize phenomena
in populations.**

SCALES AND UNITS

Scales-of-measurement are the ways that the values in a set of numbers are related to each other. The actual scale values are called *levels*. For a given scale, the increments between scale levels may all be identical, such as with heights or weights, or vary in size, such as with hurricane categories and geologic time.

Understanding scales-of-measurement is important for a couple of reasons. Using a scale that has too many divisions may lead you to be fooled by the *illusion of precision*. Using a scale that has too few divisions may *dumb down the data*. Most importantly, though, scales-of-measurement determine, in part, what statistical methods might be applied to a set of measurements. If you want to do a

*You don't need to know
my weight to a
hundredth of a gram.*

certain type of statistical analysis on a variable, you have to use an appropriate scale for the variable. You have to understand the scale-of-measurement even just to describe a dataset.

There are a variety of measurement scales and also quite a few terms used to refer to them. Table 8 lists terms describing commonly used scales of measurement. For describing a dataset, though, you usually only need to pick from three categories:

Grouping Scales. Scales that define collections in which the levels have no mathematical relationship to each other. The groups can represent categories, person and place names, numbers on an athlete's uniform instead of a name, IDs, product brands, and other sets of associated attributes. These scales are also called *nominal scales*. Data measured on nominal scales are described only by counts and statistics based on counts, like percentages.

Ordered Scales. Scales that define measurement levels having some mathematical progression or order are commonly called *ordinal scales*. Data measured on an ordinal scale are represented by integers, usually positive. Examples include: year,

Table 8. Commonly used scales of measurement.

Terms Commonly Used to Describe Scales of Measurements				
Scale	Purpose	Key Characteristic	Levels	Usage
Binary	Classification	Only two levels	Two	General, strict
Categorical	Classification	Indicates group membership	Finite and small	General, informal
Continuous	Measurement	Ratio or interval scales	Infinite	General, strict
Count	Measurement	Ordinal scale with a zero but no negative numbers	Finite	General, strict
Cyclic	Measurement	Repeating levels usually of a restricted range	Finite	General, strict
Dichotomous	Classification	Only two levels	Two	Surveys, strict
Discrete	Classification or ranking	Distinct, separate values	Finite and small	General, strict
Grouping	Classification	Indicates group membership	Finite and small	General, informal
Interval	Measurement	Scales that include decimal values without a fixed zero point	Infinite	Strict, defined by Stevens
Likert	Measurement	Ordinal, maybe representative of a spectrum	Usually 3 to 7	Surveys, strict
Location	Measurement	Scales based on coordinates or distances	Usually infinite	General, strict
Nominal	Classification	Indicates group membership	Finite and small	Strict, defined by Stevens
Numerical	Measurement	Scales based on numbers	Infinite	General, strict
Ordered	Ranking	Same as ordinal scale	Finite	General, informal
Ordinal	Ranking	Progression of integer values	Finite	Strict, defined by Stevens
Qualitative	Classification or ranking	Non-numeric scales	Finite	General, informal
Quantitative	Measurement	Scales based on numbers	Infinite	General, strict
Ratio	Measurement	Scales that include decimal values with a fixed zero point	Infinite	Strict, defined by Stevens
Restricted-Range	Measurement	Usually continuous levels between two fixed points	Depends on scale	General, strict
Time	Measurement	Scales based on time units	Usually infinite	General, strict

Description.

letter grade in school, rankings, and weight classes in sports. Counts and statistics based on medians and percentiles can be calculated for ordinal scales.

Continuous Scales. Scales that define a mathematical progression involving fractional levels, represented by numbers having values after a decimal point. These scales may be called *interval scales* or *ratio scales* depending on their other properties.

Examples of ratio scales include durations, concentrations, weights, lengths, areas, and volumes. Interval scales appear to be similar to ratio

Ordinal scales are like steps.

scales but they have no natural zero point and ratios have no physical meaning. Examples of interval scales include temperature, elevation (where the zero elevation is arbitrary), and some time scales. Any statistic can be calculated for data measured on continuous interval and ratio scales.

Four scales-of-measurement that you are likely to hear about in an introductory statistics class are *nominal*, *ordinal*, *interval*, and *ratio*. These scales are commonly referred to as *Stevens' scales* after the person who first categorized them in 1946. One way to view these differences is this:

- **Nominal (grouping) scales** are like stepping stones randomly scattered around a garden.
- **Ordinal scales** are like garden steps. You can only be on a single step, not between steps, and the steps lead progressively upward or downward. There may be many steps or just a few.
- **Ratio and interval scales** are like a garden path or ramp. You can be anywhere along the path, at high levels or low. You can move forward or back, in small or large steps.

Length is a continuous-scale measurement.

Just to be clear, measurement scales are different from *units-of-measurement*. Some pieces of information can be expressed in different units. Distance, for example, can

Description.

be measured in microns, millimeters, inches, furlongs, miles, light-years, and many other units. But, they all have the same scale properties (in this case, the properties of a ratio scale).

The key point to remember is that you can easily convert from one unit to another but not so easily from one scale to another.

There are many other scales, mostly for handling special situations, like:

Information can be measured using different scales, different units, and different devices.

- **Counts**, ordinal scale with a zero but no negative numbers
- **Restricted-Range Scales**, usually a ratio scale but with a finite range. Probability is an example.
- **Repeating-Unit Cyclic Scales**, can be any basic type of scale in which the units repeat, like days of the week or musical notes.
- **Cyclic Orientation Scales**, like degrees on a compass in which 0 degrees and 360 degrees are the same.
- **Concatenated Numbers and Tex**t, though not true scales, can often provide information that can be derived from them, like social security numbers, telephone numbers, sample IDs, date ranges, blood pressure, latitude/longitude, depth/elevation intervals, names, and addresses.
- **Location Scales** can be one-, two-, or three-dimensional, as numbers or as text.
- **Time Scales**, the quirkiest of all scales. Time (as opposed to duration) can be like an interval scale with units in hours, minutes, or seconds but with no natural zero point. They can be linear (e.g., year), cyclic (e.g., day of the week), text (mm/dd/yyyy), or ordinal (e.g., geologic time).

Analyzing these scales requires specialized approaches. Don't worry too much about them unless you really want to get into statistics.

In Stats 101, you'll have assignments in which you'll have to determine which of Steven's scales different metrics are measured on. Here are some hints:

1. If a metric can be measured with a fractional part, it's either a ratio or an interval scale.
2. If the metric has a natural zero point (the zero value for the metric is unique for all units of measurement), it's a *ratio* scale. If the zero point is different in different situations, it's an *interval* scale.
3. If the scale does not have a fractional part but does represent a progression of values, it's an *ordinal* scale.

Description.

4. If the scale does not represent a progression between the levels, it is a *nominal* scale.

VARIABLES

Variables contain the pieces of information you collect from or about each of your samples. Variable values change from sample to sample. That's why they're called variables rather than constants. If the information wasn't different for each observation, there would be no variability and no need for statistics.

There are a few things to consider about variables to be used in statistical analyses (Chapter 8).

Relevance. Variables used in statistical analyses must have some bearing on the research question.

Objectives. Prediction models should use variables and scales that are relatively inexpensive and easy to create or obtain otherwise the predictions will cost more to generate than they are worth.

I can measure temperature with either one, but which would be better?

Number. The number of variables should be kept to an absolute minimum or the analysis will become intractable. Conventionally recognized variables and scales should be used rather than creating new ones. This facilitates replication studies.

Redundancy. Some redundancy should be built into the variables used for statistical modeling if there is more than one way to measure a concept.

Scales. Quantitative variables are usually preferred over qualitative variables because they provide more scale resolution. Redundant variables having different scales are useful only if the variables are not perfectly correlated such that they add no new information to the model.

Variability. Variables should stress precision. Accuracy tends to come easy while precision is elusive. Appropriate controls to limit extraneous variability should be established involving *benchmarks*, accepted standards against which a data value is made, *processes*, activities conducted to generate data values, and *judgments*, decisions necessary by the researchers to create the data.

Measurement. It's best to use conventional, direct metrics for measuring a concept, followed by alternative metrics, indirect measures or surrogates, and lastly, newly developed measures. Measurement devices need to be simple to use, not excessively expensive, and able to be calibrated so the measurements are *repeatable* and *reproducible*.

Description.

Missing Data. Missing data is the bane of statistical analyses. There are only three strategies for dealing with the problem—delete the variable, delete the observation, or replace the missing value with a synthetic value. Not good options.

Variables are types of information. Samples are sources of information. Data are the actual pieces of information.

We are being redundant.

Why is all this important, you ask? Here's why.

All statistical calculations begin with a matrix. A matrix is nothing more than a rectangular array of numbers arranged into rows and columns. To do a statistical analysis, you typically arrange your matrix so that each row represents a different sample, each column represents a different variable, and the cells of the matrix are data. If you don't understand the difference between samples, variables, and data, you won't be able to set up your matrix properly for a statistical analysis.

Variables are the columns of a data matrix. Samples are the rows of a data matrix. Data are the cells of a data matrix.

POPULATIONS, SAMPLES, AND OBSERVATIONS

A statistical *population* is the collection of pieces of information from all the possible items that have the characteristics you are interested in. It's different from what you normally think of as a population, namely all the human inhabitants of a particular country or geographic area. A population in statistics has a different sense.

A statistical population refers to data about all the organisms or items in a group having common characteristics. A statistical population may represent all the heights of students in a school. A statistical population may represent the weights of a bottled beverage that a plant produces in a day. A statistical population may represent all the temperature measurements made on the earth's surface in a specific period of time.

The key elements in the definition of a statistical population are that the population consists of a specific kind of data from all the members of a group having some common characteristic.

We are all different in many ways but we still play together nicely.

A *sample* is a collection of elements from a statistical population. A *good* sample has the same characteristics as the population (at least the characteristics you consider to be important). Such a sample is called a *representative sample*.

Think of going to a food market that offers free samples to encourage you to buy their products. Say the market was offering a sample of pizza. The sample should include crust, sauce, and toppings such as cheese and mushrooms. If you were given only a piece of the crust, you wouldn't have a very good idea of what the entire pizza tastes like. That sample (i.e., the crust) wouldn't have the same properties (i.e., sauce, toppings) as the bigger entity (i.e., the pizza), so it would not be representative of the pizza. In other words, it wouldn't be a good sample.

Good samples are representative of the larger entity from which they are drawn. This is perhaps the most underemphasized concept in statistics. You can't make an inference from a small group of samples to the population from which the samples were taken if the samples are not representative of the population (at least on the important properties).

Now for the really confusing part. As mentioned above, a *sample* is a collection of elements from a statistical population. A sample could consist of any number of elements from the population.

But, the term *sample* has more than one meaning in statistics. A sample can also refer to a single entity from a population on which information is collected. Such an individual sample might also be called an *observation, subject, record, case, individual*, or whatever the entity is, such as a patient or student. Think of this meaning of sample as "*example*."

So, a collection of such individual *samples* is also called a *sample* because it is part of a larger collection of entities called a population. Whether the term *sample* refers to an individual entity or a collection of entities is usually taken from the context in which it is used. Confused? Don't panic. If you can handle the

I want a representative sample of this pizza.

Description.

distinctions between to, too, two, tutu, and ptew, you'll be able to handle the distinction between sample and sample.

NUMBER

Populations and samples are characterized using descriptive statistics that summarize the observed information, called data. The most fundamental piece of information about a dataset is the number of cases or records, also called the *sample size*.

In Chapter 1, one of the myths about statistics that is mentioned is that statistical analyses require many, many samples. The number of samples needed for a statistical analysis is not a simple question. One way to look at it is in terms of how much *resolution* you need. Think of the resolving power of a telescope, or a microscope, or the number of pixels in a computer image. The greater the resolution, the more detail you'll see.

Consider the images in Figure 9. You couldn't make out the image with a resolution of 9 pixels per inch and maybe not even with a resolution of 18 pixels per inch. At 36 pixels per inch, you can tell it's an image of a kitten, even if it is a bit fuzzy. At 72 pixels per inch, the image is sharp and you can tell that the kitten is **Kerpow**. Doubling the resolution again adds little to your perception of the image; it's a waste of the additional information.

81 pixels	324 pixels	1,296 pixels	5,184 pixels	20,736 pixels
9 pixels per inch	18 pixels per inch	36 pixels per inch	72 pixels per inch	144 pixels per inch

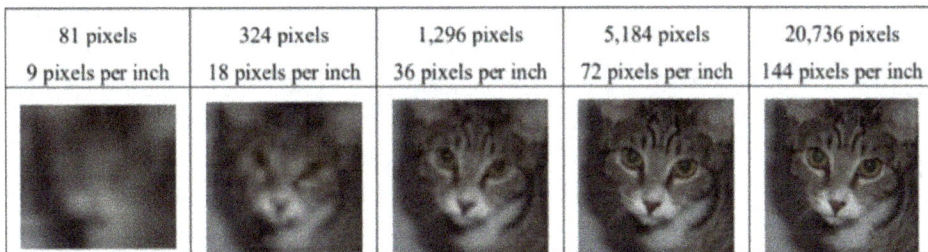

Figure 9. Resolution of a kitten.

Likewise with statistics, the greater the number of samples, the more precise your results will be, but beyond a certain point, adding samples adds little to your understanding. In fact, too many samples can have negative consequences. So, the trick is to collect the fewest samples that will achieve your objective.

Deciding how many samples you'll need starts with deciding how certain your answer needs to be given your objective. Now here's the bad news. There's no way to know *exactly* how many samples you'll need before you conduct an analysis. There are formulas for *estimating* what an appropriate number of samples *might* be and *rules-of-thumb* for when the formulas don't apply.

Say all you want to do is to collect enough samples to calculate some descriptive statistics. Maybe you want to characterize some condition, like the average weight of a litter of kittens or the average age of the players on your favorite sports team. How many samples do you need? Well if your population is small enough, like five kittens or 25 baseball players, you simply use all the members of the population.

But what if you want to calculate descriptive statistics to characterize a large population? The number of samples you'll need to describe it will depend on the *precision* you want, not the accuracy. The greater the number of samples, the more precise your estimate will be and the more it will cost to collect, process, and analyze the data.

Just as you can eat too many potato chips, you can have too many samples. A large number of samples may present challenges, like bigger efforts to process and analyze the data, less familiarity with influential data points, and too many data points to plot cleanly. Plus, collecting unneeded data points is a waste of money.

> *Samples are like potato chips. You're never satisfied with just one. Every one you take makes you want more. And, you're never sure you've had enough until you've had way too many.*
>
> C. Kufs, Making Sense of Statistical Models (course notes), 1989.

REPRESENTATIVENESS AND SAMPLING BIAS

What's really more important than the number of data points is the quality of the data points, characterized by their representativeness. *Representativeness* is a fundamental characteristic of a sample, even though it can't be measured. An individual sample must fairly represent the attributes of the population it is part of or else statistics calculated from the sample will not fairly characterize the population. Freedom from bias in sample selection helps ensure representativeness.

In an ideal world, the experimenter of a study would have the ability to select any of a limitless number of candidates. In reality, though, there are always two constraints:

🐾 The number of candidates is sometimes very limited.

🐾 Some candidates can refuse to participate.

Samples are like potato chips.

Consider the examples shown in Figure 10 of constraints on sample selection.

Some types of samples are almost entirely under the experimenter's control, such as samples of environmental media (e.g., air, water, soil), manufactured products, and

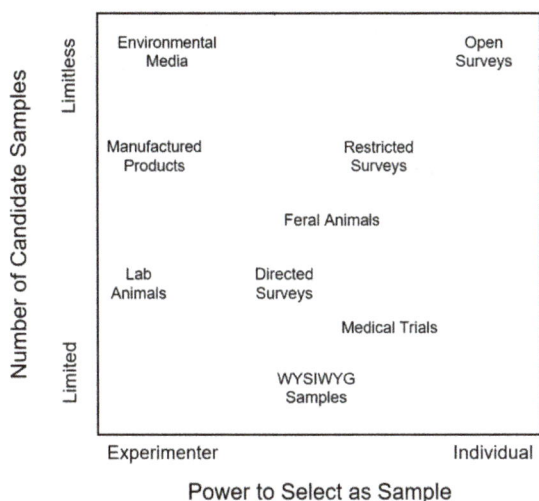

Figure 10. Constraints on sample selection.

lab animals. In contrast, feral animals have some choice in that they can avoid traps. Candidates for medical studies and surveys can choose not to participate. In open-invitation studies, experimenters give up all control over selection and allow anyone to participate. WYSIWYG (what-you-see-is-what-you-get) samples are in such limited supply that experimenters have no choice but to take whatever is available. Patients with rare medical conditions are examples.

This classification is important because of the possibility of bias. Bias can come from either the experimenter or the subject. The experimenter is the anchor of impartiality, at least in the ideal world. If there are enough candidate samples to pick from, an experimenter is supposed to select an unbiased sample. But, even the best researcher can unintentionally and unknowingly bias a selection of samples. That's why it's essential to use a sampling approach that provides some randomization.

Compared to experimenters, individuals will do whatever suits them. An open invitation to a web survey is almost guaranteed to be a biased sample, especially if it involves politics, religion, sports, or cannabis. In general, the more control subjects have in determining their participation, the more likely there is to be some selection bias.

The best way to minimize the risk of sampling bias is through the use of probability-based sampling schemes.

SELECTION

So, the first principle an experimenter needs to follow in selecting candidate samples is to avoid bias, preconceived notions, or any other expectation they might have.

There are a variety of strategies for supporting this principle. *Organizational strategies* include: selection of qualified individuals for the research team; training for members of the team on study details; preparation of research plans and schedules; oversight and approvals by committees; and reviews by collegial and independent peers. *Procedural strategies* include *standard operating procedures* (SOPs) and *quality control* (QC) measurements. *Experimental strategies* include *placebos*, *control groups*, and *blinding*.

Description.

We're samples from the litter.

The second principle that an experimenter needs to follow so that samples adequately represent the population is to ensure that every entity in the population has an equal chance of being selected. To this end, an experimenter must have a clear definition of the target population. From that, there needs to be a plan that will introduce randomization while still minimizing variability attributable to sample selection. It's a formidable task.

Sometimes, individual members of a population can't be identified so they can be selected. In these cases, a *frame* is used. A frame identifies sources for the items or individuals that may become part of a sample. For example, if every potential user of a company's new consumer product can't be identified, the company's registration lists for similar products could be used as a frame. A frame has to correspond to the focus of the study. From the population or the frame, a statistical sampling scheme is used that will provide a reasonable assurance that the samples selected are representative of the population.

PROBABILITY SAMPLING

There are a variety of sampling schemes, some of which aim to ensure that every entity in a target population has a known and equal probability of being selected. These schemes are known as *probability sampling schemes*. The most common strategies used for probability sampling are:

- Random Sampling
- Stratified Sampling
- Cluster Sampling
- Systematic Sampling

Table 9 provides examples of how these sampling schemes might be used separately or in combinations for spatial sampling, surveys, and manufacturing studies. Although these schemes are the most commonly used, there are others. There are scores of books and websites on statistical sampling that provide additional detail on these and other sampling schemes.

Table 9. Examples of commonly used sampling schemes.

Sampling Scheme	Example of a Spatial Array of 25 Samples Using the Scheme	Example of a Telephone Survey of 25 People Using the Scheme	Example of a Manufacturing Study of 25 Gizmos Using the Scheme
Systematic	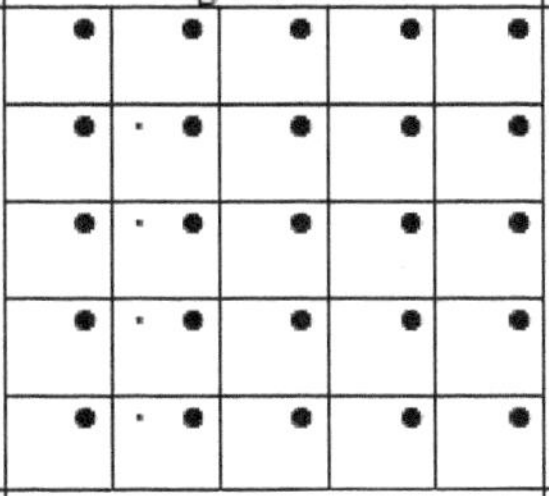	Select every 10th name from a company's five-page telephone list (frame) of 250 customers (100 past customers and 150 current customers).	Select every 10th gizmo coming off the assembly line until 25 gizmos are selected.
Random	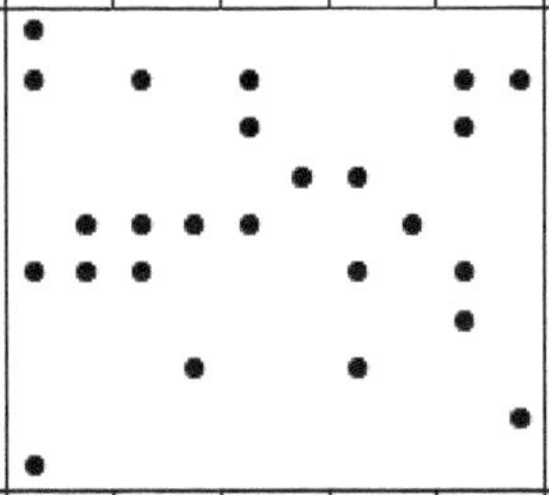	Number all 250 names on the list and use a random number generator to select 25 unique numbers corresponding to customer names.	Use a random number generator to decide how many minutes to wait between selecting gizmos.
Systematic-Random	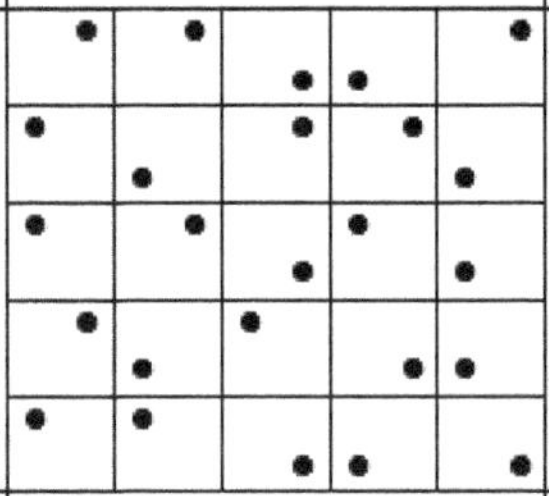	Randomly select 5 names from each of the 5 pages of the telephone list.	Select 1 gizmo from the production line every hour by using a random number generator to pick at which minute of the hour the gizmo should be selected.
Cluster	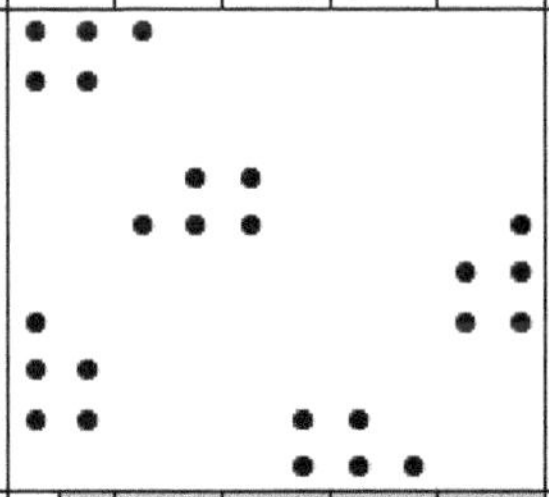	Sort the phone numbers by area code, then select a roughly equal number from each area code.	Select 8 gizmos from each of the 3 shifts of workers
Stratified-Random (40%) and Stratified-Systematic (60%)	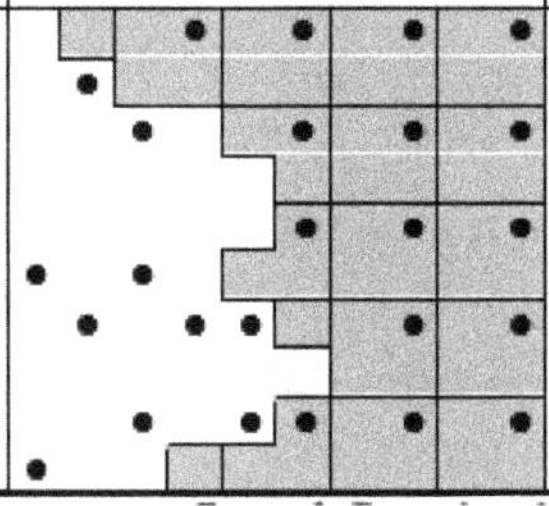	Randomly select 10 names from the 100 past customers and systematically select 15 names from the 150 current customers.	Randomly select 10 gizmos produced by experienced staff and systematically select 15 gizmos produced by newly-hired staff

RANDOM SAMPLING

Random sampling involves picking samples purely on the basis of some randomization tool, such as a table or software algorithm. Random sampling is incorporated in some form in all probability sampling schemes.

STRATIFIED SAMPLING

Stratified sampling involves subdividing the population into parts or groups, called *strata*, and then randomly selecting samples from each stratum in proportion to the number of candidate samples in the strata. You need to know a lot about your population to use stratified sampling effectively.

For example, if you were going to conduct a one-hundred-person survey of members of a gym having six hundred male members and four hundred female members, you could divide candidates for the survey by sex, and then randomly select sixty male gym members and forty female gym members. Stratified sampling is also called *proportional sampling* because the number of samples selected from each stratum is proportional to the size of the stratum.

CLUSTER SAMPLING

In *cluster sampling*, you select locations where you will sample and then randomly select individual samples at those locations. In the gym example, you might randomly select the weight room and the pool area from the ten exercise areas at the gym, and then randomly select fifty participants from each of those two areas.

Cluster sampling is also called *two-stage sampling* because there are two random selection steps, first clusters and then individuals. For comparison, *strata* are an inherent property of the population whereas *clusters* are coincidental or experimenter-imposed groupings of candidate samples.

Sample our clusters.

Composite sampling, also called *three-stage sampling*, involves taking the samples from a cluster location and compositing them by physically mixing the samples (such as with environmental media) or mathematically averaging the measurements on the samples. For example, you could collect three water samples, mix them together, and conduct just one analysis. Or, you could analyze all three samples and average the results of the tests. Compositing samples is cheaper but averaging results provides more flexibility in analyzing the data.

Description.

SYSTEMATIC SAMPLING

Systematic sampling is based on the assumption that the underlying population is randomly organized so that samples can be selected in a fixed pattern (i.e., systematically) because further randomization is not necessary. In the gym example, you might systematically select every third person to enter the gym until you had surveyed one hundred members. For this to provide a representative sample, the arrivals must be random and independent of each other, which is not very likely.

Grid sampling is a form of systematic sampling used extensively in environmental studies in which a grid is placed on a site being investigated and samples are selected in each grid cell.

If the samples are selected at the same relative location in each cell, the scheme is called a *systematic-grid sample*.

If the samples are selected randomly within each cell, the scheme is called a *systematic-random sample*.

If the grid size is designed to identify entities or properties that cannot be seen based on a probability of their occurrence, the scheme is called *search sampling* or *spatial probability sampling*.

Search sampling requires a LOT of potato chips.

NON-PROBABILITY SAMPLING

Non-probability sampling involves selecting a sample from a population that does not rely on random choice, but instead, uses subjective judgment, convenience, or other intentional criteria to pick participants. Thus, available samples do not have an equal and known probability of being selected so that there is no way to control and quantify statistical errors.

Non-probability sampling is used where random sampling can be difficult or impractical, where sample sizes are limited by cost or availability, and where focused results are needed and generalizing to a statistical population is not a goal. It is often used in preliminary studies to test measurement instruments and procedures or fine tune hypotheses before committing to large-scale studies. It is also used to target groups having specific characteristics or are hard to access, especially in qualitative and exploratory research.

Six types of non-probability sampling are:

🐾 Judgment sampling

Description.

- Surrogate sampling
- Convenience sampling
- Quota sampling
- Snowball sampling
- Self sampling.

JUDGMENT SAMPLING

Judgment sampling is also called *purposive, purposeful,* or *judgmental sampling*. It involves selecting specific entities because of some characteristics they have. The entities may have expert or specific knowledge of a study's topic, represent typical or extreme cases, unique or critical cases, or minimum or maximum variation. Examples include selecting participants for focus groups or expert panels because of their experience or knowledge, patients because of a medical condition they have, or employees based on their prior performance. This method is often used in qualitative research when there is a need to focus on a specific type of entity, experience, or characteristic.

SURROGATE SAMPLING

Surrogate sampling is also called *proxy sampling* and *double sampling*. It involves using an easy-to-select sample or metric as a substitute for a difficult-to-select sample or metric. Surrogate sampling requires the experimenter to demonstrate that the responses from the two types of samples are equivalent, thus the reason it is called *double sampling* because two entities or metrics have to be analyzed. For example, parents are often surveyed to assess their children's preferences. Field-test kits are used as surrogates for expensive laboratory tests to control costs.

We're surrogates. She's easy to catch but I'm not.

CONVENIENCE SAMPLING

Convenience sampling is also called *availability sampling.* Samples are chosen entirely because of their accessibility. It is simple and inexpensive to implement so it is often used in preliminary or exploratory research. For example, some consumer surveys target potential customers simply because they were passing by a brick-and-mortar location of a business. Sometimes the strategy has a negative context, such as when unsupervised sampling technicians collect environmental samples close to access ways rather than where a sampling plan specifies.

Description.

QUOTA SAMPLING

Quota sampling is sometimes called *consecutive sampling* and *total enumerative sampling*. It entails identifying characteristics or conditions that are important to the study and sampling entities until a predetermined number having that characteristic or condition is reached. For example, in a habitat study, a researcher may set quotas for samples of each species to be collected. In a study of a school system, a researcher may set quotas for student participants based on grade-level, sex, and so on.

I'm hiding. Who told you I was here?

SNOWBALL SAMPLING

Snowball sampling is also called *chain-referral sampling.* It involves seeking referrals from sampled entities to other entities having some similar characteristics. The strategy is used to access hidden or elusive populations, such as communities marginalized because of rare, stigmatized, or embarrassing medical conditions and behaviors or social, economic, or legal circumstances. It is most commonly used in social-science research. The same concept is used in recommend-a-friend programs.

SELF SAMPLING

Self sampling is also called *volunteer sampling.* It involves an investigator publishing an open invitation to participate in a study and allowing individuals to select themselves without any review by the investigator. Self-sampling is mostly limited to informal surveys' like those conducted on social media for entertainment.

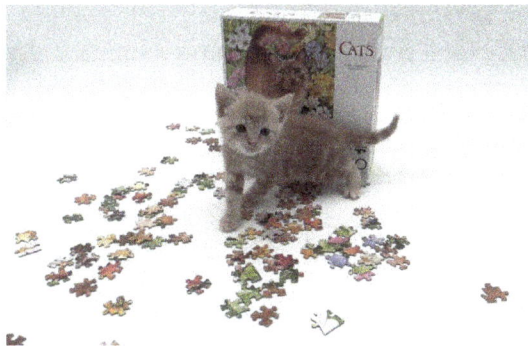

The picture on the box is the population. The puzzle pieces are samples.

Description.

DATA

Measurements about a *phenomenon* taken on *samples* from a *population* result in *data*. Everyone knows you can't do statistics without data, but that's not a problem. Data are everywhere. If data were water, we would all drown in the greatest flood in history.

A *datum* is just a piece of information, usually expressed as a number. A datum represents a specific *measurement* of a specific *phenomenon*, unlike an *anecdote*, which is a short description of one or more phenomena.

More than one datum taken together is called *data*. That's the term most people will use most of the time.

Data formatted in a similar way and stored in a computer file is called a *dataset* or *data set* depending on the spell checker. Statisticians may not take much notice of a datum but they'll definitely want to look into a dataset. After all, their mission in life is to convert data from those datasets into *information* and then *knowledge*.

We're all in this together. We must be data.

Data analysis is the process of examining *data* and *metadata* to identify patterns and relationships. Through this process, information becomes knowledge. *Knowledge* is an appreciation of what a collection of information means, in other words, how the facts are interrelated. If you combine information in just the right way, you can create something greater, like chemical compounds put together in the right way can create a cell with a nucleus in which chromosomes and mitochondria correspond to pieces of related knowledge.

If *data* are like atoms, *information* is like chemical compounds, composed of many atoms yet still a building block for something more complex. *Knowledge* is the complex objects built from the building blocks of information. *Wisdom* is when those objects combine beneficially.

Think of us as DATA. We're a challenge to manage.

Description.

COMMON TYPES OF DATA

Data can take a variety of forms. Some are readily amenable to statistical analysis and some are better suited to other methods of analysis. Some are *structured*, that is, easily formatted into the form of a matrix, Some are *unstructured,* needing extensive preprocessing before analysis. Some are *semi-structured*, needing some preprocessing but having elements that facilitate reformatting.

Table 10 summarizes twelve types of data you'll probably see if you look—the *Data Dozen.*

Table 10. The *Dirty Dozen* of data types.

Source	Data Type	Description	Examples
Device	Automatic Measurements	Information generated by devices, usually electronic or mechanical, that operate without human involvement (other than calibration and sample introduction).	Thermocouples, strain-gage scales, electronic meters, medical tests, EKGs, river discharge monitoring; IoT, satellites
Device	Manual Measurements	Information generated by devices that require human involvement to carry out the measurement.	Rulers, calipers, thermometers, balance-beam scales; medical examinations, DNA
Device	Electronic Recordings	Information stored on the internet and personal devices	Videos, audio recordings, photos, false-color images; x-rays, MRIs, CAT scans, online databases
Experimenter, Data Generator, or Data Analyst	Analog Data	Information from a source that resembles in some respect a phenomenon under investigation, i.e., a model	Experimental lab animals, wind tunnel tests; CERN
Experimenter, Data Generator, or Data Analyst	Metadata	Data about data— their origins, qualities, scales, and so on.	Time, location, and method of data generation; sampling technician ID; metric units; QA/QC information
Experimenter, Data Generator, or Data Analyst	Transformations	Information created from other information.	Percentages, sums, z-scores, ratios, and so on.
Individual or organization	Directed Responses	Information receives as the result of a specific direct inquiry.	Surveys, focus groups, interrogations, medical histories; educational testing; media interviews
Individual or organization	First Person Reports	Descriptive, qualitative information derived from a first-person encounter	Eyewitness accounts; courtroom testimony; customer comments, social media, books, blogs
Individual or organization	Secondhand Reports	Information summarized or retold by a second party based on first-person accounts.	News stories; police reports
Individual or organization	Conjectures	Information created from thought experiments rather than physical experiments.	Expert opinions; podcasts
Individual or organization	Archived Records	Information generated by an identifiable person or organization	Government records, financial data, personal diaries, logs, notes, scraped websites
Individual or organization	Unverified Reports	Information, written or retold, which cannot be disproven or verified.	Anecdotes, stories, legends; ancient texts

Description.

Data analysts use all these data types. Statisticians prefer to use data types that provide many observations so they can assess variability. Scientists and engineers may be satisfied with the results of a single, albeit well-controlled, experiment. They are deterministic breeds. Courts want every piece of evidence to be attested to by an individual, whether an eyewitness or an expert witness. They want to be able to cross-examine witnesses. Scientists prefer to leave human biases out of their data. Historians don't usually have eyewitnesses so they rely on reports, especially secondhand and even unverified reports. They'll use whatever they can find.

DATA SOURCES

Data sources and the data themselves can be classified as primary, secondary or tertiary:

- 🐾 A *primary source* creates original information called *primary data*.

- 🐾 A *secondary source* recycles *primary data* (or other *secondary data*) to create new information called *secondary data*.

- 🐾 A *tertiary source* compiles *primary data* and *secondary data* so it is more readily available.

For example, Melody likes taking pictures of her kitten, Moose. She is the *primary source* of information about Moose. Her photos are *primary data* about Moose.

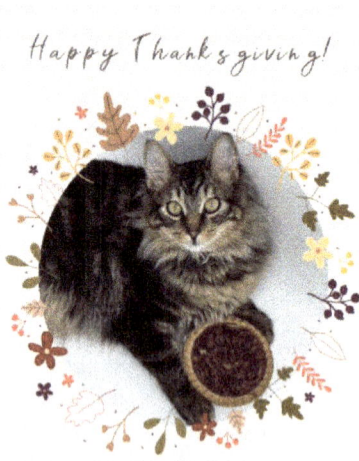

Melody's mother, Miranda, took one of the pictures of Moose and made it into a meme. Miranda is a *secondary source* of information about Moose. The meme she made is *secondary data* about Moose because it puts Melody's original picture of him into a new contextual framework.

Miranda uploaded her meme to Instagram. The Instagram link (www.instagram.com/statswithcats/) is a *tertiary source* of information about Moose because it is a compilation of memes featuring Moose and other kittens and cats.

Governments, for example, are primary, secondary, and tertiary sources of information. They are primary sources when they collect new information, such as through a census of their citizens. They are secondary sources when they analyze that census information and create projections and extrapolations. They are tertiary sources when they archive the information in documents and on websites to make it available to others.

🐈 Description.

Just because a source is classified as primary, secondary or tertiary does not necessarily reveal its quality. A primary source may provide misinformation. A secondary source may provide a biased analysis of high-quality information. A tertiary source may not be comprehensive or consistent.

Considering the source of information is sometimes as important as the information itself when the primary information is unavailable, such as in many surveys.

DATA FORMATTING AND SCRUBBING

Data can be generated in a variety of forms, including written text, spoken words, audio recordings, images and videos, device readings and transmissions, computer files, and internet downloads. As a consequence, datasets always require some processing. The two major types of data preprocessing are *formatting* and *scrubbing*.

Formatting involves creating a matrix in the software to be used for statistical analysis so that variables define columns, observations define rows, and data reside in the cells.

Data scrubbing involves correcting, replacing, or removing: invalid data; incorrectly-recorded data; incorrectly-coded data; data quality information; missing data; extraneous data; dirty data; useless data; invalid fields; out-of-spec data; out-of-bounds data; messy data; corrupted data; and mismatched data (see Chapter 9).

SPECIAL TYPES OF DATA

Certainly, different types of data are not all comparable even though they all might have value. *Big data* have a high upside for helping to solve complex problems but are challenging to manage and analyze. *Quality-control data* don't enter into a statistical analysis but they can raise red flags to other data that may not be appropriate for analysis. *Metadata* also don't enter directly into a statistical analysis but they provide important context for the data that are analyzed.

BIG DATA

What's the big deal about *big data*?

The conceptual foundations of big data began to evolve in the 1990s with advances in data storage and the explosion of data sources in the *Internet of Things* (IoT). By the 2010s, the term *Big Data* was coined to refer to datasets that were notable for their extraordinary size (*volume*), how rapidly new data were generated (*velocity*), and the many ways the data were structured (*variety*).

Before long, big data was also characterized for its quality and reliability (*veracity*) and its potential benefits for solving complex problems (*value*). These were called the *5 Vs of Big Data*. Over the next decade, five more Vs were added to the list of characteristics that define big data:

1. *Volume*, the amount of data.
2. *Velocity*, the speed at which data is generated or changes necessitating new collection and analysis.
3. *Variety*, the diversity of the formats of the data.
4. *Veracity*, the accuracy, quality, and reliability of data.
5. *Value*, the benefits the data might provide.
6. *Validity*, how appropriate the data are for the intended purpose.
7. *Volatility*, how quickly data elements age to the point that they are no longer useful and have to be removed from the dataset.
8. *Variability*, the inconsistency of the data and the presence of extreme values.
9. *Visualization*, how challenging the data are to graph and analyze.
10. *Vulnerability*, how susceptible the data are to data breaches and other security concerns.

Examples of data types that may fall in the definition of *big data* include: internet content and user logs; communications data (emails, texts, audio calls); industrial and personal IoT data; economic, financial, e-commerce and business-performance data; healthcare records and biometric data; environmental sensor data; satellite data and imagery; and government reports, records, and data.

Over time, increases in the volume, velocity, variety, and vulnerability of big data have continued to be issues, as expected. What perhaps came as somewhat of a shock initially, though, was the inadequacy of existing software to manage and analyze the data, that is, *visualization* (i.e., the characteristic of big data, not the term for statistical graphics described in Chapter 4).

I'm a really BIG kitten.

Compared to the datasets that were traditionally managed in a statistical analyses, big-data datasets required special hardware and software. Not only were the datasets large but they were often also *unstructured*, meaning that they didn't have consistent formats and required extensive processing to convert them into a matrix form suitable for analysis. Even in a structured format, big data datasets were too large to be handled by available statistical software so new strategies and software had to be developed.

Description.

With the rapid growth in data from social media, image analysis, audio and video processing, business analytics, and automated sensor data, big data has emerged as a hot topic in *data science*. That's why it's such a big deal and will get bigger in the future.

> *An anecdote is a kitten gently licking your face with a warm, wet, raspy tongue.*
>
> *Big data is a three-inch, high-pressure firehose held an arm's length away.*
>
> *They have to be treated quite differently.*
>
> C. Kufs, **Anecdotes and Big Data**, posted on 9-27-2020 at statswithcats.net.

QUALITY CONTROL DATA

The acronym *QA/QC* refers to *quality assurance* and *quality control.*

Quality assurance is a system of activities undertaken to ensure that the data will be of a level of quality appropriate for the intended use. QA components might include staffing, training, standard operating procedures, checklists, audits, and documentation.

Quality control refers to the tests and other activities undertaken to ensure that the quality specifications are fulfilled.

Did you bring the checklist?

In brief, QA focuses on the data generation process; QC focuses on the resulting data. QA information is usually included in *metadata* or other documentation of the study. QC data is usually kept close to but not in the same matrix of data to be analyzed

The most commonly used QC samples are *replicates*, *blinds*, and *blanks*.

Replicates are samples that attempt to collect the same information multiple times to evaluate consistency in the data generation process. Usually, replicate sampling involves collecting multiple physical samples for analysis by one laboratory. However, the concept can also be extended to other types of data generation. For example, soliciting the same information on a survey through the use of differently worded questions. The idea is to see if there are biases or extraneous variability that might be problematic. Replicate samples may be called *duplicates* (two samples) or *triplicates* (three samples) depending on the situation.

Blind samples are replicate samples that are given unique identifiers so they appear to be unrelated. Blinding allows a data generation process to be tested without the data generators (e.g., laboratories) knowing they are being tested. While replicates might be sent to the same or different laboratories for analysis, blind samples are usually sent to the same laboratory.

 Description.

Blank samples are specially-prepared samples used to test a single aspect of a data generation process. In sampling involving laboratory analysis, blanks might be used to assess any contamination that might have been introduced during sample collection, handling, preservation, transport, or the analysis itself.

Other QC samples are used to assess laboratory performance involving methods, reagents, glassware, and analytical hardware.

QC data are collected for the sole purpose of determining if the process of information collection is accurate and reliable. These samples are not included in a statistical analysis but are evaluated separately, usually during data scrubbing.

METADATA

Metadata are data about data. They are the information about how each datum in an analysis came to be. They may only provide convenient documentation or they may prove to be essential in deciding how to analyze the dataset. In particular, they can determine how some data points are interpreted, especially data anomalies. Examples of types of metadata for four categories of information are provided in Table 11.

I have a lot of glassware to wash.

VARIABILITY

Imagine practicing hitting a target using darts, bow and arrow, pistol, cannon, or whatever. You aim for the center of the target. If your shots all land where you aimed, you are considered to be *accurate*. If all your shots land near each other, you are considered to be *precise*. The two properties are not linked. You can be accurate but not precise, precise but not accurate, neither accurate nor precise, or both accurate and precise.

Accuracy and precision apply to both data and statistics calculated from data. If you're trying to determine some characteristic of a population (i.e., a population *parameter*), you want your statistical estimates of the characteristic to be both accurate and precise. The only way to do that is to avoid biases and minimize extraneous variability in the data.

Table 11. Examples of metadata.

Examples of Metadata for Four Types of Information				
Type of Data	Measurements	Physical Samples	Reports and Records	Direct Responses
Examples Type of Metadata	Meters, scales, tests, sensors, and other devices, education and medical testing	Medical, industrial, and environmental samples, models	Databases, news stories, books and blogs, videos, images, personal diaries and logs, Govt records	Internet posts, surveys, courtroom testimony, medical histories, customer comments
Data Identification	Measurement ID, patient or student name	Unique sample ID, patient name	Title	Title
Date and time of creation	Date and time of measurement	Date and time of sample collection	Date and time of publication	Date and time of response collection
Location of data source	Location of entity being tested	Location of entity being sampled	Link, archive, repository, library, newspaper, magazine	Link, court transcripts, medical records, business records
Sample description	Description of entity being tested	Description of media or object being sampled	Size (pages, kilobytes), number of elements (words, matrix cells), formats, image specifications	Description of entity being tested
Sample anomalies	Missing measurements, broken and miscalibrated devices, missing testing forms	Unusual color, weight, texture, smell, etc. of environmental samples, sample contamination	Database errors, news misinformation, image manipulation, plagiarism, AI	Censored posts, missing responses, missing attachments, translation needs
Sampling scheme	Probability or non-probability sampling scheme	Probability or non-probability sampling scheme	Non-probability sampling scheme	Probability or non-probability sampling scheme for surveys
Data creator	Researcher, field or lab technician, doctor, teacher, education and experience, quality assurance information	Researcher, field or lab technician, doctor, medical technician, education and experience, quality assurance information	Author, database manager, photographer, organization, individual, education and experience	Witness, interrogator, doctor, investigator, pollster, education and experience
Method of data creation	Standard operating procedures, calibration history, exceptions to procedures, chain-of-custody and security, measurement quality control	SOPs, exceptions to procedures, sample preservation, processing, and transport, chain-of-custody and security, sample quality control	Database structure, data management and analysis software, scraping, writing, downloading, photography and videography, scanning and photocopying, AI	Download and scraping, streaming, type of interview, interview script, survey questions, scanning and OCR
Definitions of metrics	Metric description and technical basis, scales and units, range of values, device sensitivity	Metric description and technical basis, scales and units, range of values, good laboratory practices, lab analysis protocols, instrument sensitivity, quality control samples, lab certification	Database structure, variable definitions, locations of archived materials	Survey questions, variable descriptions, keywords, text analysis statistics
Data analysis	Dataset formatting, scrubbing, transformations, and file names, and locations of creation notes, logs, and data archive	Dataset formatting, scrubbing, transformations, and file names, and locations of creation notes, logs, and data archive	Textual and content analysis, handwriting analysis, document authentication	Textual and content analysis, sentiment analysis, keywords, respondent authentication

Description.

To clarify, statisticians discuss variability using a variety of terms, including *errors, uncertainty, deviations, distortions, residuals, noise, inexactness, dispersion, scatter, spread, perturbations, fuzziness, differences,* and *variance.* To nonprofessionals, many of these terms hold negative connotations. But variability isn't bad … it's just misunderstood.

Variability is everywhere. It's a normal part of life. It is the spice in the soup. It is the sweet and the sour of statistics. Without variability, wines would all taste the same. All vocalists would sound alike. Every race would end in a tie. Even statistics might lose its charm.

So, a bit of variability isn't such a bad thing. Variability can reveal nuances about a phenomenon that wouldn't be imaginable in a deterministic world yet can hide those nuances when it is extraneous and uncontrolled. It needs to be understood and not ignored. And the first thing about variability that needs to be understood is what kind of variability it is.

CATEGORIES OF VARIABILITY

When you start measuring data for an analysis, you'll notice that even under similar conditions, you can get dissimilar results. That lack of precision, that variability, can be classified in a number of ways.

One way to think about data measurements as a summation of five components:

- Characteristic of Population
- Natural Variability
- Sampling Variability
- Measurement Variability
- Environmental Variability.

Characteristic of population is the portion of a data value that is the same between a sample and the population. This part of a data value forms the patterns in the population that you want to uncover.

Natural variability are the inherent differences between a sample and the population. This part of a data value is the uncertainty or variability in population patterns. In a completely deterministic world, there would be no natural variability. You would read the same value at every point where you took a measurement. But in the real world, if you made the same measurement again and again, you probably would get slightly different values. If all other types of variation were controlled, these differences would be the natural or inherent variability.

Sampling variability are differences between a sample and the population attributable to how uncharacteristic (non-representative) the sample is of the population. Minimizing sampling error requires that you understand the population you are trying to evaluate.

Measurement variability are differences between a sample and the population attributable to how data were measured or otherwise generated. Minimizing measurement error requires that you understand measurement scales and the actual process and instrument you use to generate data.

Environmental variability are differences between a sample and the population attributable to extraneous factors. Minimizing environmental variance is difficult because there are so many causes and because the causes are often impossible to anticipate or control. Environmental variability often looks like natural variability because the causes of the variability are unknown. However, it is important to

distinguish the two, at least in theory, because environmental variability might become controllable if the reasons for the variability are discovered. The causes of natural variability are always inherent, unknown, and therefore, not controllable.

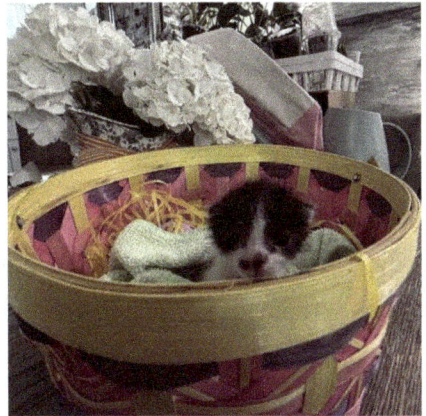

For example, say a survey were conducted with a question on consumers' overall satisfaction with a product. The "true" satisfaction might be 50% (*characteristic of the population*). There might be some variability attributable to wavering opinions held by some individuals (*natural variability*). For instance, an individual might be satisfied with some aspects of the product but not with others or they may change their mind over time.

I'm hidden like some kinds of variance. Don't forget about me.

There might be some variability attributable to who was selected to take the survey (*sampling variability*). For instance, if invitations to participate in the survey were sent to everyone who purchased the product, there might be some people who did purchase but never used the product (e.g., purchased the product for later use or purchased the product as a gift for someone else).

There might be some variability attributable to how the survey was constructed and conducted (*measurement variability*). For instance, some survey questions might have been confusing to participants (see Chapter 5).

Finally, there might be some variability attributable to something the pollster never could have anticipated (*environmental variability*). For instance, the product might have been mentioned on social media, whether as useful or unsafe. So, instead of the survey finding that 50% of users are satisfied with the product, it might find a much lower or higher percentage.

Description.

That's what statisticians mean by variability.

I know I didn't buy all these shoes for myself.

There are also other ways that statisticians classify variability based on different objectives and types of data. One classification used in *Six-Sigma statistics* (quality and management statistics) is common-cause variability vs special-cause variability. *Common cause variability* involves causes that have been observed to be inherent in a system and stable if not predictable within limits. *Special cause variability* involves unpredictable, not previously observed, significant departures from the norm, often leading to extreme values. They are attributable to unique events both within and outside of a system.

Variance is also categorized as short-term (immediate) or long-term (days and longer) variability. *Short-term variability* is usually considered relative to data-generation instruments and procedures. *Long-term variability* is related to natural changes and changes in the study environment.

When you analyze data, you usually want to evaluate characteristics of a population

How long I nap represents common-cause variability.

and the natural variability associated with that population. You want to be able to account for, if not control, common-cause variability and long-term variability. And, you don't want to be misled by any extraneous, short-term variability that might be introduced by the way you select your samples (or patients, items, or other entities) or measure (generate or collect) the data. Ideally, you won't experience uncontrolled transient events or conditions that lead to special-cause variability.

No one scheme for classifying variance will be best for all applications. Think of variance in terms of the data and the particular analyses. You'll know it's right because it will help you visualize where the extraneous variability is in the analysis and what might be done to control it.

Description.

That's why it's so important to understand the ways of variability. It's what sets statistics apart from every other approach to data analysis.

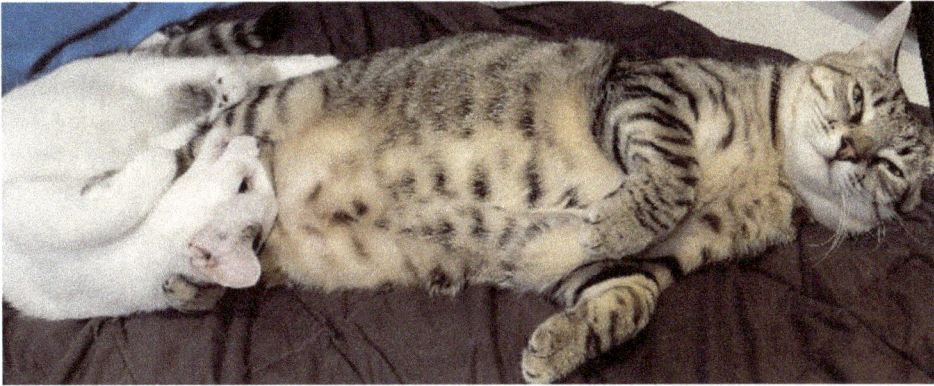

Bias and variability are like two cats that aren't related but always seem to be in the same places at the same times.

VARIABILITY VERSUS BIAS

Remember target practice? If there is little variation in your aim, the deviations from the center of the target would be random in distance and direction. Your aim would be accurate and precise.

But what if the sight on your weapon was misaligned? Your shots would not be centered on the center of the target. Instead, there would be a systematic deviation caused by the misaligned sight. Your shots would all be inaccurate, by roughly the same distance and direction from the center. That systematic deviation is called *bias*. You may not even have known there was a problem with the sight before shooting, although you would probably suspect something after all the misses.

The relationships to remember are:

Variance ↔ Imprecision
Bias ↔ Inaccuracy

Bias usually carries the connotation of being a bad thing. It usually is. It may be why statistics was mistakenly associated with lies and damn lies over one hundred years ago. But, if the systematic deviation is a good thing because it fixes another bias, it's called a *correction*. For example, you could add a correction, an intentional bias in the direction opposite the bias introduced by the misaligned weapon sight, to compensate for the inaccuracy.

So, bias can be good (in a way) or bad, intentional or not, but it's always *systematic*. On the other hand, a bias applied to only selected data is a form of *exploitation*, and is nearly always intentional and a very bad thing.

Description.

Most statistical techniques are unbiased themselves, as long as you meet their assumptions. If something goes wrong, you can't blame the statistics. You may have to look in the mirror, though.

During the course of any statistical analysis, there are many decisions that have to be made, primarily involving data. Whatever the decisions are, such as deleting or keeping an anomalous data point (*outlier*), there will be some impact on precision and perhaps even accuracy.

In an ideal world, the sum of the decisions in an analysis wouldn't add appreciably to the variability. Sometimes though, data analysts want to be "conservative", so they make decisions they believe are counter to their expectations (the *Worst-Case Fallacy*). But, when they don't get the results they expected, they often go back and try to tweak the analysis. At that point they have lost all chance of doing an objective analysis and are little better than analysts with vested interests who apply their biases from the start.

I'm a little bit accurate and a little bit precise.

Avoiding such *analysis-bias* requires no more than to make decisions based solely on statistical principles. This sounds simple but it isn't always so.

Sometimes bias isn't the fault of the data analyst, such as in the case of *reporting bias*. In professional circles, probably the most common form of reporting bias is *file drawer bias*, that is, not reporting non-significant results. Some investigators will repeat a study again and again, continually fine-tuning the study design until they reach their nirvana of statistical significance (called *p-hacking*). Seriously, is there any real difference between probabilities of significance of 0.051 versus 0.049?

But you can't fault the investigators alone. Some professional journals won't publish results that don't find significance, and of course, professionals who don't publish, perish. Can you imagine the pressure on investigators looking for a significant result for some new business venture, like a pharmaceutical? They might take subtle actions to help their cause then not report everything they did. That's a form of reporting bias.

Perhaps the most common form of reporting bias in nonprofessional circles is *cherry picking*, the practice of reporting just those findings that are favorable to the reporter's position. Cherry picking is very common in studies of controversial topics such as climate change, firearm violence, and conspiracy theories. Discussions and even academic research on such topics cherry pick egregious examples of discredited

Description.

or unlikely claims while ignoring legitimate but unexciting cases. Many political discussions on social media use information that was cherry picked.

Given that someone else's reporting bias is after-the-analysis, why is it important to other analyses? The answer is that it can misdirect future studies. There are more than a few research studies in medicine and pharmacology, for instance, that influenced the direction of funding for future research.

In general, never trust a source of statistics or a statistical analysis that doesn't report variance and sample size along with the results. Too often, it is the reporting that is misleading rather than the statistics.

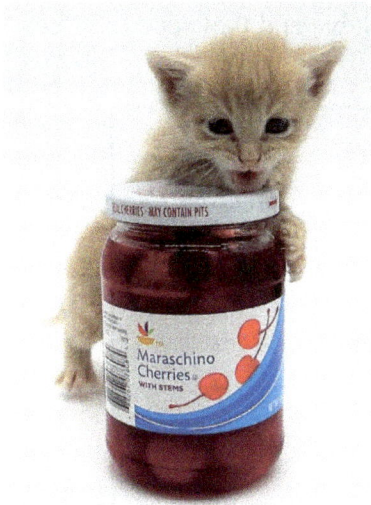

These are the cherries I picked to tell everybody about.

To sum up, a dataset should contain no bias and only natural variability. Most of the "disappointing" statistical analyses you'll see are more likely to suffer from too much variability rather than too little accuracy. To get a good result, whether in marksmanship or in data analysis, you have to control variation.

Variance doesn't go away by ignoring it.

To control variability, you have to understand it. But that's not enough. Data and variance are thoroughly intertwined. You must be proactive in planning your data collection efforts to control as much of the extraneous variability as possible.

VARIANCE CONTROL AND INFLUENCE

In addition to the sources of variability, you can think about how variability affects data in terms of:

- **Control.** The extent to which variability can be controlled so that data aren't affected.
- **Influence.** The proportion of data points that are affected by uncontrolled variability.

This is shown in Figure 11. Sampling and measurement variability usually tend to be under an experimenter's control. Sometimes environmental variability can be controlled, even if unintentionally, and sometimes it can't. These types of variability tend to affect all or most of the data. Natural variability, on the other hand, can't be controlled and it affects all data.

Biases in data affect all or most of a dataset and usually can be controlled if they are identifiable and unintentional. Intentional bias of only selected data is *exploitation*.

Description.

Random *mistakes* and *errors* may or may not be controllable and they tend to affect only a few data points.

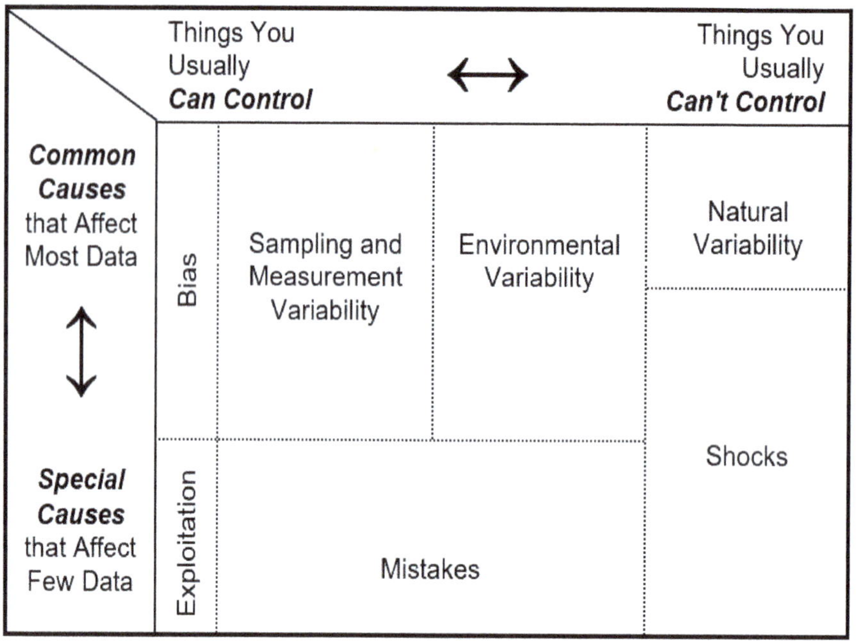

Figure 11. Causes of bias and variability.

Shocks are uncontrollable short-duration conditions or events that can influence just a few or even most of the data in a dataset. Examples of shocks include: heavy rainfall upsetting a sewage treatment plant; missing a financial processing deadline so that one month has no entry and the next has two; having a meter lose calibration because of electrical interference; mailing surveys without realizing that some have missing pages; assembly line stoppages in an industrial process; and so on.

THE THREE RS OF VARIANCE CONTROL

The fundamentals of education that we all learned in elementary school are Reading, 'Riting, and 'Rithmetic (obviously, not spellin'). With these concepts mastered, we are able to learn more sophisticated subjects like rocket science, brain surgery, and tax return preparation. Similarly, if you plan to interpret a statistical analysis, you'll need to understand the three fundamental Rs of variance control—*Reference, Replication*, and *Randomization*.

REFERENCE

The concept behind using a reference in data generation is that there is some ideal, background, baseline, standard, benchmark, or at least, generally-accepted norm that can be compared to all similar data operations or results. References can be applied both before and after data collection.

Probably the most basic application of using a reference to control variation attributable to data collection methods is the use of standard operating procedures (SOPs), written descriptions of how data generation processes should be

> *You can't understand your data without controlling variance.*
>
> *You can't control variance without understanding your data.*
>
> C. Kufs. 2011. **Stats with Cats**: The Domesticated Guide to Statistics, Models, Graphs, and Other Breeds of Data Analysis, 1st edition.

done. Equipment calibration is another well-known way to use a reference before data collection to control extraneous variability.

References are also used after data collection to assess sampling variability. This use of a reference involves comparing generated data with benchmark data. The comparison doesn't control variability, but allows an assessment of how substantial the extraneous variability is.

A more sophisticated use of a reference is to measure highly correlated but differently-measured properties on the same sample, such as measuring total dissolved solids and specific conductance in a water sample.

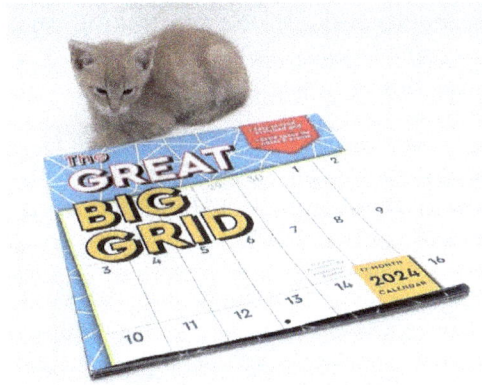

Why don't all months have the same number of days?

Deviations from a pre-established relationship may be signs of some sampling anomaly. Further, data collected on some aspect of a phenomenon under investigation can be used to control for the variability associated with the measure. Variables used solely to control or adjust for some aspects of extraneous variability are called *covariates*.

Perhaps the most well-known application of a reference is the use of *control groups*. Control groups are samples of the population being analyzed to which no treatments are applied. For example, in a test of a pharmaceutical, the test group and the control group would be identical on relevant factors such as age, weight, and so on except that patients in the test group would receive the pharmaceutical and the patients in the control group would receive a placebo.

Description.

REPLICATION

If a reference point can't be used to help control variability, it may be possible to use replication, repeating some aspect of a study as a form of internal reference.

Replication is used in a variety of ways to assess or control variability. Replicate samples or measurements are one example. Collecting two samples of some environmental media and sending both samples to a lab for analysis would be another example. Differences in the results would be indicative of measurement variability (assuming the sample of the media is homogeneous).

In addition to the data source (i.e., sample, observation, or row of the data matrix) being replicated, the type of data information (i.e., attribute, variable, or column of the data matrix) can also be replicated. For example:

- Asking survey questions in different ways to elicit the same or very similar information, such as, Did you like this …, Did this meet your expectations …, and Would you recommend this … ?
- Measuring the same property on a sample using different methods, such as pH in the field with a meter and again in the lab by titration.

We're replicates.

Replicated samples or variables require a little extra thought during the analysis. If you are looking for a fair representation of the population, a replicated sample would constitute an over-representation. Consequently, only one replicated item should be kept for analysis.

Typically, replicated samples are first compared to identify any anomalies, then, if they are similar, they are averaged. Sometimes, either the first sample or the second sample is selected instead. Samples should never be selected to use in an analysis on the basis of its value.

For replicated variables, the variables are first compared to identify any anomalies, then only one of the variables is selected to use in the analysis. Highly correlated variables will cause problems with many types of statistical analysis (called *multicollinearity*, see Chapter 8).

The concept of replication is also applied to entire studies. It is common in science to repeat studies, from data collection through analysis, to verify previously determined results. *Repeatability,* also called *direct replication*, involves repeating the original research design and procedures of a study as exactly as possible. *Reproducibility*,

Description.

also called *conceptual replication*, involves repeating a study using different designs and procedures to examine the original hypothesis.

RANDOMIZATION

Statisticians use the term *randomization* to refer specifically to the random assignment of treatments in an experimental design. But, in its common usage, randomization can involve any action taken to introduce chance into a data generation effort.

> *It is a wonderful irony of nature that introducing irregularities (randomization) into a data generation process can reduce irregularities (variability) in the resulting data.*
>
> C. Kufs. 2011. Stats with Cats: The Domesticated Guide to Statistics, Models, Graphs, and Other Breeds of Data Analysis.

Randomization is desirable in statistical studies because it minimizes (but not necessarily eliminates) the possibility of having biased samples or measurements. As a consequence, randomization also minimizes extraneous variability that might be attributable to inadvertent inconsistencies in data generation.

As with replication, randomization can be applied to both samples and variables. Samples such as study participants can be chosen at random or following a scheme that capitalizes on their existing randomness. Values for variables that are not inherent to a sample can be assigned randomly. This is done routinely in experimental statistics when study participants are assigned randomly to the treatments. Random assignments are simple to make using random-number algorithms.

If you can describe cats, you can describe datasets.

DESCRIBING DATA

Say you wanted to describe someone you see on the street. You might characterize their sex, age, height, weight, build, complexion, face shape, hair, mouth and lips, eyes, nose, tattoos, scars, moles, and birthmarks. Then there's clothing, behavior, and if you're close enough, speech, odors, and personality. Your description might be different if you're talking to a friend or a stranger, to a neighbor or law enforcement, or to someone of the same or different sex and age.

Those are a lot of characteristics and they're sometimes hard to assess. Individual characteristics aren't always relevant and can change over time. Different observers might see, or remember, them in different ways. And yet, without even thinking about it, we describe people we see every day using these characteristics. We do it mentally to remember someone or overtly to describe a person to someone else. It becomes second nature because we do it all the time.

Describing people is easy because we do it all the time. Describing datasets is even easier once you know how.

Most people don't describe sets of numbers very often, though, so they don't know how easy it actually is. Only a few characteristics have to be considered, all of which are fairly easy to assess and will never change for the dataset so long as the individual data points don't change. Once you learn how, it's hardly a challenge to get it right, unlike describing a tall, middle-aged, bald guy who is singing opera while dressed as a clown.

DATASETS

As described previously, datasets consist of four elements:

- **Variables**, in the columns of a data matrix
- **Observations**, in the rows of a data matrix
- **Data**, in the cells of a data matrix
- **Metadata**, hopefully somewhere, anywhere.

Information is data in context. Information has two parts—what the data are and what the data are about. Facts about data, such as units of measurement, are called *metadata*.

You can't have good data unless you have good metadata.

You may never encounter any big or complex datasets. Everything you see in a Stats 101 class in school will be screened by your instructor so you can use it easily. What's more, you'll probably never even see datasets behind stories in the media. Just recognize that the results of every statistical analysis is the product of far more sophisticated datasets.

Description.

There's too much to learn in introductory data analysis courses to be concerned about details such as *scraping*, *big data*, *data scrubbing*, *influential observations* and *outliers*, *replicates* and other *quality control samples*, and a host of other topics in applied statistics. They will come later if you decide to learn more about statistics and do your own analyses. That's when you might want to read **Stats with Cats**, or an introductory textbook on statistics, or spend some time searching the internet for specific topics. For now, focus on what you need. Worry about more advanced topics when you are ready.

How would you describe me?

STATISTICS

In describing a set of numbers, you'll only need to consider four attributes—*frequency*, *central tendency*, *dispersion*, and *shape*. It's that simple.

For this section, I've included example data available on the internet. It is on the topic (phenomenon) of quarterbacks (QBs) who played or are playing in the National Football League (NFL) from 1937 to 2022. The population consists of hundreds of QBs who played during that time, of which "leaders" were selected, independently by source websites, based on a variety of QB performance metrics.

Sports is a great place to start analyzing data on your own. There are a lot of data available related to sports, from player and team performances to attendance and merchandising. Many websites provide the data, often as downloadable files or tables that can be *scraped*. And, it's a lot of fun if you follow the sport you analyze. Give it a try.

Sports data and kittens are easy to find.

FREQUENCY

Frequency refers to the number of data points in a dataset. Frequency is also referred to as a *count*. If a dataset has a grouping variable, there are different frequencies for each group as well as the whole dataset.

Frequency is the primary statistic for nominal (grouping) scales and ordinal scales. The level with the highest frequency is called the *mode*. Sometimes there is no mode because none of the numbers repeat.

Description.

The mode is a curious statistic. It was first coined by Karl Pearson over a century ago in 1895. He used the term interchangeably with *maximum-ordinate*, corresponding to the vertical axis in a histogram which he also introduced. It is traditionally taught in introductory statistics courses as a measure of central tendency, but it is really a measure of frequency. While the mean, median, and mode are all equal in a Normal distribution, the mode may actually occur in a *tail* of a skewed distribution.

Modes are applicable to ordinal and even nominal scales, but can also be identified in continuous scales that have been *binned* (segmented into small ranges of the same length, as in histograms).

Mode can refer to both a single value, the most frequently occurring, and to peaks in frequency distributions, in which there may be multiple modes consisting of small ranges of values. In this latter sense, it can be considered as both a measure of dispersion and a descriptor of the shape of a distribution … but not really a measure of central tendency.

CENTRAL TENDENCY

Central Tendency refers to where the middle of a set of numbers is. It is a measure related to *accuracy*. It is used mostly for continuous (interval or ratio) scales and often with ordinal scales as well. There is no central tendency statistic for nominal-scale variables.

Accuracy is very important right now.

There are quite a few statistics that can be used to describe where the center of a dataset is including the *arithmetic mean* (the *average*), the *geometric mean*, the *harmonic mean*, *trimmed* and *winsorized means*, *weighted means*, the *median* and other statistics shown in Table 12.

The *arithmetic mean* is calculated as the sum of the values divided by the count of the values. Most people learn about the arithmetic mean before they reach high school. Arithmetic means are used for most dataset descriptions.

The *geometric mean* is calculated as the root of the product of the values where the root is based on the count (e.g., cube root for three values, fifth root for five values, and so on). The geometric mean is often used to describe growth rates.

Description.

Table 12. Common measures of central tendency.

Measures of Central Tendency		
Measure	**Description**	**Comment**
Arithmetic mean	The sum of the values divided by the count of the values	Used to describe central tendency of metrics of most quantitative scales
Centroids	The arithmetic mean of the co-ordinates of points in a shape, the geometric center of a shape or data space	Used in cluster analysis and other grouping models
Circular means	The angular mean for data distributed on circles(or spheres)	Used for averaging compass directions
Geometric mean	The *n*th root of the product of *n* data values	Used to describe growth rates
Harmonic mean	The number of data values divided by the sum of the reciprocals of the values.	Used for averaging ratios or rates
Interquartile mean	The spread of the middle 50% of the data, equal to the third quartile (75th percentile) minus the first quartile (25th percentile) of the data	Used to reduce effects of extreme values
Median	The exact center of an ordinal-scale dataset in which there are the same number of data values less than and greater than the median. If the sample size is an even number, the median is the average of the two middle data values.	For continuous scale metrics, the distribution is divided into bins
Midrange mean	The average of the minimum and the maximum	Used to reduce effects of extreme values
Moving average	The average of a specified number of data values in a set data range or time period	Used to smooth trends
Root mean square	The square root of the average of the squares of a set of values	Used when the signs of data values are not important
Trimean	The average of the first quartile (25th percentile), the third quartile (75th percentile), and two times the second quartile (median)	Used to reduce effects of extreme values
Trimmed mean	Calculated by removing the same number of values from both ends of a data distribution and averaging the remaining values. Also called a truncated mean.	Used to remove extreme values and censored data before calculating an average
Weighted mean	Averages in which individual data values have been weighted by multiplying them by some adjustment factor. If all the weights are one, the weighted mean is the same as the arithmetic mean.	Used to give more importance to certain values or to provide some desirable statistical property in an analysis
Winsorized mean	Average in the same number of values are removed from both ends of a data distribution and replaced with the next remaining values.	Used to remove extreme values and censored data before calculating an average

The *harmonic mean* is calculated as the count divided by the sum of the reciprocals of the data values. The harmonic mean is often used to calculate the average of ratios or rates.

The arithmetic mean, geometric mean, and harmonic mean are sometimes referred to as *Pythagorean means*. If all the values in a dataset are positive, the arithmetic mean will be the largest and the harmonic mean will be the smallest.

Weighted means are arithmetic (usually) means in which individual data values have been assigned greater importance by multiplying them by some adjustment factor. Sometimes, weights

That is one mean weight.

Description.

are used to provide some desirable statistical property in an analysis, like correcting for sampling bias in a survey.

The weighted mean is calculated as the sum of the adjusted values divided by the sum of the adjustments. If all the weights are one, the weighted mean is the same as the arithmetic mean.

Trimmed means are calculated by removing the same number of values from both ends of a data distribution and averaging the remaining values. Trimming is used to address extreme values and censored data. For example, in Olympic sports involving judging, trimmed means are used by excluding the highest and lowest judge scores to prevent bias.

The *median* is the exact center of an ordinal-scale dataset. There are exactly the same number of data values less than and greater than the median. The median is determined by sorting the values in the dataset and counting the values from the extremes until you find the center. If the sample size is an even number, the median is the average of the two middle data values.

The most popular central-tendency statistic for progressive-scale variables are the *median* and the *arithmetic mean*. Don't worry about modes, trimeans, winsorized means, or the circular and spherical means. Their usage is limited to special situations.

The arithmetic *mean*, or *average*, is the center of a continuous-scale, Normally-distributed dataset, even though there may not be an equal number of data values less than and greater than the mean. The mean is one of the parameters that define the Normal distribution.

If the median is equal to the mean, the frequency distribution of the data is probably *symmetrical*.

If the median is greater than the mean, the frequency distribution of the data is *skewed* to lower values. That is, the distribution has a longer tail to the left of the

I'm the median.

data peak. This is called *negative-skewed* or *left-skewed*. Concentrations of environmental contaminants and age at death are examples of negatively-skewed distributions.

If the median is less than the mean, the frequency distribution of the data is skewed to higher values. That is, the distribution has a longer tail to the right of the data peak. This is called *positive-skewed* or *right-skewed*. Income and real-estate-prices are examples of positively-skewed distributions.

Sometimes medians and means show great disparities. For example, the mean income of American households is much higher than the median income because there are many more low-income households than there are high-income households. In cases such as this, it is important to evaluate the frequency distribution of the values.

Table 13 shows central tendency statistics calculated for ratings of 214 leading quarterbacks from 1937 to 2022. (QB rating is based on the number of passes completed per attempt, yards per attempt, touchdowns per attempt, and interceptions per attempt.) QB rating is on a scale of 0 to 158.3.

Table 13. Mean career passer ratings of NFL quarterbacks.

Career Ratings for 214 NFL Quarterbacks 1937-2022	
Measure of Central Tendency	**Value**
Arithmetic mean	78.87
Geometric mean	78.21
Harmonic mean	77.54
Interquartile mean	78.79
Median	78.50
Midrange mean	77.90
Mode (4 of 214 values, 1.9%)	80.40
Root mean square	79.51
Trimean	78.64
Trimmed and winsorized means	
Trimmed average after removing 2 data points (1%)	78.88
Winsorized average after removing 2 data points (1%)	78.86
Trimmed average after removing 6 data points (3%)	78.91
Winsorized average after removing 6 data points (3%)	78.92
Trimmed average after removing 11 data points (5%)	78.88
Winsorized average after removing 11 data points (5%)	78.99
Weighted means	
Where the most important data are weighted 1 and the rest of the weights are between 0 and 1	
Weighted for the year of the draft (2021=1; 1937=0.087)	81.25
Weighted for the draft number player was picked at (#1=1; #333=0.003)	78.99
Weighted for the number of seasons played (26=1; 2=0.12)	78.27
Weighted for the number of teams played for (1=1; 9=0.11)	80.34
QB ratings can range from 0 to 158.3. Values in this dataset range from 52.2 to 103.6. Data from pro-football-reference.com and drafthistory.com.	

Values in the dataset range from 52.5 to 103.6. Measures of the central tendency of the QB ratings range from 77.54 (harmonic mean) to 81.25 (weighted mean based on the year a QB was drafted) compared to the arithmetic mean of 78.87.

Which statistic for central tendency is most correct will depend on the nature of a dataset and the planned use of the statistic.

DISPERSION

Dispersion refers to how spread out the data values are on progressive scales. It is a measure related to *precision*. There are quite a few statistics that can be used to describe the dispersion of a dataset including the *range*, the *standard deviation*, the *variance*, the *coefficient-of-variation,* and other statistics shown in Table 14. Measures of dispersion tend to be more susceptible to changes in a dataset than are measures of central tendency.

The *range* of a dataset is simply the *maximum* value minus the *minimum* value. It is used frequently in data scrubbing to check for out-of-range errors, computational inefficiencies, and the need for data *transformations*.

Table 14. Measures of dispersion.

Measures of Dispersion		
Measure	Description	Comment
Coefficient of variation	The standard deviation divided by the mean	Can be used to assess the form of a data distribution
Distance standard deviation	The root of the sum of the variances of the xyz coordinates in a spatial dataset	Used in analyzing spatial statistics
Interquartile range	The difference between the 75th and 25th percentiles	Used to build box plots, identify outliers, and evaluate data distributions.
Minimum and maximum	The lowest and highest values in a dataset	Describes span of data, useful for data scrubbing
Mean angular deviation	The average of the deviations from the mean of a set of angles	Used for describing the dispersion of circular scales
Mean (or average) absolute deviation	The average of the absolute (unsigned) differences between each data point and the mean or median.	More resilient to outliers than the standard deviation.
Mean squared error	Average squared difference between predicted and actual values	Used in assessing the precision of models.
Median absolute deviation	The median of the absolute deviations from the data's median.	More resilient to outliers than the standard deviation.
Quartile coefficient of dispersion	The interquartile range divided by the sum of the 75th and 25th percentiles	More resilient to outliers than the standard deviation.
Quartile Deviation:	The average of the differences between the first quartile and the second quartile, plus the third quartile and the second quartile.	More resilient to outliers than the standard deviation.
Range	The difference between the maximum and the minimum	Describes span of data, useful for data scrubbing
Relative mean difference	The average magnitude of the difference between each data point and the mean, relative to the mean itself.	Similar to the coefficient of variation with the mean of the differences replacing the standard deviation
Variance and Standard deviation	Variance is the average of the squared differences between each data point and the mean. Standard deviation is the square root of the variance.	Variances are additive in statistical models. Standard deviations use the same units as the original data.

Description.

The *standard deviation* is probably the most commonly used dispersion statistic in data description. It is also used in most statistical testing to find differences in data populations. It is one of the parameters of the Normal distribution. The *variance* is the square of the standard deviation

The standard deviation is calculated by:

- Subtracting the mean of a dataset from each value in the dataset
- Squaring each subtracted value
- Adding all the squared values
- Dividing the sum of the squared values by the number of values in the dataset (if you're describing a sample) or by the number of values in the dataset minus 1 (if you're describing a population). The difference in the formulas for standard deviations of samples (divisor of n) and populations (divisor of n-1) has to do with statistical comparisons, a big topic in Stats 101.

In mathematical notation:

$$\text{Standard deviation} = \sum (x - \mu)^2 \div (n-1)$$

Standard deviations are in the same units as the original data, so examining them is usually more intuitive than examining variances. Variances have the nice statistical property of being *additive*. You can add up the variances of individual groups to get the total variance. This is useful if you're interested in looking at segments of your population, for example in advanced statistical procedures like ANOVA.

Dispersion can be confusing.

The *coefficient-of-variation* is the *standard deviation* divided by the *mean* times 100%. It is sometimes called the *relative standard deviation*. The coefficient of variation is often used to compare variables and decide if data transformations might be needed.

Frequency can also be used to show how data values are spread. Frequency-based measures of dispersion are usually expressed in *quartiles* (the four scale values that represent 25% fractions of data), *quintiles* (the five scale values that represent 20% fractions of data), and *deciles* (the ten scale values that represent 10% fractions of data).

Unlike measures of central tendency, which all look for the same property of a data distribution, measures of dispersion explore different aspects of the concept of data dispersion. The standard deviation and the absolute deviation are based on differences between data points and a measure of central tendency. Frequency-based measures of dispersion look more at the shape of a data distribution.

The most popular dispersion statistic for progressive-scale variables is the *standard deviation*. Don't worry about the other measures of dispersion as their usage is limited to more advanced situations.

Table 15 shows dispersion statistics calculated for the ratings of 214 leading quarterbacks from 1937 to 2022. The dispersion of the QB ratings range from 6.5 (median absolute deviation) to 13.0 (interquartile range) compared to the standard deviation of 10.15. Which statistic for dispersion is most appropriate for a given situation will depend on the nature of a dataset and the planned use of the statistic.

SHAPE

Shape refers to the frequencies of values in a dataset at each level of the scale. There are a number of methods that can be used to assess a variable's frequency distribution.

STATISTICS

One method to assess a dataset's shape is to compare the mean to the median. If they are about the same, the distribution is probably *symmetrical*. If the mean is less than the median, the sample is *left skewed*. If the mean is greater than the median, the sample is *right skewed*.

If the *coefficient of variation* is substantially greater than 1, the distribution may be *truncated* on the left and *elongated* on the right, such as with a *lognormal distribution*.

Skewness and *kurtosis* are statistics that provide information about a variable's frequency distribution. They are power functions, so they are

Table 15. Measures of dispersion for career quarterback ratings.

Career Ratings for 214 NFL Quarterbacks 1937-2022	
Measure of Dispersion	**Value**
Minimum	52.20
Maximum	103.60
Range	51.40
Mean absolute deviation	8.02
Median absolute deviation	6.50
Interquartile range	13.00
Quartile deviation	6.50
Quartile coefficient of dispersion	8.2%
Coefficient of variation	12.9%
Standard deviation	10.15
Variance	103.10

QB ratings can range from 0 to 158.3. Values in this dataset range from 52.2 to 103.6. Data from pro-football-reference.com and drafthistory.com.

even more challenging to calculate manually than the standard deviation. However, they are easy to interpret. The farther they are from zero (sometimes 3 is used for kurtosis), the more likely that there are deviations from the Normal distribution.

They do have a tendency to make small departures from Normality look large, though.

Skewness measures distribution symmetry. A value of zero indicates that the distribution is symmetric. Negative values indicate a long tail on the left side of the distribution (i.e., *left skewed*). Positive values indicate a long tail on the right side of the distribution (i.e., *right skewed*). Values outside the range of about ± 3 suggest that asymmetry may be a problem.

The *kurtosis* is sometimes referred to as the *peakedness* of a distribution, but it's really more of a measure of the proportion of samples in the tails of a distribution versus the proportion in the middle of the distribution. The kurtosis of a standard Normal distribution is 3, so some software subtracts 3 from the kurtosis value so that the statistic is centered on zero. Positive kurtosis values indicate a distribution has a high peak and short tails, and is termed *leptokurtic*. Negative kurtosis values indicate a distribution has a flat peak and long or thick tails, and is termed *platykurtic*.

There are also a variety of statistical tests that are used to evaluate whether a dataset comes from a population that is Normally distributed. Testing for Normality is a fairly advanced topic that often isn't even taught in Stats 101. Don't worry about it unless you plan to go further in statistics.

GRAPHS

Dataset descriptions are often augmented by graphs that depict the shape of a data distribution. They are particularly revealing because you can see how the data compares to a Normal distribution. The three most commonly used statistical graphics for describing the shape of data distributions are histograms, box-and-whisker plots, and probability plots.

For ordinal scales, the preferred graph is a *histogram*, which shows the number (frequency) of data values for each level of the scale. For continuous scales and ordinal scales with many levels, small increments of the data scale, called *bins,* are created to summarize the data frequencies.

Yup. The shape of these data looks Normal to me.

Figure 12 is a histogram showing data frequencies for the 214 quarterback ratings in six bins, each consisting of ten scale values on the horizontal (black) axis. The blue bars indicate the number of ratings on the left (blue) vertical axis corresponding to each of the binned rating intervals. The red line on the histogram indicates the percentage of the total number of ratings on the right (red) vertical axis. Overall, the histogram suggests that the ratings approximate a Normal distribution based on its appearance.

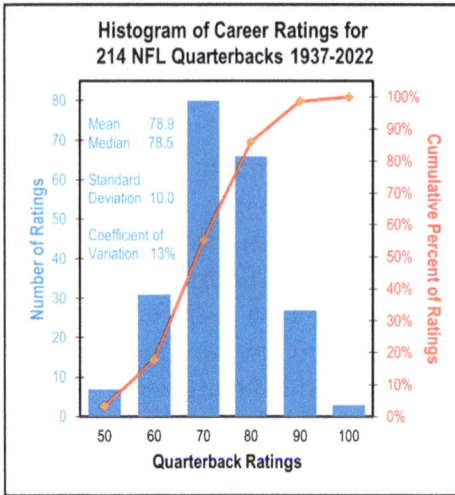

Figure 12. Histogram of quarterback ratings using six bins.

There are hundreds of distribution models, but the most commonly used is the Normal distribution. The normal distribution model has two parameters—the mean and the standard deviation. That's why those two statistics are used so often.

One issue with histograms is that the distribution shapes they present are not unique. Histograms for the same data will look different depending on the bin intervals that are selected. Figure 13 is a histogram showing the distribution of the same ratings data using twelve bins instead of six. This histogram also suggests that the ratings approximate a Normal distribution, it just looks different.

Box-and-whisker plots show a data distribution as if you were looking at a histogram from the top down. They show the *box* bounded on the bottom by the *25th quartile* of the data and the top by the *75th quartile* of the data. The height of the box represents the *interquartile range*.

The lower cross bar on the *whisker* (the vertical line running through the box) represents the 25th quartile of the data minus 1.5 times the interquartile range. The upper cross bar on the whisker represents the 75th quartile of the data plus 1.5 times the interquartile range. Data points that appear as markers above or below the whiskers represent extreme values, possibly outliers. In some box diagrams, the ends of the whiskers represent the data minimum and maximum instead of quartiles ± 1.5 times the interquartile range.

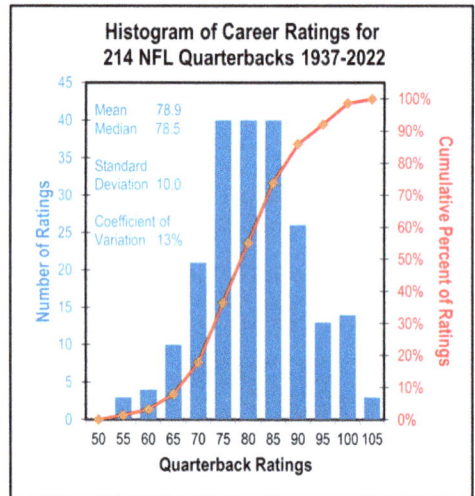

Figure 13. Histogram of quarterback ratings using twelve bins.

Description.

Inside the box, the mean is represented by an **X** and the median is represented by a horizontal line. Box plots can also be enhanced to show frequencies and other measures of central tendency and dispersion.

Figure 14 is a box plot of the QB rating data. It shows a symmetrical data distribution with the mean and median being nearly equal. There is one possible outlier at the lower end of the distribution.

A third type of statistical graphic used to show distribution shape is a *probability plot*. Probability plots show the values of the dataset versus their position in a Normal (or other) distribution. In a probability plot, deviations from a straight line suggest departures from Normality.

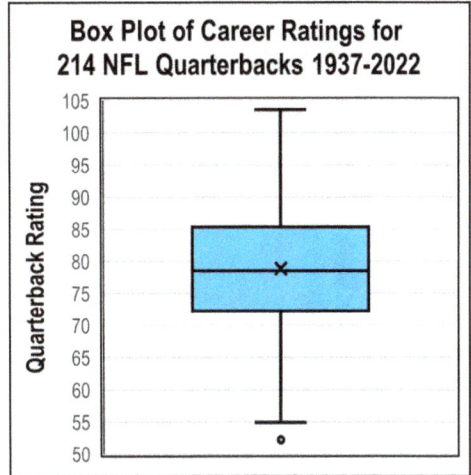

Figure 14. Box plot of career QB ratings.

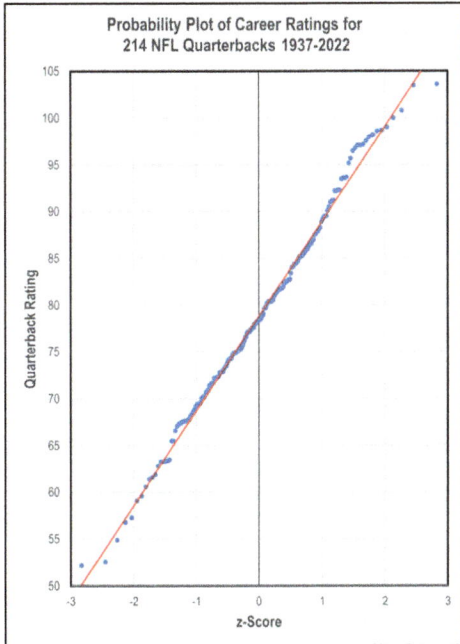

Figure 15. Probability plot of the career quarterback ratings.

Figure 15 shows a probability plot of the example data. The blue circles represent individual quarterback ratings. The red line represents a Normal distribution. The probability plot also suggests that the distribution of quarterback ratings approximates a Normal distribution with some deviations in the tails. This is a common occurrence. Unfortunately, the tails are where the significance levels of statistical tests are decided, so such deviations from a Normal distribution are problematic (Chapter 6).

The next three figures show plots for quarterback career passing yards instead of quarterback ratings. While the quarterback ratings appear to be approximately Normal, passing yards are left-skewed to lower values. This is because passing yardage is unbounded on the upper end while quarterback rating scale is limited in practice if not bounded. In fact, passing yardage is one of the quarterback performance metrics included in the rating model. The metrics for the rating model—

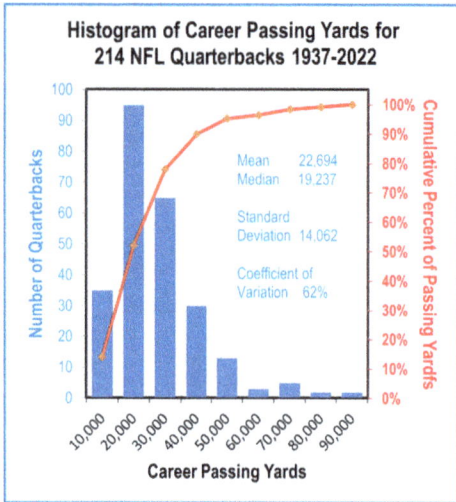

Histogram of Career Passing Yards for 214 NFL Quarterbacks 1937-2022

Mean 22,694
Median 19,237

Standard Deviation 14,062

Coefficient of Variation 62%

Figure 16. Histogram of career quarterback passing yards.

passes, yards, touchdowns, and interceptions—are all normalized by dividing them by the number of attempts, thus limiting their range.

The histogram (Figure 16) for the total career passing yards clearly shows asymmetry in the data frequency distribution. The median is much less than the mean compared to the dataset of the quarterback ratings in which they are virtually the same. The coefficient of variation is high at 62% though not over 100%, which might still indicate a lack of Normality.

The box plot (Figure 17) for passing yards shows the data asymmetry even more clearly than the histogram. It also suggests that the highest six data points might be considered to be outliers.

Finally, the probability plot (Figure 18) for passing yards shows how much the data deviate from a Normal distribution, especially in both tails. Conducting a statistical test (Chapter 6) on this dataset would require special consideration of this circumstance.

Comparing shapes of continuous-scale data to mathematical models (equations) of frequency distributions is like comparing a person to some well-known celebrity. The two individuals wouldn't be identical but sometimes they would be similar enough to warrant a second look. In other cases though, the discrepancies are so striking that the notion of any similarity can be dismissed outright. That's how statistical graphics of distribution shape can focus subsequent analysis plans.

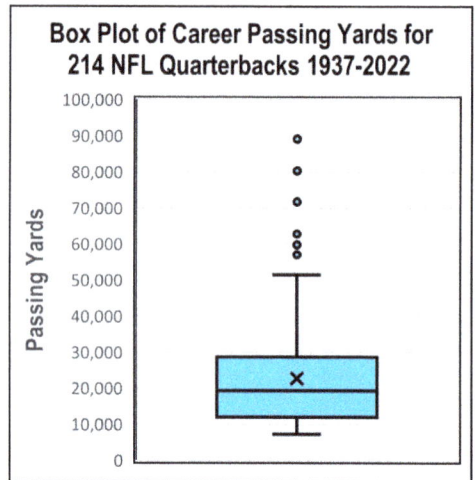

Box Plot of Career Passing Yards for 214 NFL Quarterbacks 1937-2022

Figure 17. Box plot of career quarterback passing yards.

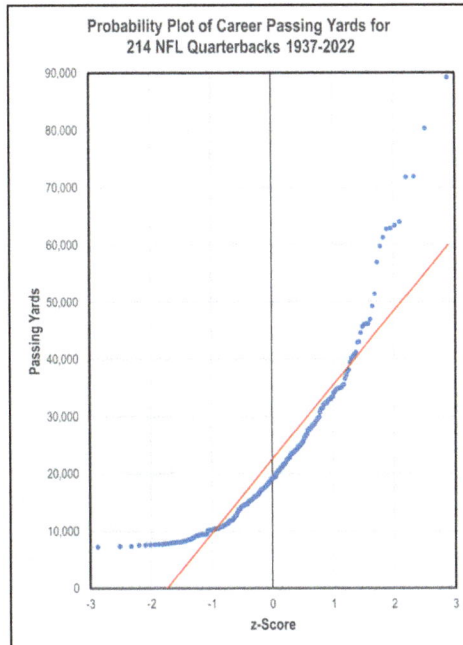

**Figure 18. Probability plot
of career quarterback
passing yards.**

WHAT TO LOOK FOR IN DATA DESCRIPTIONS

Some activities are instinctive. A baby doesn't need to be taught how to suckle. Most people can use an escalator and open a door instinctively. The same isn't true of playing a guitar, driving a car, or understanding statistics. But, at least for describing a dataset, it's pretty straightforward.

Here are the five basic elements to look for—four statistics plus metadata:

- **Frequency**. Number of samples, mode.
- **Central Tendency**. Mean and median.
- **Dispersion**. Minimum, maximum, standard deviation, coefficient of variation.
- **Distribution Shape**. Histogram, box-and-whisker plot, probability plot.
- **Metadata**. Description of phenomenon, aim of research. description of population (size, key common characteristics), sampling strategy, description of metric (name in dataset, definition, relevance, planned use), measurement (device/process, creator), scale, unit, data source and reliability, variability

Description.

control measures, data scrubbing actions (errors, missing data, outliers, other issues), number of other metrics in dataset, and so on.

What you might notice from this is that it's not the statistics that make data description challenging, it's the metadata.

Table 16 summarizes the statistics most often used to describe a dataset for the most common types of scales. For example, a nominal-scale dataset would be described by providing counts or percentages of observations in each group. An ordinal-scale dataset would be described by providing counts or percentages for each level, the median and percentiles, and ideally, a histogram. A continuous-scale dataset would be described by providing the closest distribution model and estimates of its parameters, such as "Normally-distributed with a mean of 10 and a standard deviation of 2." Continuous-scale datasets can be described so succinctly because the

They figured this out all by themselves.

distribution-shape specification contains so much of the telling information.

The statistics are easy, so here are a few hints for assessing the validity of the metadata behind data descriptions.

DATA SOURCE

Data used in a statistical analysis can either be generated by the experimenter (the *primary source*) or acquired from another source (a *secondary* or *tertiary source*). Each has its own potential issues.

ACQUIRED DATA

Often, statistical analyses are conducted on data that have been acquired from one or more sources outside of the analyst's control. This is not in itself a bad thing. Some of the most revealing analyses come from datasets that combine information that originated from different sources. Think of all the analyses that incorporate information from the U.S. Census Bureau on the country's population. The important things to consider are the reliability of the sources and the contextual compatibility of the variables.

SOURCE RELIABILITY

Data acquired from outside sources, usually online sources, may not describe how the data were generated, so the reliability of the source is the first consideration. Government and academic organizations that specifically compile data and package

Table 16. Statistics most often used to describe a dataset for the most common types of scales.

SCALE			CHARACTERISTIC	STATISTIC
GROUPING SCALES	NOMINAL		FREQUENCY	COUNTS OR PERCENTAGES
			CENTRAL TENDENCY	
			DISPERSION	
			SHAPE	
PROGRESSION SCALES	ORDINAL		FREQUENCY	COUNTS, PERCENTAGES, MODE
			CENTRAL TENDENCY	MEDIAN, MEAN
			DISPERSION	RANGE, QUARTILES AND OTHER PERCENTAGES
			SHAPE	HISTOGRAM, DOT PLOT, STEM-LEAF DIAGRAM
	RATIO		FREQUENCY	COUNTS, PERCENTAGES, MODE
			CENTRAL TENDENCY	MEAN, MEDIAN
			DISPERSION	STANDARD DEVIATION, QUARTILES AND OTHER PERCENTAGES
			SHAPE	PROBABILITY PLOT, SKEWNESS, KURTOSIS

them for dissemination are the best sources. Sources that provide data without adequate metadata and documentation are the worst.

Errors in datasets built from published data take a variety of forms. First, everything that can happen in the creation of a locally-created dataset can happen in a published dataset, so there could be just as wide a variety of errors. Reputable sources, however, will scrub out *invalid data*, *dirty data*, *out-of-spec data*, *corrupted data*, and *extraneous data*. Most will not address *missing data*. None will deal with *messy data*. Missing and messy data are the responsibility of the data analyst.

Datasets that have been combined from multiple sites or even different files from the same site can suffer from two types of problems—*integration process* and *contextual differences*. Neither of these issues is always readily apparent.

We're getting too tired to look any more.

ACQUISITION PROCESS

Acquiring data from the internet is sometimes as easy as clicking on a button. The datasets are all compiled and prepared for download. I love these sites.

 Description.

Government agencies are excellent sources of data.

Sometimes, however, data have to be copied-and-pasted or *scraped* from web sites or reports. These processes can sometimes introduce errors. This happens more often than is supposed in cases in which the process is highly automated and there is little effective data validation before analysis. Beware of scraping without adequate data scrubbing.

Some errors in a dataset may occur when the acquired datasets are combined. The integration process of putting the datasets together usually aims to add either variables or observations. There are several terms that refer to this process. *Merging* datasets using a common variable *key* adds variables to existing observations. *Concatenating* datasets usually involves adding observations. *Appending* may add variables to existing observations or observations to existing variables. *Aggregating* involves mathematically consolidating either observations (e.g., *replicates*) or variables (e.g., *transformations*).

Sometimes the process, automated in most software used in data management, isn't consistent, complete, or appropriate for one reason or another, usually because of irregularities in the dataset. Examples of such errors are mismatched observations or variables, the inclusion of replicates and other quality-control samples, and glitches from extraneous formatting.

It is the responsibility of the data analyst to check for errors that might have been generated, but this isn't always done, at least not thoroughly. As a result, calculated statistics may be incorrect, often apparently at random. Look at the number of samples included in each calculation. They should all be the same or there should be a good reason why they aren't.

Be careful when you're gathering data together from different sources. They might not all be comparable.

It goes without saying that merging data from different sources can be satisfying yet terrifying, like bungee jumping, cave diving, and registering for Stats 101.

Description.

COLLECTED DATA

Data collected by the experimenter doing the analysis and reporting have the benefit of having a single, known person or organization controlling the process. There are challenges, however.

Every measurement can be thought of consisting of three elements:

- ❧ *Benchmark*. The accepted standard against which a data value is made. Scientific instruments, meters, rulers, scales, comparison charts, and survey question response options are all examples of measurement benchmarks.
- ❧ *Processes*. Repetitive activities that are conducted as part of generating a data value. Equipment calibration, measurement procedures, and survey interview scripts are examples of measurement processes.
- ❧ *Judgments*. Decisions made by the individual to create the data value. Examples of measurement judgments include reading instrument scales, making comparisons to visual scales, and assigning correct codes for data entry.

For any particular data type, all three of these elements change over time. *Benchmarks* change when new measurement technologies are developed or existing meters, gauges and other devices become more accurate and precise. Standardized tests, like the SAT, change to safeguard the secrecy of questions. Likewise, *processes* change over time to improve consistency and to accommodate new benchmarks.

Judgments improve when data collectors are trained and gain work experience. Such changes can create problems when historical and current data are combined because variance differences attributable to evolving measurement systems can produce misleading statistics.

Be sure you read the scale right.

For data collected by the original researcher, look for quality control that considers the *benchmarks*, *processes*, and *judgments* used to generate the data and the *reference*, *replication*, and *randomization* used to minimize extraneous variability. Look for indications that the process involved sound scientific practices, including hypotheses stated before data collection, statistical sampling, data scrubbing, and peer-reviewed analyses. Specific items of concern are detailed in the following sections.

Description.

DATA CONTEXT

Datasets having different contexts may not be compatible for analysis. Inconsistencies in such datasets may be attributable to differences in data definitions, differences in the conditions under which the data were generated, differences in business rules and data administration policies, and differences related to the passage of time.

For example, metrics may have been measured using different devices or procedures, like pH from a titration, meter, or indicator paper. They may have the same variables but come from different populations or have been generated using different sampling strategies.

These differences can be overtly stated in metadata or buried deep in the way the creation of the data evolved. In either case, the errors aren't always visible in the actual data points; they have to be discovered. And even if you discover inconsistencies, you may not be able to fix them. Percentages and other calculated variables are often red-flags for hidden inconsistencies. Be sure it's clear what the bases for the percentages are.

INCONSISTENT DEFINITIONS

When you combine data from different sources, or even evaluate data from a single source, be sure you know how the data metrics were defined. For example, some counts of students in college might include full-time students at both two-year and four-year colleges. Other counts may exclude two-year colleges but include part-time students.

Say you're analyzing the number of diabetics in the U.S.. The first glucose meter was introduced in 1969, but before 1979, blood glucose testing was complicated and not quantitative. In 1979, a diagnosis of diabetes was defined as a fasting blood glucose of 140 mg/dL or higher. In 1997 the definition was changed to 126 mg/dL or higher. Today, a level of 100 to 125 mg/dL is considered prediabetic.

Data definitions make a real difference. So, if you're analyzing a phenomenon that uses some judgment in data generation, especially

This is me buried in my metadata.

phenomena involving technology, be aware of how those judgments might have evolved.

Description.

INCONSISTENT CONDITIONS

In addition to different data definitions, the context under which a metric was created may be relevant. For example, the Major League Baseball (MLB) record for most home runs in a season has been held by eight men. The four who have held the record the longest being: Babe Ruth (60 home runs, 1919 to 1960); Roger Maris (61 home runs, 1961 to 1997); Ned Williamson (27 home runs, 1884 to 1918); and Barry Bonds (73 home runs, 2001 to 2019). The other four record holders held their record for fewer than five years each.

During those time periods, there have been changes in rules, facilities, equipment, coaching strategies, drugs, and of course, players. It would be ridiculous to compare Ned Williamson's 27 home runs to Barry Bonds' 73 home runs.

There are many examples of data being generated under different conditions. Consider how perceptions of race and gender might be different in different sources, say a religious organization versus a federal agency. Even surveys by the same organization of the same population using different frames can produce different results. Be sure you understand the contexts data have been generated under when you look at files that have been combined.

INCONSISTENT TIME

Time is perhaps the most challenging framework to match data on. In business data, for example, relevant parameters might include: fiscal and calendar year; event years (e.g., elections, census, leap years); daily, monthly, and quarterly cutoff days, and seasonality and seasonal adjustments. Data may represent snapshots, statistics (e.g.,

Time scales can be frustrating to deal with.

moving averages, extrapolations), and planned versus reprogrammed values. In addition, sometimes, the rules change over time. The first fiscal year of the U.S. Government started on January 1, 1789. Congress changed the beginning of the fiscal year from January 1 to July 1 in 1842, and from July 1 to October 1 in 1977. Time is not on your side no matter what Mick Jagger says.

DATA NUMBER AND QUALITY

SAMPLE SIZE

The number of observations used to describe a dataset is important for several reasons. First, resolution will be better if more observations are used to evaluate a variable's central tendency, dispersion, and frequency distribution. Second, sample size is a good statistic to check to see if any observations have been inadvertently omitted. Third, sample sizes are a good way to evaluate the importance of groupings of observations. So, sample size plays a role in identifying data errors, evaluating subgroups of data, and providing adequate resolution for calculated statistics. It is the first statistic you should look for in reviewing a data analysis.

VARIABLE RELEVANCE

It's good to examine whether the variables evaluated in a study are sufficient to answer the study's research question. Measuring the wrong things or not measuring all the right things may seem ridiculous but it is not uncommon. The wrong fraction of a biological sample could be analyzed. The wrong specification of a manufactured part could be measured. And in surveys, the demographic defining the frame could be off-target.

Sometimes collecting supporting data is neglected, like recording ambient conditions in environmental sampling or questioning respondent demographics in surveys. This issue can occur for an individual data point or for the whole dataset. Identification of these issues can be challenging if not impossible if you're not knowledgeable of the subject area.

DATA QUALITY

Data quality involves sound sampling and measurement to ensure that the data are representative of the population. The best way to evaluate this is to look for processes that involve the three Rs—reference, replication, and randomization. If metadata aren't provided to demonstrate the data generation process, question whether something is being forgotten, ignored, or hidden.

DATA ISSUES

Every dataset has issues that the data analyst has to deal with before conducting the statistical analysis. For example, *replicates* and other quality control samples have to be removed. *Missing data* has to be examined to identify any patterns that might influence analysis interpretation.

Censored data are reported as "less than" (<) or "greater than" (>) some limit of measurement usually because the metric can't be quantified accurately with the available measurement instrument. Very low concentrations of chemicals in environmental or biological samples, for example, are censored when the instrument can detect the pollutant but not quantify its concentration. Multimeters for electrical testing have upper limits on the voltage they can measure thus censoring higher values.

Sometimes entries for censored data can be problematic because the < and > qualifiers are interpreted by software as text, thus upsetting a variable's number formatting.

That must be an outlier.

The same is true when text abbreviations like NA or ND are used to designate missing values.

There are a variety of ways to address censored or missing values either by replacing affected data points or by using statistical procedures that account for censored data. Nevertheless, censored and missing data are a nightmare for applied statisticians because there is no consensus on the best way to approach the problem in a given situation.

Outliers are anomalous data points that just don't fit with the rest of the measurements in the dataset. They present big challenges in data analysis. You might think that outliers are easy to identify and fix, and there are many ways to accomplish those things, but there is enough judgment involved in those processes to allow damning criticism from even untrained adversaries. That is a nightmare for an applied statistician. They can be 100% in the right yet still be made to appear as a con artist.

POOR QUALITY DATA

Some data are, or at least appear to be, of low quality. They might be impossible or unlikely values that are outside the boundaries of the metric scale. Examples include pH outside of 0 to 14, an earthquake larger than 10 on the Richter scale, a human body temperature of 115°F, negative ages and body weights, and sometimes, percentages outside of 0% to 100%. Look for data *minimums* and *maximums* if they are reported.

Some data are accurate but not precise. That is a nightmare for statisticians because statistical tests rely on extraneous variance to be controlled. You can't find significant differences between mean values of a metric if the variance in the data is too large. A large variance in a metric of a dataset is easy to identify just by calculating the standard deviation and comparing it to the mean for the metric (i.e.,

Description.

the *coefficient of variation*). There are methods to adjust for large variances, but the best strategy is prevention.

All of these data quality issues should be documented in a data analysis report.

DESCRIPTION FOCUS

Data descriptions might be approached differently depending on whether they are focused on population characteristics, snapshots in time, changes in conditions, trends and patterns, or anomalies.

POPULATION CHARACTERISTICS

Most descriptions of data are probably focused on characterizing

I'm being an outlier.

populations. For continuous scales, look at the *median* and the *mean*. If they're close, the frequency distribution is probably symmetrical. You can confirm this by looking at a histogram or the skewness. If the *standard deviation* divided by the mean (*coefficient of variation*) is over 1, the distribution may be *lognormal*, or at least, asymmetrical. *Quartiles* and *deciles* will support this finding.

Compare the measures of central tendency and dispersion. If the dispersion is relatively large, statistical testing may be problematic. For grouping (nominal scale) variables, look at the frequencies of the groups. You'll want to know if there are enough observations in each group to break them out for further analysis.

SNAPSHOT

What do the data look like at one point? Usually, it's at the same point in time but it could also be a common condition, like after a specific business activity, or at a certain temperature and pressure. Snapshots aren't difficult. You just decide where you want a snapshot and record all the variable values at that point. The thing you look for in a snapshot is something unexpected or unusual that might direct further analysis. Consider whether a data anomaly represents an *outlier* or a *rare event*. Snapshots can also be used as a baseline to evaluate change.

CHANGE

Change usually refers to differences between time periods but, as with snapshots, it could also refer to changes in some common conditions. Change can be difficult, or at least complicated, to analyze because you must first measure the variables at the baseline time (or condition) and the subsequent time (or condition), and then

Description.

calculate the change you want to explore. When calculating changes, be sure the intervals of the changes are consistent.

Look for changes that are very large in either the negative or positive direction. Check to see if the percentages of change are consistent for all variables. Consider what some reasons for the changes might be.

Calculate the mean and median changes. If the indicators of central tendency are not near zero, you might have a trend. Verify the possibility of a trend by plotting the change data. You might even consider conducting a statistical test to confirm that the change is different from zero.

If you pause a video, you get a snapshot.

TRENDS, PATTERNS, AND ANOMALIES

Finally, look for trends, patterns, and anomalies in the data. This is best done graphically with *scatter plots.* Trends, patterns, and anomalies are described further in Chapter 4 (Graphs) and Chapter 7 (Data Relationships).

SCALES AND STATISTICS

VARIABLE SCALES

First, review the measurement scales of the variables so you know what analyses would be applicable to the data. Also, look for nominal and ordinal scale variables to consider how the observations might be segmented.

If it looks like different scales are being used for a variable, consider if this is an uncorrectable issue. For example, a variable might appear to use both ordinal and ratio scales, which might not be a *fatal flaw* if the ratio measurements can be rounded. A nominal-scale metric can appear to be measured on an ordinal-scale if numbers are used to identify the categories. Time and location scales can also be problematic. Compared to fixing metrics with inconsistent units, fixing metrics with inconsistent scales can be challenging.

Data for a metric should all be measured and reported in the same units. Sometimes, measurements can be made in both English and metric units but not converted when included into a dataset. Sometimes, an additional metric is included to specify the unit, however, this can lead to confusion.

A famous example of confusion over units occurred when NASA lost the $125 million Mars Climate Orbiter in 1999. Government engineers based their designs on metric units while contractor engineers provided crucial data in English units. The discrepancy wasn't noticed until well after the spacecraft crashed.

It was also the reason that the Air Canada Boeing 767 Flight 143, the Gimli Glider, ran out of fuel in 1983. Because of a fuel gauge malfunction, a dipstick measurement of fuel in the aircraft's tank had to be converted from centimeters to liters to kilograms. However, the density value for the jet fuel used in the conversion was in pounds/liter instead of kilograms/liter so the weight of fuel on the plane was underestimated by almost half.

Consistency of units is not a new issue. Experts believe that the Swedish warship Vasa sank on its maiden voyage in 1628 because it was top-heavy and asymmetrical making it heavier on the port side. The asymmetry has been attributed to carpenters from Sweden using 12-inch Swedish-foot rulers while carpenters from Holland used 11-inch Dutch-foot rulers.

Fixing metrics that have inconsistent units is usually straightforward if the problem is recognized in time.

STATISTICS

Be sure that the description provides statistics for the *number of samples*, *central tendency*, *dispersion*, and *distribution shape*. Also, be sure the particular statistics are appropriate for the variable scales. Look for any comparisons to common baselines or previous values for context.

This is the closest I'm going to get to swimming.

If the analysis involves groupings within the dataset, start by looking at the statistics for the entire dataset. Divide and aggregate groupings only after you have a feel for the global situation. The reason for this is that the number of possible combinations of variables and levels of grouping variables can be large, even overwhelming. Each one could be an analysis in itself if there are enough observations. Like peeling an onion, explore one layer of data at a time until you get to the core.

SUMMARY

Don't rely on any single number or graphic for describing the key statistical criteria—frequency, central tendency, dispersion, and distribution form—of a dataset. Statistics are like presidential advisors. If a president has no preconceived notion about an issue, he will listen to the opinions of his advisors and then make a decision. If he has an intuitive feeling about an issue, he will listen to the opinions of

Description.

his advisors and then send the advisors that don't agree with him off on a tour of New Jersey's landfills.

Data analysis should be viewed the same way. Sometimes a particular statistic or statistical test doesn't agree with other analyses. It happens all the time. Understand the consensus of what the statistics are saying and send the exception to the landfill.

Don't be discouraged if you find the amount of information in his chapter overwhelming. There is a LOT of it for sure, but it's all foundational information. You'll read about the same concepts of *population*, *data*, *scales*, *sampling*, *variance*, and so on throughout the rest of **Stats with Kittens**, Stats 101, **Stats with Cats**, and whatever other instructional material you might encounter. You're over the hump.

And as promised, describing a dataset isn't that hard compared to describing a person.

Was that not-as-bad or much-worse-than I imagined?

Description.

A picture is the best description.

Description.

CHAPTER 4. GRAPHICS.
Seeing Data Is Believing Data.

INTRODUCTION

Everybody learns about graphs in elementary school, even continuing through high school. So, what more does everyone need to know? Why is the topic part of learning about statistics in everyday life?

It's because the pie charts, bar charts, and scatter plots you learned about are but a frost layer on the iceberg of statistical graphics.

You probably don't even realize how many types of statistical graphics you see every day that go far beyond the graphs you learned in school. There are graphs for illustrating data properties, mixtures, distributions, relationships, and concepts. There are graphs for specific data types, graphs for repeating data, and graphs that combine graphs. There are graphs that show flow, interconnection, organization, or concepts instead of data. There are even graphs that we all use every day but don't think of as graphs. There are just too many different types of graphs to count.

I can do almost all of these charts in Excel.

Nobody knows every kind of graph, even professional statisticians. Nevertheless, it's important to know how to look at graphs in general and decide how they might be illustrative or misleading.

That's what this chapter is about.

THE JARGON OF STATISTICAL GRAPHICS

First, let's get past the basic jargon of plots, charts, graphs, and diagrams. All of these terms are defined as visual representations of data. All are used synonymously. All are used as both nouns and verbs. All have other meanings.

To split hairs:

- *Plots* tend to place more emphasis on individual data points.
- *Charts* tend to involve lines and areas more than individual points.
- *Graphs* tend to be more mathematically complex than charts and plots.

Graphics.

😼 *Diagrams* tend to be more artistic and fill the entire plot area.

Most specific kinds of visual representations of data are called *plots* or *charts*, and to a much lesser extent, *diagrams*. Diagrams tend to imply some drawing rather than plotting, like Venn *diagrams* and brainstorm *diagrams*. The term *graph* is used mostly in a general sense. A few statistical graphics are referred to as *maps* without being linked to locations, like *mind maps*, *heat maps*, and *tree maps*. Some graphics have unique names without using the four main terms, like *histograms*, *timelines*, and *word clouds*.

Not everyone would agree with these distinctions, of course. That being said, you can usually refer to visual representations of data by any of the four main terms, especially as a novice, though you may be called out by a smart-aleck critic. If you're referring to a specific kind of visual representation of data, one of the four terms might be preferred, for example, pie charts, scatter plots, block diagrams, and stream graphs.

All of these statistical graphics are sometimes referred to as *visualizations.* In general, **visualizations explore**. They are created to make sense of data visually and to explore data, often interactively. The process of visualization is by nature iterative in the sense that different options are tested for interpretability through the use of software.

Visualization usually starts with the default settings of the software, then explores different types of graphs and groupings of data, and finishes with labeling and even artistry. The processes used to create visualizations can be applied efficiently to almost any dataset.

Another term, *infographics,* is used to refer to combinations of visualizations, text, images, and artwork designed to inform or persuade general audiences. **Infographics explain**. They are specific, elaborate, attractive, and self-contained. Every infographic is unique and must be designed from scratch for visual appeal and overall reader comprehension. There is no software for automatically producing infographics the way there is for visualizations.

Visualizations tend to be more objective than infographics and are better for allowing audiences to draw their own conclusions, although the audience needs to have some skills in data interpretation. Data visualizations do not contain infographics.

Infographics are better than visualizations for guiding the conclusions of an audience. They are more subjective

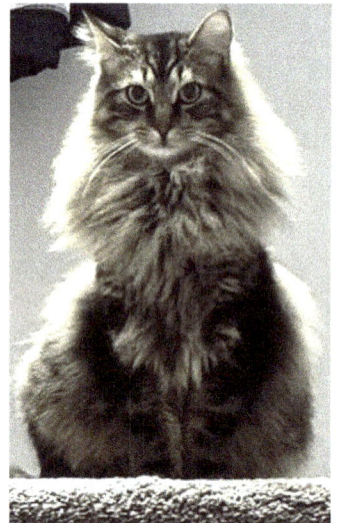

So, visualizations provide a visual depiction while infographics provide a data story. Sounds simple enough to me.

🐈 Graphics.

than visualizations. Infographics may or may not contain visualizations. Both can be static or animated. Both require a knowledgeable person to create them.

Table 17 summarizes characteristics of visualizations and infographics. Unfortunately, the two terms are not always used consistently. They are even used interchangeably by some people. Don't worry about it. Just remember:

Visualizations explore. Infographics explain.

Table 17. Characteristics of visualizations and infographics.

	Visualization	Infographic
Objective	Analyze	Communicate
Audience	Some data analysis skills	General audience
Components	Points, lines, bars, and other data representations	Graphic design elements, text, visualizations
Source of Information	Raw data	Analyzed data and findings
Creation Tool	Data analysis software	Desktop publishing software
Replication	Easily reproducible with new data	Unique
Interactive or Static	Either	Static
Aesthetic Treatment	Not necessary	Essential
Interpretation	Left to the audience	Provided to the audience

Understanding statistical graphics requires appreciating the *Three Fs of Statistical Graphics:*

- 🐾 *Foundation.* The fundamental type of graph selected on the basis of the *aim* and *focus* of the graph and the characteristics of the variables to be plotted.
- 🐾 *Framework.* The technical specifications of the graph: the plot area; the variables (*data series*); the points to be plotted, and the *axes*.
- 🐾 *Facade.* The labeling and other details that make the chart easier to understand or more appealing to audiences.

Graphics.

These concepts define every individual statistical graphic. They are what differentiate data presented in a STEM journal from the same data depicted in a MSM (mainstream media) story.

THE FOUNDATION OF GRAPHS

Before creating or even perusing any statistical graphic, you first have to understand a bit about the dataset. This is, of course, a chicken-egg dilemma because graphics are a fundamental way of understanding data.

Second, you have to have some notion of what the graphic aims to accomplish. Is it to show characteristics of the dataset or relationships between groupings in the data or between different metrics measured on samples? Whether you're creating the graphic or just studying it, know where the finish line is.

Finally, if you are creating the graphic, you'll need appropriate software. No software or internet application can create every possible statistical graphic. There isn't even a list of every kind of statistical graphic that software might provide. And even if there were, the list would become outdated quickly as new types of graphics are devised. Don't worry about these issues.

Chicken versus egg? There's no dilemma. Lizard eggs were around long before chickens.

There are plenty of common statistical graphics for any need you might have that can be created with readily-available software.

So, with data and a purpose, you can at least begin looking at statistical graphics. The place to start is the graph's *aim* and *focus*.

Aim is the reason the data are being plotted, that is, what the plot is expected to reveal, such as:

- Data frequency and distribution
- Variability and anomalies
- Relative proportions of the components of a mixture
- Properties or values of variables
- Properties or values of samples or sample groups
- Trends, patterns, or other relationships among variables.

Focus is the concept of what gets plotted. There are three choices:

- Plots of all the individual data points, or if there are too many, a random sampling of the points. This is what is usually done, at least to start.

Graphics.

- Plots of values that represent groups of data points. Usually means or medians are plotted, but the representative statistic could just as well be standard deviations or another descriptive statistic. The groups are usually a nominal-scale variable, but could be an ordinal-scale variable if there are only a few groups.
- Plots of a selected few data points as icons that show the characteristics of several variables. This option isn't used often because most statistical analyses focus on the big picture rather than individual data pieces. Perhaps as a consequence, these kinds of diagrams aren't supported by as many readily-available, easily-usable software as other types of diagrams are.

The focus of a diagram will be on either the *variables* or the *observations*. If the aim of the graph is to show how a number of *observations* are related to each other on the basis of one or more variables, one plot is used for each variable or set of variables showing all the observations. *Scatter plots* are used in this way.

How am I supposed to see the big-picture from all these pieces?

If the aim of the graph is to show how a number of *variables* are related to each other for a very small number of observations, one plot is used for each observation or group of observations. Icon plots (*radar plots*, *polygon plots*, *star plots*) are used for this purpose. HINT: most of the time, the focus will be on showing the relationships between observations rather than relationships between variables.

In creating a graph, the first thing that is done is determining what kinds of graphs could be drawn by answering these questions:

- **What is the aim of the graph?** Will the primary aim be to show: data frequency and distribution; proportions of the components of a mixture; properties or values of a variable; or trends, patterns, or other relationships among variables.
- **Is your focus on variables or samples?** Will the focus be on showing how a number of samples are related to each other on the basis of one or more variables *or* on showing how a number of variables are related to each other for a very small number of samples?
- **Will individual points or group means be plotted?** How many data points have to be plotted? Will the points be shown individually or will the averages of groups of data points be shown (this is useful when there are a large number of data points)?

Graphics.

 How many axes will be needed? The axes to be used have to be determined based on the aim and focus, the number of variables, whether the variables are measured on the same or different scales, and whether the scales are discrete or continuous.

Once those questions are answered, the basic kinds of graphs appropriate for a dataset can be determined. Consider the same questions whether you are studying a graph created by someone else or creating your own.

This is easy but yet more demanding than in Stats 101 where the instructor tells everybody what to do.

THE FRAMEWORK OF GRAPHS

The *framework* of a graph refers to the elements that make a graph useful for the audience. The graph wouldn't exist if these parts were not included. They're also the parts that can make a graph misleading if they are not specified correctly.

PARTS OF GRAPHS

The *Framework* of a graph consists of the specifications of the *plot area*, the variables (*data series*), the *axes*, the *points* to be plotted, and some *labeling*.

PLOT AREA

The *plot area* is the space at the center of the graph where data are plotted. In two-dimensional graphs, the plot area is usually square, a rectangle longer in the vertical direction, or a rectangle longer in the horizontal direction. The choice of the dimensions should consider the specific data, how that type of data is customarily graphed, the medium where the graph will be displayed, and of course, the audience. The *plot area* is bounded by the *axes*.

Considering the data, the dimensions of a graph should have the longer axis where there are more *meaningful units*. Meaningful units are the number of units in the data range that provide precise, clear-cut, reliable information. Units on a graph can be divided by eye, perhaps by halves and even quarters. However, if the smallest increment the instrument gathering the data can detect is larger than that, further visual subdivision is

Data stay in the plot area.

meaningless. So, multiply the number of meaningful units by the data range for each metric and decide which has the most meaningful units. Make the axis for that metric the longer axis.

Graphics.

If gridlines are provided to help viewers tie data points back to the axes, they shouldn't be too intrusive. They should be colored as a light gray rather than black. Also, if the horizontal and vertical axes are measured in the same units, the gridlines should be approximately square. If the graphic is a geographic map, the axes must be scaled identically.

Considering the presentation, graphs meant for computer screens should make the horizontal dimension longer. Graphs meant for books should make the vertical dimension longer. This certainly is not a hard-and-fast rule.

There are two exceptions to these guidelines. First, time axes should almost always be placed on the horizontal axis. Second, a *dependent variable* (the variable being analyzed in an experiment, also called the *outcome variable*) should always be placed on the vertical axis.

Plot areas should generally not be filled or shaded unless there is a compelling reason for doing so. Visual clarity is the primary reason for shading the plot area.

DATA SERIES

Data series are the variables plotted on a graph. Each data series must be associated with only one set of axes although several data series can be associated with the same axes. The scale of the axis must be compatible with the scale of measurement of the data series. A graph can have more than one data series.

AXES

Of all the elements of a graph's framework, the *axes* are the most challenging. There are many things to specify and no rules on the best ways to do them. There are some conventions. Dependent variables are plotted on the vertical axis. Time is plotted on the horizontal axis. Maps have to have exactly the same number of data units per inch of axis on both axes. After that, it's how you want to present the data.

The number of axes that are needed will depend on how many variables have to be plotted and their scales of measurement. Some people prefer plotting several data series on the same graph even if an additional axis has to be used. They prefer packing as much information into the graph as they can. Other people prefer plotting each data series separately on their own graphs. They prefer to keep the graph simple and uncluttered. There is no one correct way of assigning axes.

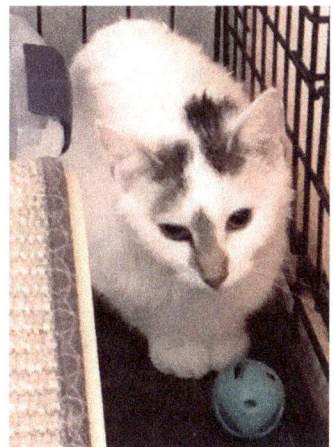

I think you're trying to pack too much stuff in this space.

One misleading notion about axis scaling is that all axes should begin at zero, perhaps a product of Huff-induced

Graphics.

paranoia from the 1950s. This is taught in introductory math classes by teachers wanting to avoid trying to explain *resolution, precision,* and *meaningfulness* as it applies to data units. It's like telling a child not to talk to strangers. That's fine for a five-year-old but by the time they grow up, they'll have to talk to many strangers, like blind dates, stadium beer vendors, and IRS auditors.

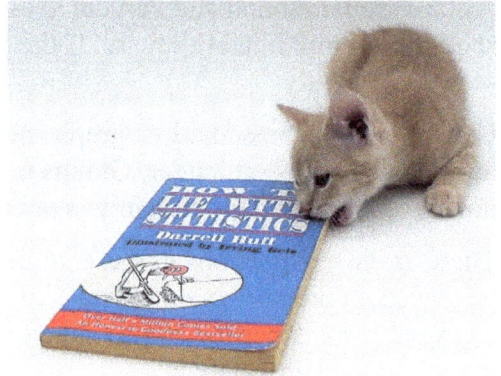

Unfortunately, some people never learn how axes should be scaled. Using a zero point for a time axis or for interval scales like temperature makes no sense. Starting an axis at zero is problematic for scales with negative numbers and scales with little

Don't believe everything you read.

variation across the axis range. On the other end of the spectrum, expanding an axis scale to show minute, meaningless differences leads to the fallacy of *false precision.*

DATA POINTS

Data points are a fundamental part of graphs that you might think are not controversial at all. They do, however, sometimes present problems. If the points used on the graph are too big, they can obscure other observations and be misleading, especially on maps. If there are many observations to plot, even tiny data points can turn into graphic mud.

Sometimes the data on a line graph or scatter plot will look sparse but only because observations with the same values for both axes overplot each other. This is especially true when one or both of the axes use ordinal scales.

If the data represents spatial locations, consider how large the data points would appear in the real world. A data point that seems to be a reasonable size on a graphic may represent a huge area that a sample would have been taken from.

So, in creating a graph, start with small simple dots to show the locations of data points relative to the axes. Look for overlap that might require frequency bubbles or plotting group means. If data density is not an issue, decide if you want to show more information through *embellishments*, like *data icons* and *labels*. That's called the graph's *Facade.*

LABELING

The most important *Framework* elements of a graph that require labels are the axes. Each axis must have an explanatory title and values for the scale. *Tick marks* are usually helpful also but must be compatible with the precision of the metric. How

Graphics.

many values are labeled is a matter of preference but the labels should be easily readable.

Other labels that are useful for most graphs are a *title*, a *legend* identifying the data series, and a text box identifying the *source of the data*. Some graph creators prefer to label individual points and lines rather than having a legend.

FRAMEWORK CHALLENGES

Some large datasets present special challenges in creating graphs.

TOO MANY OBSERVATIONS TO PLOT

There aren't many instances in which statisticians get to complain about having too many samples, but trying to graph thousands of data points is one of those occasions. This has become a more common issue since the 2000s when *Big Data* came to be recognized. There are three options for plotting large datasets:

- Take a random sample of the data and plot those points.
- Graph the means of data groups.
- Create a bubble plot where the bubble sizes represent the numbers of samples.

In the first case, the graph should have a notation on it explaining how many observations were available, how many were selected, and how the selection was carried out. In the second case, the number of observations in each group should be indicated, usually by a textual notation on the graph. On bubble plots, a visual reference for frequency can be linked to bubble size (i.e., area or diameter).

TOO MANY VARIABLES TO PLOT

Having many variables to explore simultaneously is a common challenge. If there aren't too many data points, a matrix plot might be appropriate. Matrix plots contain small scatter plots of a set of variables, taken two at a time, arrayed as a matrix. Because of their small size, the scatter plots turn to mud if there are a lot of data points. Matrix plots can be good for *variable triage* by allowing the many variable relationships to be viewed quickly so that the ones with the strongest trends and patterns can be identified.

ADDITIONAL SCALES NEEDED

Sometimes there are challenges when several variables have to share axes on the same graph. For example, accountants need to graph dollar amounts for large and small budget categories. Meteorologists need to plot ambient temperatures measured at several elevations. Public health officials have to plot hospitalizations from large and small populations. In all these examples, the same metric could be plotted on one axis even though the values of the groups are very different.

Graphics.

There are four ways to address the need for additional axis scales: add new axes; convert variable units; shift variable units; and change variable scales.

ADD NEW AXES

Most graphing software allows the addition of a second y-axis. This allows two variables to be scaled independently. Some software also allows the addition of a z-axis. This allows three variables to be scaled independently. Unfortunately, some people have difficulty understanding data presented in three dimensions, so this shouldn't be the first course of action.

CONVERT VARIABLE UNITS

If two variables measure the same concept except in different units, the variables can be converted to a common unit. The conversion might involve familiar units, such as converting cubic feet of a liquid to gallons, or specialized relationships, such as converting inches of snowfall to inches of rainfall.

SHIFT VARIABLE UNITS

There are just too many of us.

If two variables use the same unit of measurement but have values on very different ranges of the scale, the variables can be multiplied or divided by constants so that all the variables fall in the same range. Factors of ten are used commonly for this purpose, although the constants could just as well be 5 for one variable and 20 for another.

CHANGE VARIABLE SCALES

Variables can be *normalized* by subtracting the minimum value from each observation and then dividing by the data range. This rescales the range to 0 to 1.

Variables can be *indexed* to more representative values by dividing each measurement by values that facilitate comparisons. Good examples are economic metrics that are indexed to the Gross Domestic Product (GDP).

If variables are normally-distributed, they can be converted to z-scores by subtracting the mean from each value and then dividing by the standard deviation. 95% of the values will then fall into the range -2 to +2. This is called *standardization*. This scale change does not affect patterns of relationships between variables even though the data values change.

Graphics.

Figure 19 shows data from an example presented in Chapter 6. The original data have a range of 56.9 to 75.2. The *normalized* data have a range of 0 to 1. The *indexed* (to 50) data have a range of 1.14 to 1.50. The *standardized* (to z-scores) data have a range of -1.75 to 1.72.

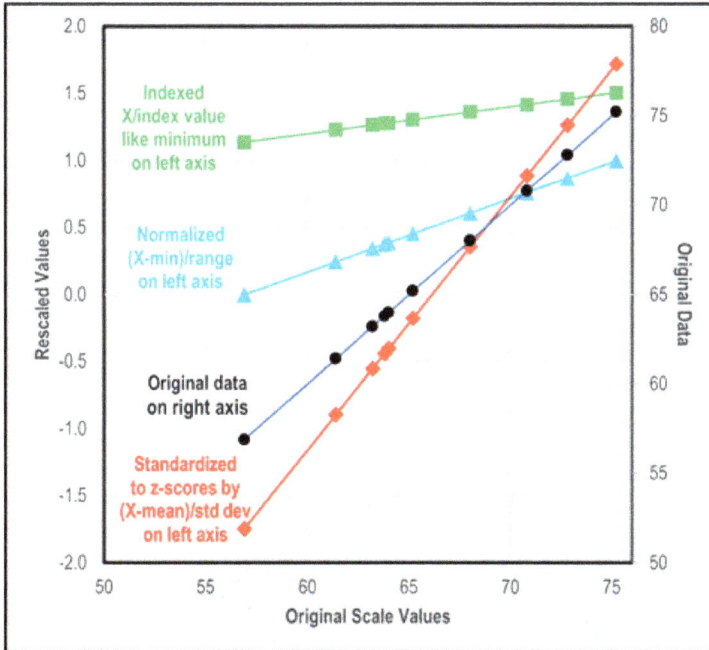

	Rescale		Data									Range
Original data	from chapter 6	56.9	61.4	63.2	63.8	64	65.2	68	70.8	72.8	75.2	18.30
Normalized	(X-min)/range	0.00	0.25	0.34	0.38	0.39	0.45	0.61	0.76	0.87	1.00	1.00
Indexed (to 50)	X/index value	1.14	1.23	1.26	1.28	1.28	1.30	1.36	1.42	1.46	1.50	0.37
Standardized	(X-mean)/std dev	-1.75	-0.90	-0.55	-0.44	-0.40	-0.18	0.35	0.88	1.26	1.72	3.46

Figure 19. Normalized, indexed, and standardized data transformations.

All of these scale transformations compress the original data range while preserving the order of the original data points. This would allow the transformed data to fit on a graph axis more efficiently. The scales all have perfect linear correlations with the original data so trends and patterns in the original data are maintained in the transformed data.

This is just an example. Results for an actual situation requiring a scale change would be much more impressive.

Graphics.

THE FACADE OF GRAPHS

The *Facade* of graphs is all about *embellishments* that are added to either convey additional information, draw viewer attention, or make the graph easier to interpret. The key to using the facade of graphs effectively is understanding the *audience*.

WHO'S LOOKING

We all see statistical graphics every day, at school, in the media, and on the internet. Each of those content providers needs to appeal to a different audience. Academic graphics can be unusual, complex, and sterile because students have to study them to get good grades. The mainstream media has to keep their graphics simple yet appealing for non-technical readers. The internet has everything. If you plan on providing or reviewing a statistical graphic, you should understand who the intended audience is before you create or criticize.

ART OF THE CHART

There are competing philosophies of graphing, divided to some extent by perceptions about the audience for a graph. The philosophy of many art directors of newspapers and magazines is to keep the graph simple, interesting, and attractive in order to engage the reader. Look no further than *USA Today*, *Newsweek*, or *Time* to see three-dimensional exploded pie charts and bar charts made of little soldier icons, dollar bills, or some other cutesy graphic.

I found the key!

In contrast, Edward Tufte, an expert in informational graphics, espouses a philosophy that assumes the audience is knowledgeable and interested. He maintains that graphs should provide only as much information as needed, and do it as efficiently as possible. Tufte's guidelines include:

- 🐾 The dimension of a chart must not be greater than the dimension of the data. For example, if you're plotting two variables on a Cartesian (rectangular) graph, don't add an extra axis (dimension) for depth. It may be visually appealing but it's scientifically misleading.

- 🐾 Data must be presented in context. You shouldn't show just part of a dataset. Label everything you need to make sure the data are presented accurately and meaningfully.

🐈 Graphics.

- Maximize the *data density* and the *data-ink ratio*. Put enough data in your graph to make it worthwhile. Eliminate everything on the chart that isn't data or contributes to the interpretation of the data.

- Eliminate *chart junk*, the unnecessary pictures, dimensionality, grid lines, fill patterns, and other objects that clutter a graph while adding no scientific value.

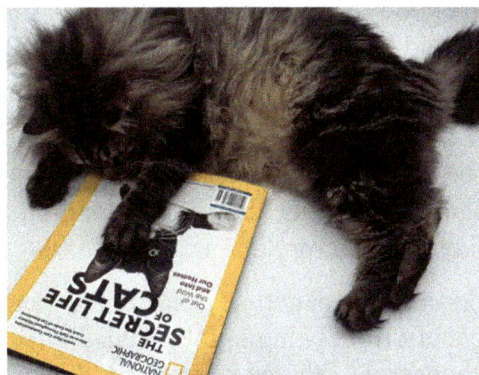

I love their charts and maps.

Tufte believes he has the audience's attention while the art directors believe they have to compete for it. Then there are authors like David McCandless who look at presenting data from an artistic perspective. Graphics that follow this perspective are truly works of art even though the graphs are based on data and aimed at specific audiences. All of these graph developers make valid points. They simply have different perspectives, different audiences, different aims, and different data.

CHART ENHANCEMENTS

Chart embellishments are not the cursed objects that Tufte makes them out to be so long as they serve a useful purpose. If they provide additional relevant information, they are useful. If they make the graph easier to interpret, especially for readers who need some assistance, they are useful. If they are eye candy meant to draw viewer attention, maybe they shouldn't be added to the graph. There are better ways to peak reader attention than some other quaint graphics. Here are three ways that charts can be usefully embellished.

STRUCTURAL LINES

Structural lines can be useful or superfluous, depending on your perspective. Some people like black, 1-to-2 point *borders* around the plot area and the whole chart because the lines draw the eye into the data presentation.

Some people like a few *gridlines* to help them track data points back to axis values. Such gridlines should be subtle, perceptible but not obtrusive, like 0.5-to-0.75 point, light-gray lines.

Axis lines are usually black and the same or a bit heavier than the border of the plot area.

Tick marks to subdivide scale values are helpful so long as there aren't too many and they occur at *meaningful* intervals. For example, tick marks at multiples of 2, 5, or 10

Graphics.

are good for most axes. Tick marks for time axes should be based on days, weeks, months, or years, whatever is relevant.

If there are multiple y-axes to scale multiple data series, the axes and scale values can be made the same *color* as the data point identifiers to facilitate interpretation.

Sometimes, patterns appear in the data that would benefit from being highlighted. *Divider lines* or *shapes* and *fills* can be used for this purpose. Again, they should be subtle so they don't distract from the data points.

I'm embellished.

POINT IDENTIFIERS

Point identifiers are necessary, even if they are just simple dots. They can often be made more useful with *embellishments* that provide additional information.

Data points can be identified in many ways, usually using different *shapes*. The shapes can be enhanced by using different *sizes* and *colors* to show group memberships. Care should be taken to ensure that the shape sizes aren't so large that they misrepresent the position of the data point on the graph. This is especially true for maps where the size of a data point represents a physical space.

The selection of colors for data points should be sensitive to any color blindness some readers may have. Points can also be represented by *clipart* that provides relevant information on group membership, like logos for businesses, schools, sports teams, and so on. Points can also be *labeled*, although positioning labels so they don't overlap can be challenging.

Icon plots are scatter plots in which the point identifiers are replaced by miniature versions of multivariable graphs that provide additional information about the data point. They are visually complex but can display more dimensions of information than other charts.

The graphs in icon plots could be *pie charts*, *bar charts* (horizontal and vertical), *area charts*, lines (*sparklines*, *profile lines*), radial plots (*radar plots*, *star plots*, *sun ray plots*, *polygon plots*), and even more complex icons like *Chernoff Faces*. Examples of these icons are shown in Figure 20.

Because humans are accustomed to distinguishing subtle differences in human faces, zany statistician Herman Chernoff reasoned that representing data values using facial characteristics would facilitate identifying patterns. He developed a procedure in 1973 to replace the conventional axes for icons by such characteristics as head size and shape, and the positions, lengths, and angles of the eyebrows, eyes, nose, ears,

Graphics.

Kitten or cat? Moose, on the left, is a one-year-old kitten. Marbles, on the right, is a four-year-old cat.

and mouth. Hence, Chernoff faces … proof that the disco era of the 1970s was a lot more fun than anyone wants to remember.

Chernoff faces aren't used often in statistical graphics because they are not familiar to most audiences and they require specialized software to create. But boy, are they fun.

PATTERN IDENTIFIERS

If there are trends, patterns, clusters, or anomalies in a data plot, it may be useful to add embellishments that highlight the interpretation. This is mainly to aid non-technical audiences.

Trend lines are the most commonly used embellishment. They can show how data compare to linear or curvilinear models.

Context points or *lines* can be used to show data for some benchmark condition. An example would include using a context point or line to show a group mean alongside the raw observations.

Connector lines are used to associate data groups, often in bar charts. *Divider lines, shapes,* and *fills* with different color schemes can also highlight group membership.

Flow lines, particularly on a scatter plot, are used occasionally to indicate how a particular observation had different values at different times or under different conditions.

Text boxes can be used on complex graphs to clarify interpretations for non-technical audiences.

The epitome of a chart embellishment is the use of *motion*. Motion charts involve creating a series of charts for different times or conditions and compiling them into an animated gif or a video for presentation. Motion charts are laborious to create but they are very compelling for depicting a data interpretation.

There's no pattern here. Maybe it's because they're frogs and not data.

 Graphics.

Pie-chart icons show proportions as angles of the circle. It's difficult to identify patterns if there are more than a few variables. If you can't use color, you're at an added disadvantage.	
Sun ray icons can show actual values as well as percentages. Each variable is assigned a ray of the sun as an axis. The rays are set apart at equal angles so information is conveyed using shape rather than color or pattern.	
Star plot icons are similar to sun ray icons except that the axes (sun rays) outside the data values are omitted so every ray may be a different length. This difference allows patterns to be identified more easily at the expense of some perception of the values of the variables.	
Polygon icons take the objective of star plots one step further by eliminating all axes and filling in the resulting shape. Polygons show patterns very well but provide little sense of the values of the variables.	
Bar chart icons have the same problem as pie chart icons; as the number of bars increase, it becomes harder to distinguish patterns in the data especially if you can't use color.	
Profile icons make pattern detection easier by eliminating the bars and replacing them with the shape the bars formed. This difference is analogous to the difference between star icons and polygon icons.	
Sparklines are profile icons with the frame and the fill eliminated. They emphasize data patterns over data values.	
Chernoff face icons replace conventional graph axes by facial characteristics. It's easy to detect differences between faces but difficult to interpret patterns in the differences.	

Figure 20. Icons used to compare variables for individual data points.

Graphics.

SIX TYPES OF GRAPHS

In the 1950s, you only had to know about a few kinds of charts, mainly pie charts, bar charts, and scatter plots. In Huff's 1954 book **How to Lie with Statistics**, almost all of the graphs he presented as examples were bar/area charts, line charts, or scatter plots. Not even any pies. Furthermore, they were all hand drawn.

Back then, early mainframes did enable knowledgeable users to create rudimentary statistical graphics, though it required programming. Now, most everyone has access to a computer and software for graphics and statistical analysis. You can even find websites that will create fairly advanced statistical graphics for you. As a consequence, the number and variations of types of statistical graphics has skyrocketed.

Stop making me talk about Huff's book.

There are now more kinds of graphs than most people, even data analysts, would ever need to know about. Some are used specifically in particular disciplines, like finance or geography. Some are used for particular applications, like *Six Sigma* or *geostatistics*. But most graphs can be used for any type of data. Furthermore, while some diagrams are fairly unique, a good many are just variations of basic graphs.

Pie charts, bar charts, and scatter plots are still the primary types of statistical graphics that are taught in elementary schools. Calendars, maps and organization charts aren't even recognized by many people to be types of statistical graphics.

You won't see many of the types of graphs presented in this chapter in most Stat 101 courses. You will see them in advanced college classes, though, and not just STEM classes either. Even if you don't create them yourself, you'll see them everywhere on the internet and in the media. Be prepared.

Perhaps what distinguishes many graphs from each other isn't just their purposes, it's the statistical scales that the graph can accommodate.

For example, almost all *concept diagrams* use qualitative information. *Rose diagrams* are meant for repeating scales like compass directions; that's why they use a circular format. *Control charts* use some measure of time on the horizontal axis. *Pie charts* must use data measured in percentages that sum to 100%. *Radar plots* and other multivariable charts require the scales of the variables to be standardized so that all variables use a common scale. Both axes of *maps* have to be identically

Graphics.

scaled and formatted. And as described in Chapter 3, histograms are designed for interval-scale data but can be used for continuous-scale data by creating data *bins*.

With so many types of graphs and so many possible uses, it's tough even for professionals to keep them all straight.

There's really no foolproof classification system for statistical graphics let alone a comprehensive system for picking the perfect chart to use for a particular purpose and data set. Nevertheless, to try to simplify the process, consider these six categories of commonly used graphs based on their primary usage:

- **Property Diagrams**, used mainly to display statistical properties.
- **Mixture Diagrams**, used mainly to display the proportions or amounts of data groupings.
- **Distribution Diagrams**, used mainly to display data frequencies and the shapes of data distributions.
- **Relationship Diagrams**, used mainly to display the relationships between data points.
- **Concept Diagrams**, used mainly to display non-numerical information.
- **Combination Diagrams**, two or more types of charts that are incorporated in the same graphic, used for *data storytelling*.

Some graphs can serve multiple purposes and often do, so there is considerable overlap in these categories. For example, box plots show statistical properties like central tendency and dispersion but they also illustrate distribution shapes. Bar charts show mixtures but can also show statistical properties. Venn diagrams can show conceptual relationships or mixtures.

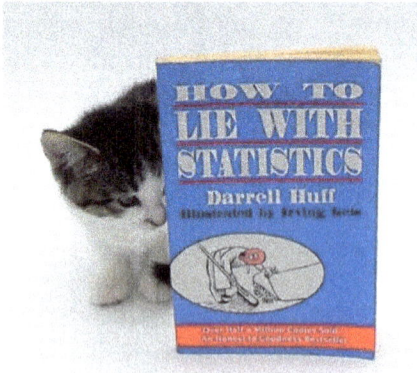

It would have been a thicker book if he had more charts.

Statistical graphics are complicated. If they weren't, every instructor would teach the same information in the same way to everybody. So, don't worry too much about how a diagram might be classified, focus on how the diagram could be used or is being used in a particular situation.

Here are twenty-six examples of the five basic types of charts plus three examples of combination charts. That's a lot more than Huff had to deal with in 1954.

1. PROPERTY DIAGRAMS.

Property diagrams aim to show a statistical property, usually measures of central tendency, dispersion, or frequency. Typically, a property diagram will show many properties of a dataset, often in different ways, at the same time. Box plots, described

Table 18. Examples of graphs used to show statistical properties.

Types of Diagrams Used to Show Properties														
		AIM — Purpose of Plot						FOCUS — What's Plotted				Usual Layout	Axis Scales		
Diagram	Description	Properties central tendency & dispersion	Distributions frequencies & shapes	Mixtures amounts & proportions	Relationships trends & patterns	Concepts non-numerical information	Time or Duration	One Data Point	Individual Data Points	Grouped Data	Nonnumeric Information	Line / Rectangle / Triangle / Cube / Circular / Ad Lib	Horizontal Axis	Vertical Axis	Other Axes
Autocorrelation plot	Correlation between the values of variable at different time points			X	X		X		X	X		Rectangle	C	C	NA
Box & whisker plot	Bar graph showing descriptive statistics	X	X						X	X		Rectangle	N	C	NA
Bullet graph	Bar charts with extra visual elements	X		X	X		X		X	X		Rectangle	N	C	NA
Calendar	Matrix of dates						X		X			Rectangle	O	O	NA
Candlestick, stock, high-low-close chart	Box plots showing stock data over time	X	X		X		X		X	X		Rectangle	C	C	NA
Combination chart	Different chart types shown on the same diagram	X	X	X	X		X	X	X	X		Rectangle	C	C	NA
Contour map	Scatterplot with lines showing equal values				X				X			Rectangle	C	C	O
Dot matrix chart	Data quantities represented by circles in a matrix		X	X					X	X		Rectangle	O	O	O
Funnel diagram, population pyramid	Bar chart centered on a point and sorted vertically		X	X					X	X		Rectangle	C	N	NA
Heatmap	Matrix showing data values using colors		X	X					X	X		Rectangle	O	O	NA
Maps	Diagram showing geography and other information				X	X			X		X	Rectangle	C	C	NA
Periodogram	Plot of signal amplitude vs frequency	X			X		X		X			Rectangle	C	C	NA
Radar chart	Multivariable radial chart				X			X				Circle	NA	NA	C
Rose diagram, radial bar chart	Bar chart displayed on a circle for repeating scales	X	X				X		X			Circle	NA	NA	C
Span chart	Bar chart of data ranges	X	X		X		X		X	X		Rectangle	C	O	NA
Spatial diagram	Diagrams showing data using geographic maps				X	X			X	X		Rectangle	C	C	NA
Spiral plot	Radial bar chart for repeating data	X	X				X		X			Circle	NA	NA	C
Star plot	Multivariable radial chart				X			X				Circle	NA	NA	C
Variogram	Plot of spatial variance vs distance between sample pairs			X	X				X	X		Rectangle	C	C	NA

Layout: Line=one axis; Rectangle=2 axes at 90°; Triangle=three axes at 60°; Cube=four axes at 90°; Circular=Multiple axes arranged radially; Ad Lib=No set layout.

Axis Scales: C=continuous ratio & interval scales; O=ordinal scales; N=nominal grouping scales; NA=no scale. Horizontal & vertical axes can be switched in some graphs.

in Chapter 3, show central tendency with the mean and median, dispersion with the maximum and minimum, interquartile range, and extreme values, and distribution shape by the way it is constructed. It can also be enhanced to show frequency by adding the width of the box as a graphical element.

Table 18 highlights nineteen types of graphs used to display information about statistical properties.

BAR CHARTS

Bar charts are the most commonly used graphic to display quantities, especially by groups, as well as mixtures and distributions. By the time you finish high school you've probably have seen dozens of bar charts and maybe even created some yourself as homework assignments.

Graphics.

Bar charts have an ordinal or continuous-scale variable on the vertical axis and a nominal or ordinal-scale variable on the horizontal axis. The axes are reversed in *column charts*.

The quantitative axis in bar charts usually represents *frequency*, either the number or percentage of observations. Data are represented by bars although other shapes are used occasionally. The shapes are usually two-dimensional but may be depicted in three-dimensions, mainly to infuriate statisticians who see no need for the artistry.

Variations of bar charts include: *area charts*, *side-by-side bar charts*, *stacked-bar charts*, *radial bar charts*, *Pareto charts*, and quite a few others.

Bars representing different data groups can be arranged *side-by-side* or *stacked*. A bar chart sorted by frequency is called a *pareto chart*, which requires that the horizontal axis uses a nominal scale. 100% Stacked bar charts can also show proportions. The bars can even be connected in *area charts*.

When a bar chart has two quantitative axes, one axis represents frequency and the other usually represents ordinal-scale groupings. In a histogram, for example, the horizontal axis represents the scale of the variable being depicted. If that variable is measured on a continuous scale, the scale of the bar chart axis has to be converted to ordinal-scale groupings, called *bins*. Picking an appropriate nominal-scale to represent a continuous scale is a challenge usually involving trial-and-error.

Bar charts are fun to make.

Figure 21 is a bar chart depicting the number of wins for 32 NFL teams during the 2021-2022 regular season. The bar chart was created for presentation on social media, so it has a lot of embellishments to attract attention. The audience was assumed to consist mainly of NFL fans who may follow just one team or many. No statistical knowledge was assumed. Ease of understanding was a primary concern.

Depicting the number of wins could also have been accomplished using other types of graphs, such as a scatter plot, but the bars provided several advantages over scattered data points. Mainly, the bars provide a focus on the organization of the teams. This was a consideration related to the *Foundation* of the graphic.

The bars depicting the number of wins are horizontal with the scale positioned at the top of the plot area. This quantitative scale was placed horizontally instead of the more commonly used vertical placement of bars to simplify formatting. The 32 teams required more space than the maximum 17 wins, thus creating a more compact

Graphics.

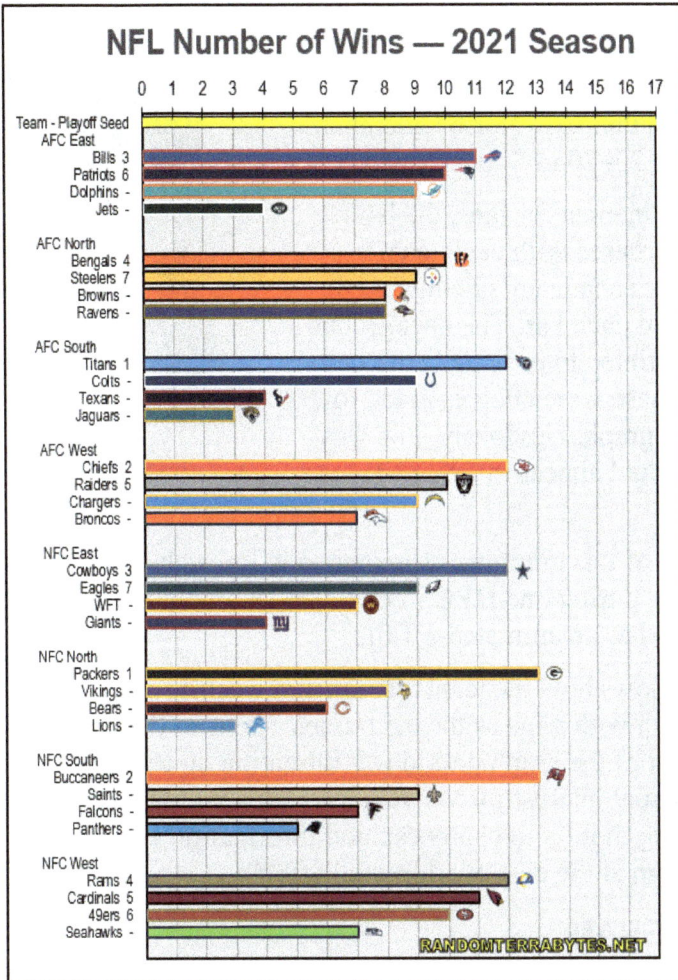

**Figure 21. Number of wins for
32 NFL teams during the
2021-2022 season.**

format. Furthermore, time metrics are conventionally represented on horizontal axes, formatted left-to-right from oldest to youngest. These were considerations related to the *Framework* of the graph.

A perfect season would have been 17 wins although no team had more than 13 wins during 2021-2022. Still, the horizontal axis was scaled to 17 to represent the entire season and convey how far teams fell short of a perfect season with no losses. The Packers and the Buccaneers each had the most wins, 13 (76%). The Jaguars and the Lions each had the fewest wins, 3 (18%).

Graphics.

The vertical scale represents the 32 teams grouped into their two Conferences and eight Divisions. The NFC-West Los Angeles Rams (12 wins) defeated the AFC-North Cincinnati Bengals (10 wins) 23–20 in Super Bowl LVI.

To provide visual interest for fans, the bars were filled and bordered with each team's colors and icons of each team's helmet were placed at the end of each bar. The background of the plot area is tinted to provide better color discrimination. These were choices made for the *Facade* of the graph. Obviously, this was not a chart that would appear in a STEM journal.

What's your favorite team? Ours is the Lions. We're little lions.

Eighteen versions of this graph were created, one for each week of the season (each team does not play during one BYE week). The 18 graphs were compiled into a single *motion chart* as an animated-gif file.

Running the gif shows how the teams fared from week-to-week (i.e., the bars increasing in length with wins as the gif frames advance) during the season. The motion component of the graph provides information on changes in the number of wins as well as visual interest for viewers over what would have been provided by a static chart. Adding motion to charts can add information as well as being entertaining, though at the cost of additional effort.

FUNNEL DIAGRAM

Funnel charts show data frequency, as a number or a percentage, as bars similar to bar charts. The differences are that the bars are arranged vertically, centered horizontally on a point, and sorted in some meaningful manner.

In Figure 22, the horizontal axis of the example funnel chart shows the percentage of game victories of NFL teams during the 2021 season. The axis isn't displayed because it shifts for each bar, so only the relative lengths of the bars are important.

The vertical axis shows the 32 teams in the NFL. The bars are sorted vertically by the team's success in the 2021-2022 playoffs. The Rams are at the top of the funnel having beaten the Bengals, listed second, in the Super Bowl. The 49ers and the Chiefs are next, having advanced to, but lost, their Conference playoff games. The next four teams lost their Divisional playoff games and the following six teams lost their Wild Card playoff games. The remaining 18 teams did not make the playoffs and are sorted by their win percentage from the regular season.

Graphics.

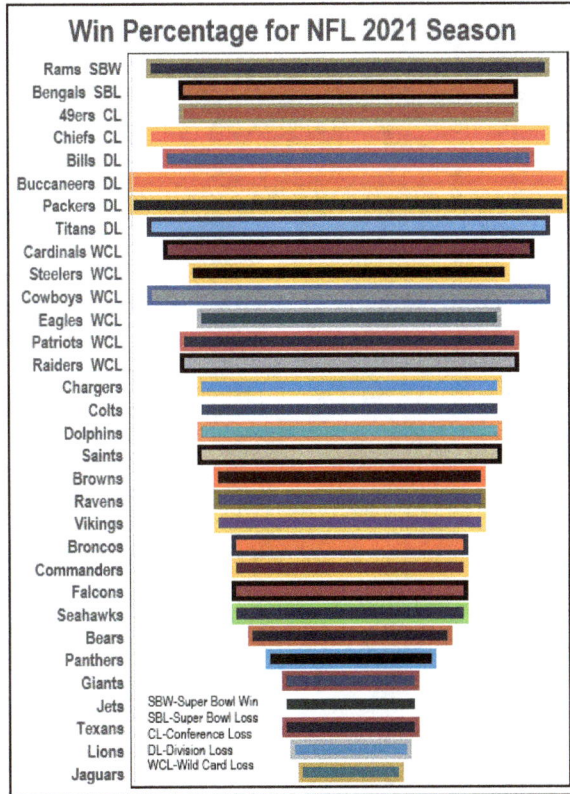

**Figure 22. Funnel diagram showing
NFL win percentages for 2021.**

The bars are decorated with each team's colors to add visual interest for hard core
NFL fans.

One popular version of a funnel chart is a *population pyramid*. Population pyramids
are used to display and compare demographic frequencies often involving group
comparisons. Depending on how a pyramid is constructed, it can represent two back-
to-back histograms showing the frequencies of two binary groupings.

HEAT MAPS

Heat maps are statistical graphics that show information on two variables using
colors. The maps take the form of a matrix with the axes using either nominal or
ordinal scales or continuous scales that have been *binned*.

Figure 23 is an example heat map in which the horizontal axis displays four metrics
for how the 32 NFL teams performed during the 2021 season. The variables
represent: the number of attempts to pass versus rush (run) on plays; the net yards
gained; the number of first downs gained; and the number of touchdowns scored; as

Graphics.

2021 Team	playoffs	Passing vs Rushing Ratios				Win %
		Attempts	Yards	1st Downs	TDs	
Rams	☺	0.55	0.84	0.84	1.00	0.71
Bengals	☺	0.37	0.76	0.72	0.44	0.59
Buccaneers	⊖	1.80	1.00	0.98	0.48	0.77
Packers	⊖	0.43	0.56	0.71	0.67	0.77
Chiefs	⊖	0.57	0.63	0.77	0.46	0.71
Cowboys	⊖	0.47	0.54	0.76	0.57	0.71
Titans	⊖	0.87	0.12	0.27	0.05	0.71
Bills	⊖	0.52	0.37	0.46	0.30	0.65
Cardinals	⊖	0.29	0.48	0.42	0.11	0.65
49ers	⊖	0.13	0.38	0.32	0.12	0.59
Patriots	⊖	0.20	0.31	0.20	0.06	0.59
Raiders	⊖	0.62	0.87	0.79	0.26	0.59
Steelers	⊖	0.72	0.64	0.87	0.45	0.56
Eagles	⊖	0.00	0.00	0.00	0.00	0.53
Chargers	☹	0.69	0.74	0.79	0.40	0.53
Colts	☹	0.15	0.05	0.04	0.13	0.53
Dolphins	☹	0.49	0.63	0.85	0.29	0.53
Saints	☹	0.09	0.21	0.27	0.49	0.53
Browns	☹	0.17	0.08	0.16	0.08	0.47
Ravens	☹	0.28	0.21	0.18	0.11	0.47
Vikings	☹	0.45	0.52	0.59	0.79	0.47
Broncos	☹	0.29	0.31	0.26	0.14	0.41
Commanders	☹	0.25	0.26	0.24	0.25	0.41
Falcons	☹	0.56	0.75	1.00	0.31	0.41
Seahawks	☹	0.30	0.27	0.34	0.26	0.41
Bears	☹	0.24	0.25	0.31	0.10	0.35
Panthers	☹	0.42	0.33	0.28	0.01	0.29
Giants	☹	0.52	0.38	0.55	0.33	0.24
Jets	☹	0.69	0.56	0.77	0.19	0.24
Texans	☹	0.40	0.65	0.68	0.55	0.24
Lions	☹	0.49	0.39	0.52	0.34	0.21
Jaguars	☹	0.64	0.41	0.57	0.04	0.18

Blue cells are predominately PASS. Red cells are predominately RUSH.

Figure 23. Heat map of four statistics for team offenses during the 2021 NFL season.

well as their win percentage in the far right column. All of the values were standardized to z-scores and normalized to a range of 0 to 1.0.

The vertical axis shows the 32 teams sorted by their playoff success as in the example of the funnel chart.

The colors for the performance metrics range from blue, indicating a higher ratio of passing over rushing, to red, indicating a higher proportion of rushing over passing.

The colors for the win percentage range from green to gold. It would come as no surprise to hard core NFL fans that the most successful teams in the playoffs favored passing over rushing.

Graphics.

This heat map was created in Microsoft Excel by assembling a matrix of the data and using conditional formatting to assign the colors. The procedure was straightforward and easy, except perhaps for rescaling the data.

Other commonly used property diagrams are *matrix charts* in which the cells of the matrix may include text, graphics, or other charts, not just colors. A *calendar* is a common form of matrix chart.

RADAR CHARTS

Matrices come in many useful and different forms.

Radar charts are different from most statistical graphics in several ways. The data are presented in a radial format rather than the more common rectangle format. The charts are always based on more than two variables, usually three to seven so they don't become too cluttered. They depict data for several metrics but only for one sample. And, they are scaled so the more important element in interpreting them are the shapes produced by the scaled data.

Usually many radar charts are created to compare subjects in a dataset. In Figure 24, radar charts depict the career achievements of nine NFL Hall of Fame quarterbacks. The seven axes of the radars, appearing as spokes from the center of the charts, represent the variables. In this example, the seven variables are: the number of games the quarterback played; the percentage of passes completed; the number of yards gained by passing; the number of touchdowns thrown; the number of interceptions thrown (fewer interceptions rated higher); the number of times sacked (fewer sacks rated higher); and the number of yards the quarterback ran for. All of the data were standardized to z-scores and normalized to a range of 0 to 4.5. Icon colors reflect the colors of each quarterback's primary team.

Nine quarterbacks were selected for this figure based on their differently shaped radars. Johnny Unitas and Terry Bradshaw were "well rounded" in the sense that their radar icons are roughly circular, compact, and centered on zero.

Steve Young and Joe Montana, both quarterbacks for the San Francisco 49ers, were similar in their performances only Young was known for his rushing while Montana was not. Jim Kelly (not shown) was more like Montana.

Peyton Manning was the premier passer shown by his icon being large and shifted to the right along the axes for percentage, yards, and touchdowns. Dan Marino (not shown) was a bit like that.

Graphics.

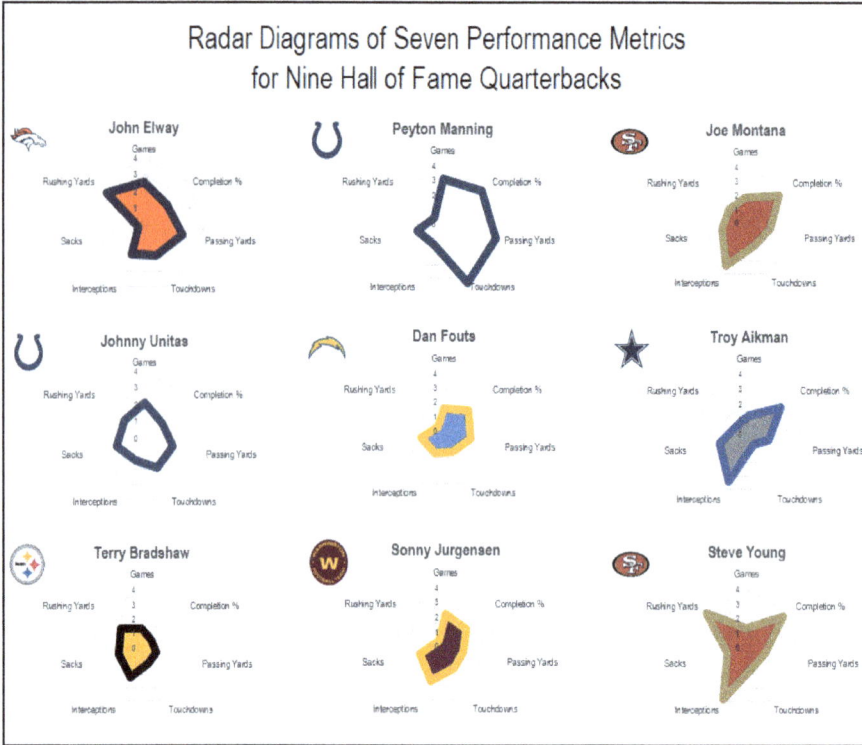

Figure 24. Radar charts showing career performance of nine NFL Hall of Fame quarterbacks.

John Elway and Fran Tarkenton (not shown) had similar career performance characteristics. Sonny Jurgensen, Dan Fouts, and Ken Stabler (not shown) were similar based on their radar icons.

That's how you interpret radar charts.

Similar types of radial plots displayed as icons include *sun ray charts*, *star charts*, and *polygon charts*. Rectangular plots that can be used as icons include *pie charts*, *bar charts*, *area charts*, *profile charts*, and *sparklines*. And for a wild and crazy time, search the internet for *Chernoff faces*.

SPATIAL DIAGRAMS

Spatial diagrams show data on entities associated with specific locations. Such data may be spatially independent or spatially dependent. *Spatially-independent* data have no correlation with data from locations around it. Examples include population demographics from bordering territories. *Spatially-dependent* data are correlated with nearby measurements of the same variable based on how far apart the locations are. Spatially dependent metrics are called *regionalized variables*. Topography, for example, is a regionalized variable.

Graphics.

Everybody has seen a map, the most common kind of spatial diagram. Maps are used to present a wide variety of geographic, geologic, biologic, meteorologic, and natural resources information, not to mention cultural information like political boundaries, roads, and population demographics.

For example, Figure 25 is a map of the U.S. showing the percentages of NFL players by State. These data are spatially independent because there is no correlation between the States.

In contrast, Figure 26 is a contour map of the topography of the Big Island in Hawaii featuring Mauna Kea (northern peak) and Mauna Loa (southern peak). Kilauea is the active volcano located on the southeastern flank of Mauna Loa.

This model is a lot better than a world map. It's not flat.

Topography is a regionalized variable because the elevation of any point will be similar to the points around it. (This map was created on the website contourmapcreator.urgr8.ch/.)

Contour maps are really just x-y-z scatterplots in which the x and y are location coordinates and the axes are identically scaled and formatted. The lines on a contour map are called *isopleths* or *isolines.* They represent locations having equal values for a specific regionalized variable. Contour maps are used commonly in the Earth and environmental sciences to show spatial patterns of regionalized variables, like concentrations of ore deposits or groundwater contamination.

Spatial diagrams can also represent subsurface areas either parallel to the surface (*subsurface maps*), or perpendicular to the surface (*cross-sections*), or both (*block diagrams*). These diagrams are mostly used to show information about geology or natural resources.

2. MIXTURE DIAGRAMS.

Mixture diagrams aim to show quantities and proportions. The most commonly used mixture diagrams are *bar charts*, which are also used to show properties, and *pie charts.*

Table 19 highlights eleven types of graphs used to display information about mixtures.

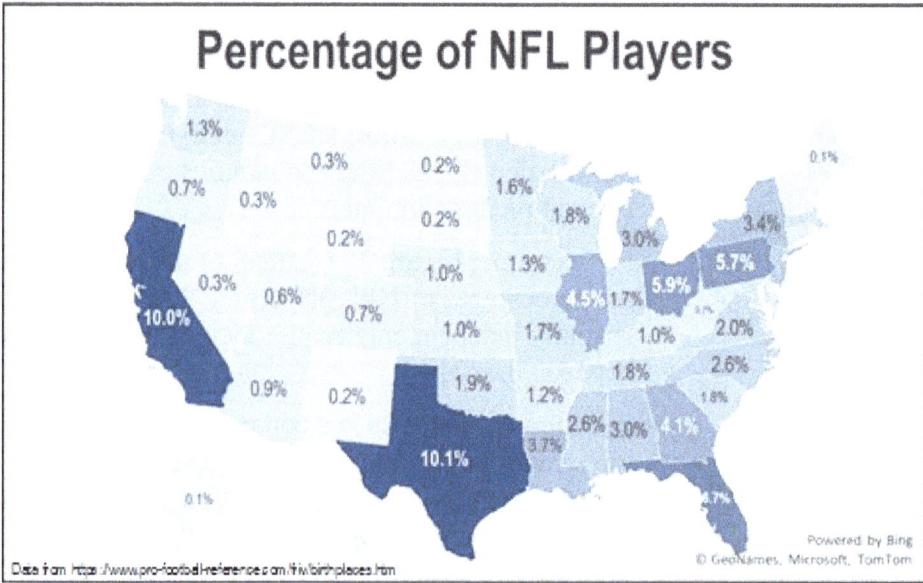

Figure 25. Percentages of NFL players by state.

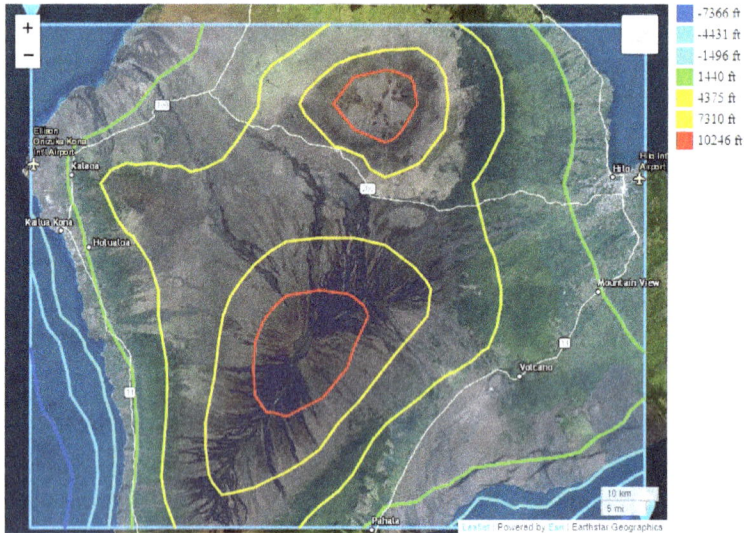

**Figure 26. Contour map of the topography
of the Big Island of Hawaii.**

PIE CHARTS

Pie charts are the most commonly used graphic to display **proportions** of data groups. They show the proportions as slices of a circle. There is no axis for scale; proportions are implied by areas of the circle, like slices of a pie. *Donut plots* are pie

Graphics.

Table 19. Examples of graphs used show mixtures.

Types of Diagrams Used to Show Mixtures

Diagram	Description	AIM — Purpose of Plot: Properties (central tendency & dispersion)	Distributions (frequencies & shapes)	Mixtures (amounts & proportions)	Relationships (trends & patterns)	Concepts (non-numerical information)	Time or Duration	FOCUS — What's Plotted: One Data Point	Individual Data Points	Grouped Data	Nonnumeric Information	Usual Layout	Horizontal Axis	Vertical Axis	Other Axes
Area graph	Bar chart with areas instead of bars			X	X		X			X		Rectangle	N	C	NA
Bar chart	Data quantities represented by bars			X	X		X					Rectangle	N	C	NA
Donut chart	Pie chart with a hole in the center			X					X	X		Circle	NA	NA	C
Mosaic, percent-stacked-bar chart	Quantities by sized colored areas			X	X				X	X		Rectangle	N	C	NA
Pareto chart	Bar chart with category frequencies sorted high to low			X	X				X	X		Rectangle	N	C	NA
Pie chart	Circular graph of proportions of quantities			X					X	X		Circle	NA	NA	C
Side-by-side bar chart	Bar charts with grouped data shown horizontally			X	X				X	X		Rectangle	N	C	NA
Stacked-bar chart	Bar charts with grouped data shown vertically			X	X				X	X		Rectangle	N	C	NA
Treemap	Rectangular representation of a hierarchical structure			X	X				X	X		Rectangle	N	N	NA
Waterfall chart	Bar chart for changes over time or between groups			X	X		X		X			Rectangle	C	C	NA
Word cloud	Word count proportions			X	X	X					X	Ad Lib	NA	NA	O

Layout: Line=one axis; Rectangle=2 axes at 90°; Triangle=three axes at 60°; Cube=four axes at 90°; Circular=Multiple axes arranged radially; Ad Lib=No set layout.

Axis Scales: C=continuous ratio & interval scales; O=ordinal scales; N=nominal grouping scales; NA=no scale. Horizontal & vertical axes can be switched in some graphs.

charts with the center removed. Sometimes the pies are extruded into a third dimension and tilted to engage non-technical audiences. For some reason, these are not called cake diagrams. Sometimes slices are pulled away (*exploded*) from the rest of the pie to highlight the segment.

Despite their negative perception, pie charts are easy for most people to understand, and consequently, are ubiquitous in the general media.

These are really the best kind of pie.

Pie charts must be based on proportion data that sum to 100%. If a pie chart is going to be wrong, it is usually because the percentages were calculated incorrectly.

There shouldn't be too many sections, *slices*, in a pie chart because they inhibit comprehension. If the labels on the data sections are too crowded to read, there are probably too many sections.

Graphics.

Figure 27 is an example of a pie chart. It shows the percentages of players in the NFL Hall of Fame by the positions they played when they were active. Green slices represent offense players and red slices represent defense players.

There is another example of a pie chart in Chapter 5 (average percentages of voters and nonvoters).

MOSAIC CHARTS

Mosaic charts, also called *proportional area charts,* are like pie charts in that they depict differences in group sizes, except that they are square or rectangular. Like pie charts, they rely on visual perceptions of area rather than scales.

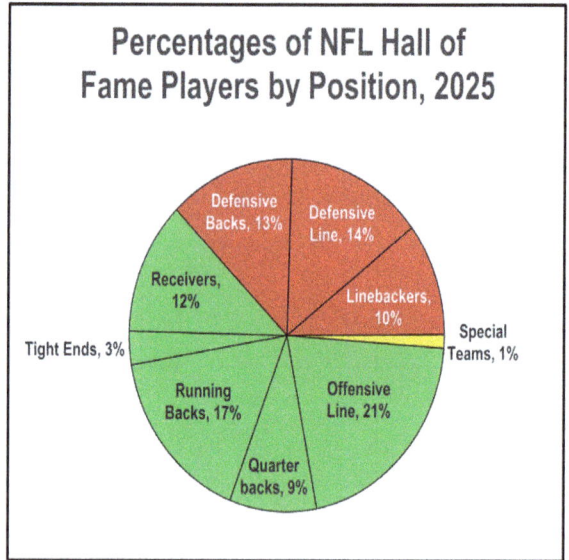

Figure 27. Pie chart of NFL Hall of Fame players by position.

Mosaic charts are also similar to *heat maps* in that they show information in rectangular arrangements of cells using colors. The two terms, mosaic charts and heat maps, are sometimes used interchangeably but they are quite different. Heat maps have defined rows and columns so that the cells are similar sizes. In mosaic charts, however, the rows and the columns are not uniform so that the cells are not necessarily similar sizes.

Mosaic charts are also called *Mekko charts* or *Marimekko charts*, or more commonly, *percent-stacked-bar charts*. They are sometimes used as alternatives to pie charts. Mosaic charts may be more difficult for some people to understand than pie charts because it is more challenging to compare the sizes of differently sized rectangles than slices of a circle. Airline baggage handlers and people who prepare Chewy packages for shipment don't have this problem.

Figure 28 is an example of a mosaic chart for the average weights of position groups on an NFL team. In the figure, you might be able to tell that the average offensive lines on teams weigh about the same as the average defensive lines, but not that

I like pies better than pie charts but they're both fun to play with.

Graphics.

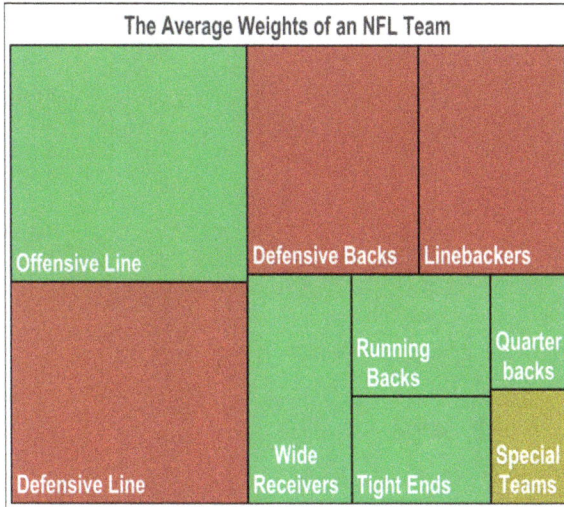

Figure 28. Mosaic chart for the average weights of position groups on an NFL team.

offensive lines are almost 200 pounds heavier. You might be able to tell that the average defense without the line is heavier than the average offensive without the line, but not that it is over 400 pounds heavier. Such things are interesting to know.

(Data for this mosaic chart came from hortonbarbell.com/average-height-weight-of-nfl-players-by-position/, bleacherreport.com/articles/1640782 -the-anatomy-of-a-53-man-roster-in-the-nfl, and denversportsradio.com/ average-weight-of-nfl-punters/.)

WORD CLOUDS

While most people are familiar with bar charts and pie charts, they may not be familiar with *word clouds*. Word clouds are used to represent how many times specific words appear in a passage of text. The relative frequency of the words is shown using text size (and sometimes colors and fonts) in a graphic.

The use of text formatting to characterize the relative frequency of words dates back to the 1970s but grew to prominence in the 2010s.

The advantages of word clouds are that the concept behind the graphic is simple and there are websites where word clouds can be created easily for free. The main disadvantages are that word clouds are only semi-quantitative and involve more artistry than statistical rigor. It's impossible to estimate the number of times a particular word appears in a passage from the word cloud. Furthermore, creators can manipulate what words appear and how they are counted in order to make the graphic more attractive. Marketers love them; academics hate them. Still, word clouds are a useful statistical graphic to know about.

Figure 29 is a word cloud based on the 15,000 words in this chapter about statistical graphics and formatted as a cat. It was created at

Every member of a team has a role to play. I am the quarterback and they are my offensive line.

Graphics.

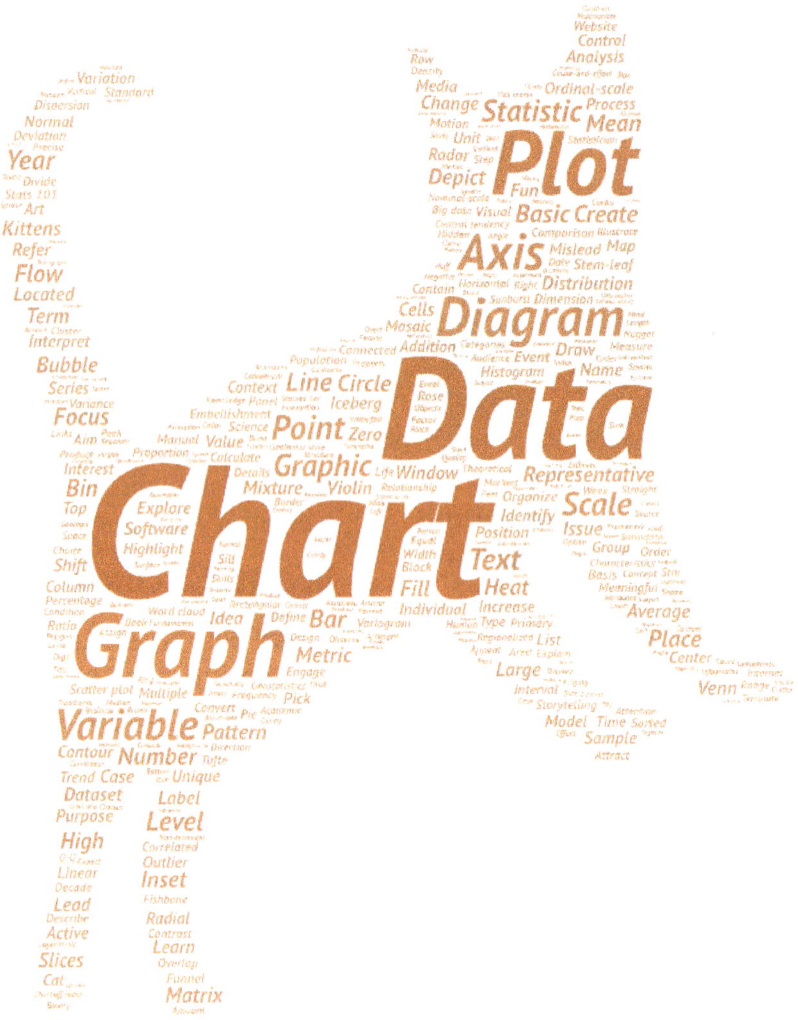

**Figure 29. Word cloud based on the
16,000 words in this chapter.**

wordart.com. There's another word cloud based on all of **Stats with Kittens** at the
beginning of the Glossary.

3. DISTRIBUTION DIAGRAMS.

Distribution Charts show the number of data points at scale values or intervals thus
depicting the central tendency, dispersion and shape of the dataset. The three most
commonly-used statistical graphics for describing datasets are *histograms*, *box plots*,
and *probability plots*. Other distribution charts include *stem-leaf diagrams*, *dot plots*,
density plots, and *violin charts*.

Graphics.

Table 20. Examples of graphs used to display information about data distributions.

Types of Diagrams Used to Show Distributions		AIM — Purpose of Plot						FOCUS — What's Plotted				Usual Layout	Axis Scales		
Diagram	Description	Properties: central tendency & dispersion	Distributions: frequencies & shapes	Mixtures: amounts & proportions	Relationships: trends & patterns	Concepts: non-numerical information	Time or Duration	One Data Point	Individual Data Points	Grouped Data	Nonnumeric Information	Line / Rectangle / Triangle / Cube / Circular / Ad Lib	Horizontal Axis	Vertical Axis	Other Axes
Density plot	Smoothed version of a histogram	X	X	X					X	X		Rectangle	C	C	NA
Dot plot	Histogram with dots instead of bars	X	X	X					X	X		Rectangle	O	C	NA
Histogram	Bar chart with interval counts	X	X	X					X	X		Rectangle	O	C	NA
Probability plot	Data z-scores vs Normal distribution probabilities	X	X						X			Rectangle	C	C	NA
Q-Q plot	A probability plot of the quartiles of two distributions	X	X						X			Rectangle	R	C	NA
Stem-leaf diagram	Histogram using numbers	X	X	X					X	X		Rectangle	O	O	NA
Surface plot	3D grid of data		X	X	X				X	X		Rectangle	C	C	C
Violin plot	Box plot with frequencies	X	X	X					X	X		Rectangle	C	C	NA

Layout: Line=one axis; Rectangle=2 axes at 90°; Triangle=three axes at 60°; Cube=four axes at 90°; Circular=Multiple axes arranged radially; Ad Lib=No set layout.

Axis Scales: C=continuous ratio & interval scales; O=ordinal scales; N=nominal grouping scales; NA=no scale. Horizontal & vertical axes can be switched in some graphs.

Table 20 highlights eight types of graphs used to display information about statistical distributions.

Distribution Charts tend to be more complicated than property charts or mixture charts. The things you'll want to look for in these graphs are:

- **Shape.** Are there an equal number of values in each interval (a uniform distribution), or is there a cluster of data around a particular value (a unimodal distribution), or is there more than one cluster of data modes?

- **Symmetry.** Do the data fall symmetrically around the distribution's center? Symmetry is the property measured by *skewness*.

- **Central Tendency.** Where is the center of the distribution? This is the property measured by the *median*. If your frequency distribution isn't symmetrical, your calculated *mean* may be misleading.

- **Dispersion.** How spread out are the data? Dispersion is the property measured by the *standard deviation*. If your frequency distribution isn't unimodal and symmetrical, your calculated standard deviation may be misleading.

- **Outliers.** Are there any data points located far away from the rest? *Outliers* will affect virtually any statistic you calculate.

- **Model.** If you know what mathematical models of data frequencies look like, you can get a sense of which model might be a good approximation for your data. If you can identify an appropriate theoretical distribution, like the

Graphics.

Normal distribution, you can use that distribution's equation to make predictions about the population your sample is from. This makes statistics a lot of fun.

There are several examples of distribution graphs in this Chapter and Chapter 3.

BOX PLOTS

One type of graph that can display data properties is a *box-and-whisker plot*, usually just called a *box plot*, as discussed in Chapter 3. Box plots are different from most statistical graphics in that they present descriptive statistics rather than individual data points. They also rely on lines, areas. and shapes rather than just points to convey information.

Multiple box plots are frequently combined to show differences in groups of data. Figure 30 is an example of eight box plots showing average NFL quarterback ratings (also called passer ratings because they focus only on a quarterback's pass attempts, completions, passing yards, touchdown passes, and interceptions) by decade from the 1940s through the 2010s.

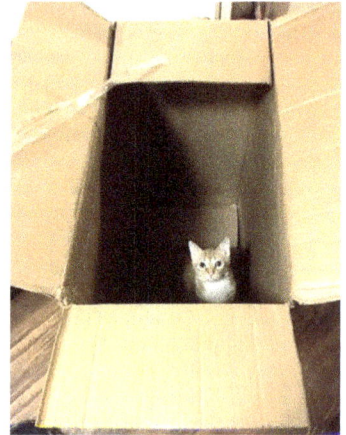

I've got a big box to hold all of my big data.

The box plots suggest that the average ratings have been increasing since the 1970s when the formula for the ratings was first developed. That trend may be attributable to more skilled players but it may also reflect better recordkeeping and more data than from earlier decades.

A popular variant of box plots are *candlestick charts*, which are used to analyze stock market information. They resemble box plots but use the open and close prices to define the box instead of the 25% and 75% quantiles.

HISTOGRAMS

The most readily-available type of graph for frequency distributions is the *histogram*. As discussed in Chapter 3, histograms are a popular variant of bar charts in which the vertical axis represents the number of observations in data ranges represented on the horizontal axis.

Histograms are good for visualizing the form of data distributions and for comparing them to a theoretical model like the *Normal distribution*. They're acceptable for looking at dispersion and outliers but not quite so good for looking at central tendency. In contrast, box plots are good for looking at central tendency, dispersion, and outliers, but not as good for distribution shape. Histograms are available on websites and even low-end statistical software as well as spreadsheet programs like Microsoft Excel.

PROBABILITY PLOTS

The best visual comparison of a continuous-scale variable to some theoretical distribution is a *probability plot*. As discussed in Chapter 3, probability plots use ranked data plotted as a cumulative frequency versus what would be expected from a theoretical distribution, usually the *Normal distribution*.

Statistical software is usually needed to create probability plots. However, interpreting the plots is relatively easy. Plotted data points should fall on the

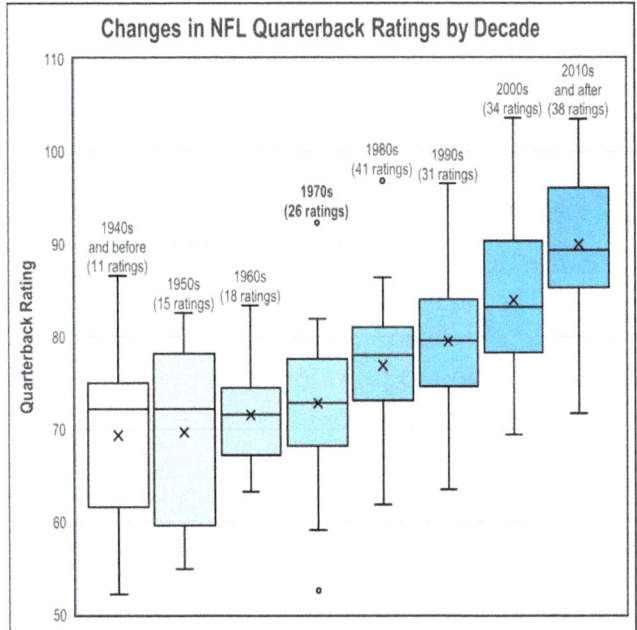

Figure 30. Box plots showing average NFL quarterback ratings by decade from the 1940s.

straight line representing the theoretical distribution. The farther away the data points are from the line, the poorer the fit.

Like box plots, there are many variations of probability plots. The term probability plot usually refers to what is called a *Q-Q plot* or *quantile-quantile plot* because data are grouped into regular intervals along the variable scale called *quantiles*.

While probability plots usually compare a data distribution to the Normal distribution, that's not always the case. Probability plots can compare any two distributions, either empirical data or theoretical models.

Histograms are a kind of bar chart.

There are also important variations of probability plots. *Detrended probability plots* plot the differences between two distributions on the y-axis instead of the values of the distribution. The deviations should plot along a horizontal line at 0.0 on the y-axis. If there are any patterns in the plotted points, excessive scatter, or shifts from the zero line, there are likely to be important differences

Graphics.

between the distributions. These plots make it easier to notice differences between the data distribution and the theoretical distribution because the line is horizontal rather than at a 45-degree angle.

You have to be aware of exactly what kind of probability plot you're looking at.

STEM-LEAF DIAGRAMS

Stem-leaf diagrams are like histograms except that the bars are replaced by data values. The values are split into two parts, the *stem* consisting of the beginning digit of a group of values and the *leaf* consisting of the remaining digits of the values. Stem-leaf diagrams provide more information than histograms because it's possible to interpret what the actual data values are.

Stem	Leaf
5	2 3 5 7 7 9
6	0 1 1 2 2 3 3 3 3 3 4 6 6 7 7 7 7 8 8 8 8 8 8 8 8 9 9 9 9 9
7	0 0 0 0 0 1 1 1 1 2 2 2 2 2 2 2 2 2 3 3 3 3 3 3 3 4 4 4 4 4 4 4 4 5 5 5 5 5 5 5 5 5 5 5 5 5 5 5 6 6 6 7 7 7 7 7 7 7 7 8 8 8 8 8 8 8 8 8 9 9 9 9 9 9 9
8	0 0 0 0 0 0 0 0 0 0 0 1 1 1 1 1 1 1 2 2 2 2 2 2 2 2 2 2 2 2 3 3 3 3 3 3 4 4 4 4 4 5 5 5 5 5 5 5 5 6 6 6 6 6 6 6 7 7 7 8 8 8 8 8 8 9 9 9
9	0 0 0 1 1 1 1 2 2 2 2 4 4 4 4 5 6 7 7 7 7 7 8 8 8 9 9 9
10	0 1 4 4

**Figure 31. Stem-leaf diagram of
quarterback ratings from the 1980s.**

Figure 31 is a stem-leaf diagram (of quarterback ratings from the 1980s) in which the stems appear in the column on the left of the figure and the leaves are listed consecutively in the column on the right. Thus, the first row of the diagram shows the data values: 52, 53, 55, 57, 57, and 59. This graphic was created using Edward Furey's "Stem and Leaf Plot Generator" at calculatorsoup.com/ calculators/ statistics/stemleaf.php.

If the diagram is drawn vertically, the stems are arranged on the horizontal axis and the leaves are stacked above. The data values must be typed in a monospaced font in which each number occupies the same amount of space.

Dot plots are like histograms except that the bars are replaced by dots or other icons. This approach provides more information than histograms because it's easier to see how many data points are included. Dot plots are used mostly for small data sets.

*These are M&Ms
not a dot plot.*

VIOLIN PLOTS

A violin plot is a combination of a box plot and a *kernel density plot*. The width of the violin represents the number of data points at each data value. A violin plot

Graphics.

shows the data frequency as a smooth curve rather than the bars shown in histograms. As a consequence, it is easier to identify patterns in violin plots than in histograms because there is no need for bins.

Figure 32 is an example of a violin plot for career rating of NFL quarterbacks from the 1980s. In the figure, the width of the violin showing the data frequency suggests that the distribution of ratings is a bit asymmetrical and that there may be multiple peaks in the data.

Violin plots are challenging to create because of the complicated kernel density calculation. There are, however, a few websites that will perform the calculations and create the chart. This figure was created at www.statscalculators.com/.

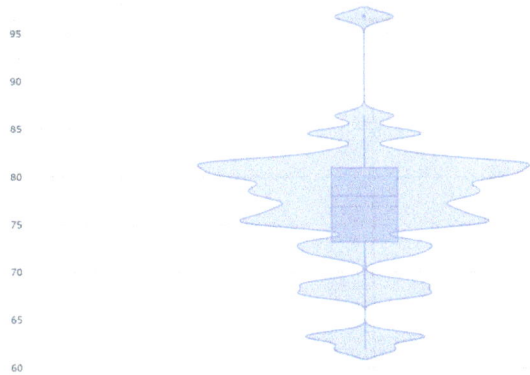

**Figure 32. Violin plot for career rating of
NFL quarterbacks
from the 1980s.**

4. RELATIONSHIP DIAGRAMS.

Relationship diagrams show data points for two or more variables at a time so that any patterns or trends can be identified.

The most commonly-used statistical graphics for exploring data relationships are line charts and scatter plots for multiple data points and radar charts for single data points and multiple variables.

Table 21 highlights ten types of graphs used to display information about statistical relationships. Relationships that can be explored include:

- **Patterns**—*shocks*, *steps*, *shifts*, *cycles*, and *clusters*
- **Trends**—*linear*, *curvilinear*, *nonlinear*, or *complex* trends involving different dimensions (temporal, spatial, categorical, hidden, or multivariate).

These are discussed further in Chapter 8.

LINE CHARTS

In *line charts,* the horizontal axis represents an ordinal-scale variable and the vertical axis represents either an ordinal or a continuous-scale variable. Often, the horizontal axis represents time. Points are often but not always connected by lines. Because of the ordinal axis in a line graph, the variation of data points representing groups can be estimated and displayed at each level on the axis.

Graphics.

Table 21. Examples of graphs used to display relationships.

Types of Diagrams Used to Show Relationships

Diagram	Description	Properties central tendency & dispersion	Distributions frequencies & shapes	Mixtures amounts & proportions	Relationships trends & patterns	Concepts non-numerical information	Time or Duration	One Data Point	Individual Data Points	Grouped Data	Nonnumeric Information	Usual Layout	Horizontal Axis	Vertical Axis	Other Axes
Bubble, balloon chart	Scatterplot with bubbles or balloons depicting a third dimension				X		X		X	X		Rectangle	C	C	C
Control chart	Line charts with calculated data limits	X			X		X		X	X		Rectangle	C	C	NA
Icon plot	Scatterplot with graphs as data point icons				X			X	X	X		Rectangle	C	C	C
Line chart	Scatterplot with one ordinal axis	X			X		X		X	X		Rectangle	C	C	NA
Matrix diagram	Matrix showing relationships between components or variables			X	X	X			X	X	X	Rectangle	N	N	NA
Motion chart	Chart showing incremental changes over time or other conditions	X	X		X		X		X	X		Rectangle	C	C	NA
Polygon plot	Radial line chart				X	X	X		X	X		Circular	C	C	C
Scatterplot, x-y plot	Data plotted on two continuous-scale axes	X	X	X	X		X		X	X		Rectangle	C	C	NA
Ternary plot	Triangular scatter plot	X		X	X				X	X		Triangle	C	C	C
x-y-z plot, 3D plot	Scatterplot with three axes	C	C	C	C		C	C	C			Cube	C	C	C

Layout: Line=one axis; Rectangle=2 axes at 90°; Triangle=three axes at 60°; Cube=four axes at 90°; Circular=Multiple axes arranged radially; Ad Lib=No set layout.

Axis Scales: C=continuous ratio & interval scales; O=ordinal scales; N=nominal grouping scales; NA=no scale. Horizontal & vertical axes can be switched in some graphs.

Figure 33 is a line chart of the career passer rating (on the continuous-scale vertical axis) of 29 HoF quarterbacks versus the number of seasons they played in the NFL (on the ordinal-scale horizontal axis). These quarterbacks are identified by **green** diamond-shaped markers. In addition, 18 active or recently retired quarterbacks with the potential of being selected to the HoF are identified by **blue** circular-shaped markers.

The markers for all of the quarterbacks are labeled with the quarterback's name color coded in the same manner. The potential HoFers are also labeled with their record of Super Bowl appearances and whether they are active or retired.

There are also three reference lines indicating the average ratings for the 29 HoFers, the 18 potential HoFers, and 71 quarterbacks that have played during the 2000s.

The *Foundation* of the plot, its *aim* and *focus*, is to show how quarterbacks have evolved over time. The *focus* is on both individuals and groups. Individual Hall of Fame players (HoFers shown in **green**) are shown along with modern players (shown in **blue**). They are both compared to group statistics for HoFers, selected modern players, and all starting quarterbacks since 2000.

The metrics for the comparison are the number of seasons played (on the horizontal axis) and the NFL quarterback rating (on the vertical axis), which includes the

Graphics.

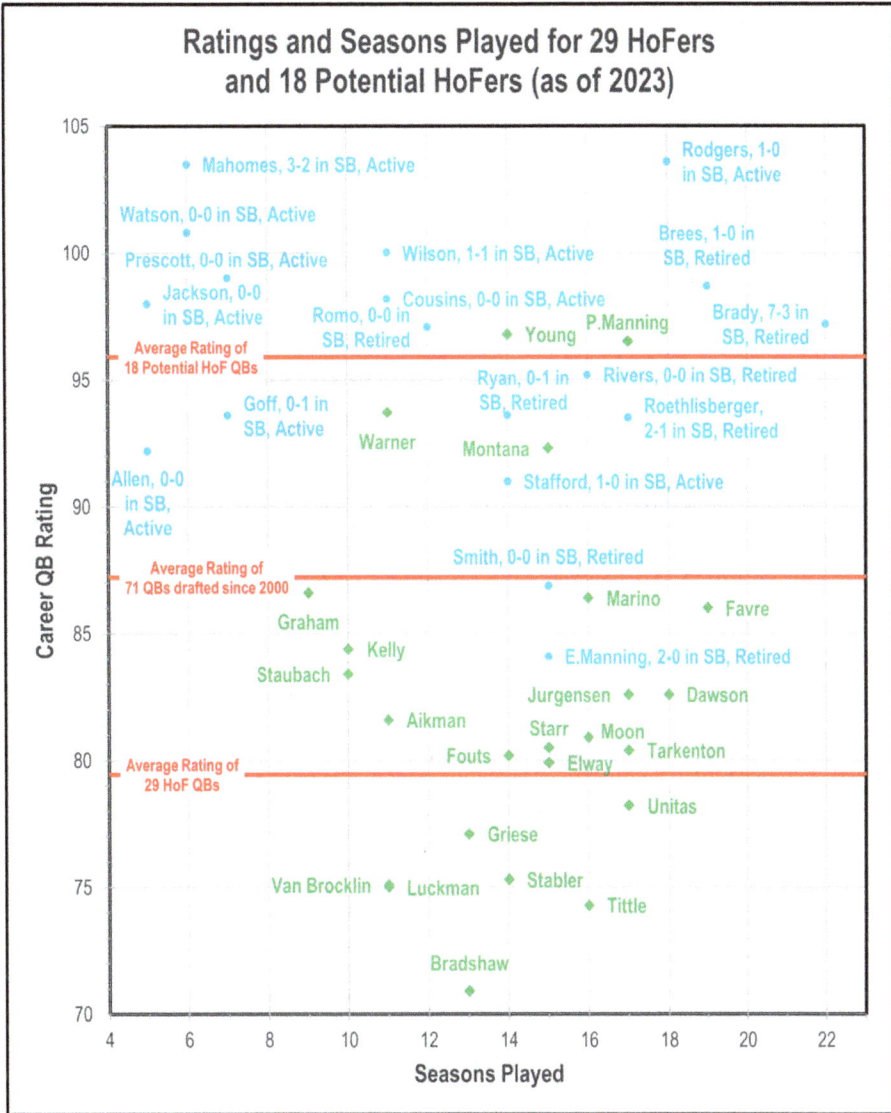

**Figure 33. Line chart of the career passer ratings
of quarterbacks versus the number of seasons
they played in the NFL.**

traditional measures of quarterback proficiency (i.e., passing attempts, completions, yards, interceptions, and sacks).

These metrics aren't the only factors considered in an individual's eligibility for the Hall of Fame, there are many others. For example, a quarterback's win-loss record in the Super Bowl is another. Still, the graph provides a reasonable initial comparison.

Graphics.

The aim of this graph could have relied on either a *line graph* or a *scatter plot*, but the seasons-played metric is measured on an ordinal scale making a line plot the better choice.

The *Framework* of the plot involves consideration of the axes, in particular, the vertical axis for quarterback rating. The range of the quarterback-rating metric is 0 to 158.3, about 158 divisions. However, the range of the data is 70.9 to 103.6, about 33 units. So what should the range of the axis be?

Some people would argue that the axis should cover the complete range of the metric "for perspective." This is often true such as in Figure 21, the bar chart of NFL team wins during the 2021 season.

My passer rating would be great if I could just get a grip on the ball.

Others would prefer that the axis cover only the data range with a small buffer on the ends. This plot uses the second strategy with the axis ranging from 70 to 105, 35 units. This approach prevents 80% of the axis from being blank, which would result from using the full data range. It's a reasonable approach that doesn't interfere with a reader's ability to interpret the graph and doesn't hide or mislead.

The *Facade* involves all the markers, colors, labeling (name, HoF or potential HoF, Super Bowl record, active or retired), and reference lines needed to tell the data's story. It makes for a colorful and engaging graphic.

The plot shows that, in general, quarterbacks are becoming more proficient in their craft. The average passer rating for the 29 HoF quarterbacks is 16.5 units lower than for the 18 HoF hopefuls. There is little overlap between individual HoFers and potential HoFers. Furthermore, the number of seasons played doesn't factor into those findings. Modern players aren't likely to have lower ratings as they play longer. It will be interesting to watch how many of these potential HoFers will get elected once they become eligible.

If you are interested, there are other examples of line charts in Chapter 1 (Google searches for statistical jargon) and Chapter 5 (Google searches for survey terms, household ownership of telephones).

SCATTER PLOTS

Scatter plots may appear identical to line charts but scatter plots have two continuous-scale axes while line plots have one ordinal-scale axis and one ordinal-

Graphics.

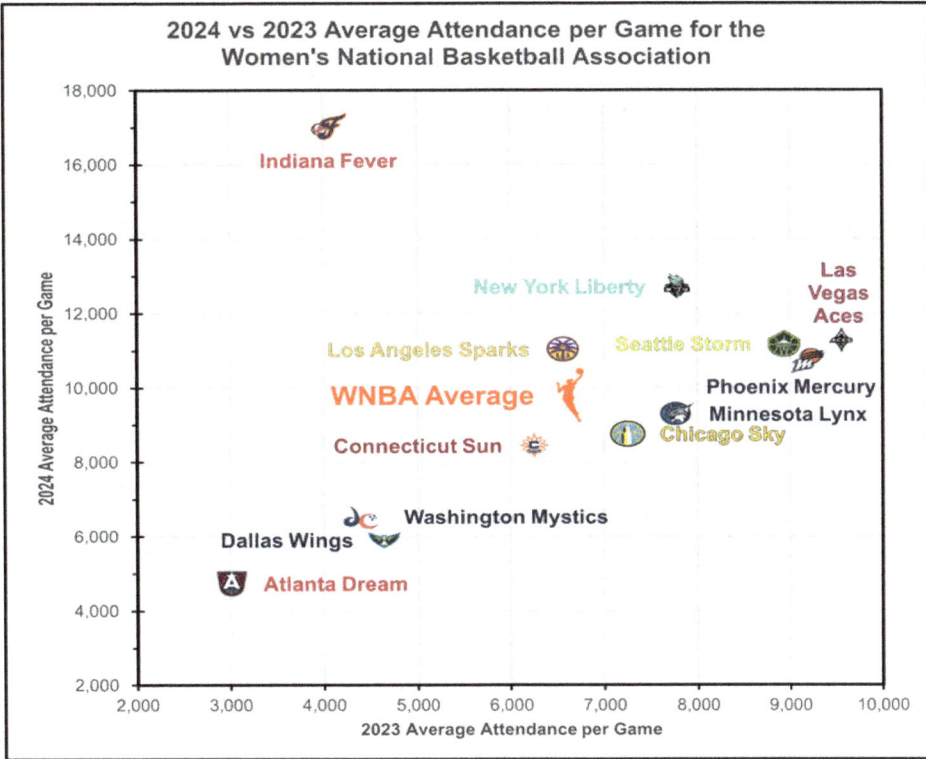

Figure 34. Scatter plot of the average attendance per WNBA game during 2023 and 2024 seasons.

scale or continuous-scale axis. Scatter plots are used commonly in statistical analyses involving data relationships.

Figure 34 is a scatter plot of the average attendance per game during 2023 and 2024 for the twelve teams in the Women's National Basketball Association. The plot shows that overall attendance at WNBA games increased by 48% from 2023 to 2024, with most of the teams seeing 16% to 68% more fans.

There is one major exception, however. The Indiana Fever had a 319% increase in attendance from 2023 to 2024. That's why the team's attendance plots in the upper left corner of the graph far away from the rest of the teams. They went from the low end of team attendance to the highest. The team is what is called an *outlier* (first mentioned in Chapter 1 and throughout the rest of

*Caitlin sure is bringing in
a lot of money for the WNBA.*

 Graphics.

Stats with Kittens). Without the Fever, WNBA attendance increased only 34%.

If you follow the WNBA, you'll recognize that this is what is referred to as the *Caitlin Clark effect,* based on the extraordinary talent, charisma, and celebrity of **Caitlin Clark**, the number 1 overall draft pick in 2024 by the Fever. Attributing the Caitlin Clark effect as the reason the Fever's attendance is an outlier is generally acknowledged by both fans and the media. Deciding what to do about the effect isn't, at least for the WNBA and some players and pundits.

Why is this a scatter plot instead of a line graph? Because the axes represent averages, which are measured on a continuous scale. Both axes are based on continuous scales, hence, a scatterplot is appropriate rather than a line plot.

More than one variable can be included in a scatter plot, which sometimes requires the inclusion of a second, differently-scaled axis. Data are usually represented by simple points. The points may also be connected by straight or curved lines, or even have a trend line placed through them. Data points are also sometimes represented by statistical icons (e.g., *radar charts*) that depict additional information about the point. Scatter plots that use statistical icons instead of simple points are called *icon plot*s.

Other examples of scatter plots appear in Chapters 2, 4, 5, and 7.

BUBBLE CHARTS

Bubble charts are scatter plots in which markers for the data points are replaced by circles (or spheres) indicating the value of a third variable. The bubbles don't have axes but their size, by either width or area, is proportional to the value of the third variable. *Balloon charts* are a variation of bubble charts.

Figure 35 is a bubble chart of yards allowed by team defenses (vertical axis) versus yards gained by team offenses (horizontal axis) for NFL teams during the 2021 regular season. The area of the bubbles is proportional to the win percentage minus 0.5. **Green** bubbles indicate more wins than losses, a winning season. **Red** bubbles indicate more losses than wins, a losing season.

From the graph, it looks like most of the teams that gained at least 250 yards per game (4,200 offense yards for the season) had a winning season regardless of how good their defense was. The two teams that played in the Super Bowl, the Rams and the Bengals, had average defenses and their offenses were only about 15% better than average, but they reached the pinnacle of success in 2022.

Of course no data set is without some curiosities, and in this case it's the Buffalo Bills. Their bubble appears at the bottom-center of the graph. The team's offense was average but its defense was 600 yards better than the next closest teams. That's about a quarter of the total data span. It's an *outlier* for sure. Football analysts believe their defensive performance was exceptional because of elite starting players with exceptionally skilled back-ups, a knowledgeable and resourceful coaching staff, a

Graphics.

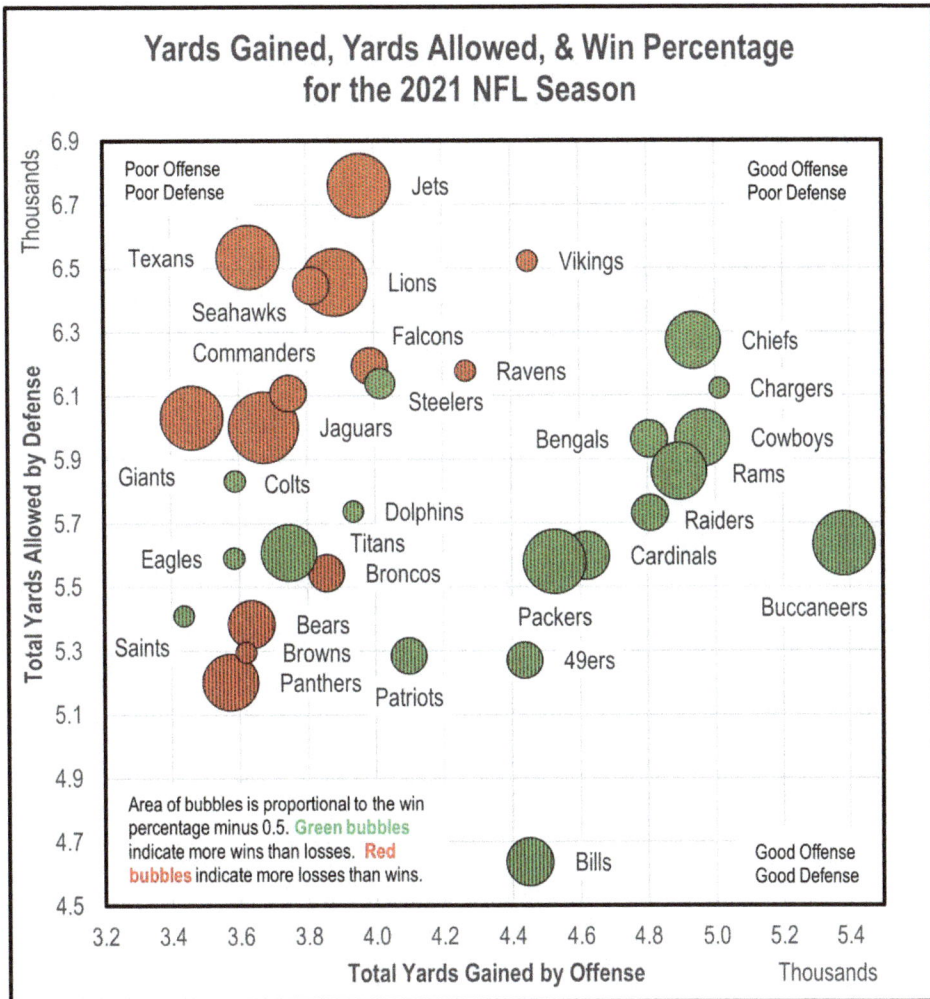

Figure 35. Bubble chart of defensive yards allowed versus offensive yards gained for NFL teams during the 2021 regular season.

well-structured defensive scheme, and a positive team chemistry. Unfortunately, they made the playoffs but lost in the Divisional round.

As enlightening and fun as statistics are, they can't capture all the unforeseeable happenings that go into some human endeavors.

WATERFALL CHARTS

Waterfall charts are all about change. They are used to show how data values increase or decrease from an initial point.

Graphics.

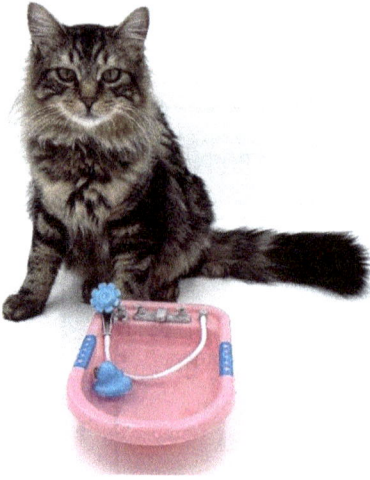

This is my indoor waterfall.

The horizontal axis is usually an ordinal variable, most often time, but could also represent nominal or ordinal-scale variables. The vertical axis is not fixed, or even shown on the chart for that matter.

Changes in data values are represented by bars that successively change position in either an upward (positive) or downward (negative) direction. Each bar is essentially its own *floating axis*. The bars in a waterfall chart are often also color-coded to represent either positive or negative changes.

Figure 36 is a waterfall chart representing changes in the number of U.S. Federal employees from 1960 to 2024. The graph begins in 1960 when the government employed 2,381,750 individuals. Employment during this period peaked at 3,195,667

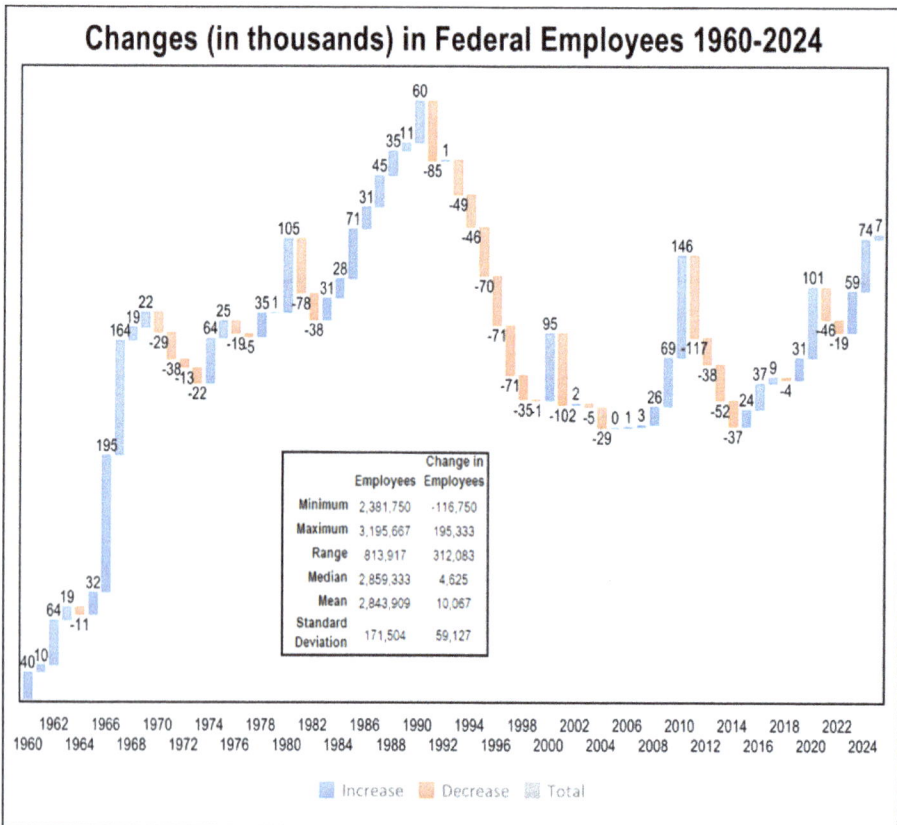

Figure 36. Waterfall chart representing changes in the number of U.S. Federal employees.

 Graphics.

in 1990. Large changes around years ending in zero were years in which the Census Bureau hired additional staff to complete their once-a-decade census.

VARIOGRAMS

If you plan to go into a STEM field, especially mining engineering or environmental science, you'll see *variograms* as part of preparing geostatistical contour maps. They depict the *spatial correlation* structure of a *regionalized variable*. Variograms are used to estimate the parameters—*nugget, range,* and *sill*—for *kriging*. Whew!

A *regionalized variable* is a variable in which the measurements at a location are similar to measurements that are located nearby. This similarity among samples or measurements located near each other is called *spatial correlation*. The analysis of spatial correlation, usually involving the creation of *contour maps*, is called *geostatistics*.

A *variogram* (*semivariogram* to geostatistical purists) is a plot of the spatial variance (on the vertical axis) versus the distances between pairs of sampling points (on the horizontal axis).

Geostatistics involves two parts—*variogramming* and *kriging*. Variogramming is the process of creating a *variogram* of a set of data for a *regionalized variable*. *Kriging* is the process of creating a contour map of the measurements of the regionalized variable based on the variogram.

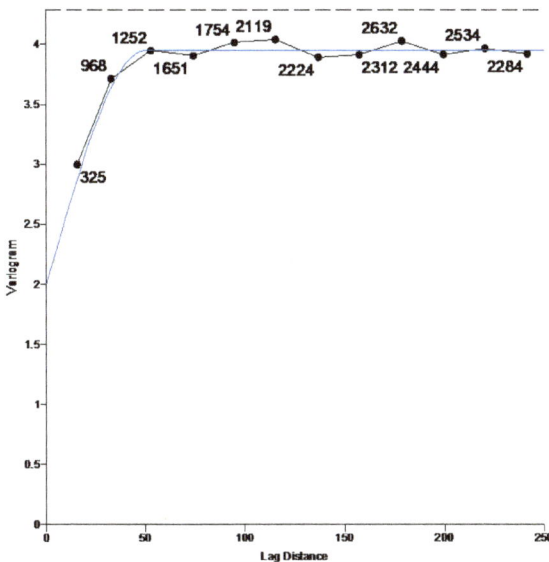

Figure 37. Variogram of the concentrations of an environmental contaminant.

The spatial correlation of a regionalized variable is characterized by three parameters that are derived from a variogram. The *range* is the distance at which measurements of a regionalized variable are no longer correlated. The *sill* is the spatial variance at the range and at distances between locations larger than the range. The *nugget* is the spatial variance of co-located samples or measurements. Nuggets are analogous to the natural variability in metrics that are not regionalized variables. These terms come from the mining industry where geostatistics was first developed. Kriging was named after Danie Krige (1919-2013), a South African statistician and mining

Graphics.

engineer who pioneered the field of geostatistics.

Figure 37 is a variogram of the concentrations of an environmental contaminant in the sediments of a lake. In the variogram, the *nugget* is 2 and the *sill* is about 4 (read from the vertical axis). The *range* is about 50 feet (read from the horizontal axis). The numbers besides each data point are the number of data pairs used to calculate the spatial variance.

These parameters and the measurements of the contaminant concentrations were then entered into specialized software for geostatistical analysis to produce a contour map by kriging.

Of course it's more complicated than that, but at least you'll recognize variograms and geostatistics if you see them.

5. CONCEPT DIAGRAMS.

There is a wide variety of diagrams for showing ideas, hierarchies, processes, and other non-numeric information. Most of the diagrams are drawn manually or from templates. There are no specific data formats. The information can range from verifiable knowledge to informal opinions.

Table 22 highlights twenty-one types of graphs used to display information about concepts.

CONCEPT MAPS

A *concept map* is a general term for any kind of diagram that shows relationships between ideas or concepts rather than numerical data. They are usually created manually because every concept map is unique in both its layout and content. Different types of concept maps include *brainstorm diagrams* and *mind maps.*

There are examples of concept maps in Chapter 6 (phases of statistical testing), Chapter 8 (categories of models), and Chapter 9 (elements of applied statistics).

CONTEXT DIAGRAMS

Context diagrams provide background information related to some idea or event to support discussions about the idea or event. Context diagrams take many forms, usually involving text. There are dozens of examples throughout **Stats with Kittens** and **Stats with Cats**.

Figure 38 is a context diagram used as part of an article on how sociological generations (e.g., Boomers, Millennials, Zoomers) behave differently because of their different life experiences. Individuals in each generation grew up experiencing different world events, employment, entertainment, and technology, leading to different viewpoints and goals. The graph was meant to draw a reader's attention to what they experienced directly and what they learned about the past.

Graphics.

Table 22. Examples of graphs used to display concepts.

Types of Diagrams Used to Show Concepts		AIM Purpose of Plot						FOCUS What's Plotted				Usual Layout (Line Rectangle Triangle Cube Circular Ad Lib)	Axis Scales (Continuous Ordinal Nominal Not Applicable)		
Diagram	Description	Properties central tendency & dispersion	Distributions frequencies & shapes	Mixtures amounts & proportions	Relationships trends & patterns	Concepts non-numerical information	Time or Duration	One Data Point	Individual Data Points	Grouped Data	Nonnumeric Information		Horizontal Axis	Vertical Axis	Other Axes
Arc diagram	One-dimensional network diagram				X	X			X		X	Line	NA	NA	O
Brainstorm diagram	Drawing linking ideas				X	X			X		X	Ad Lib	NA	NA	NA
Cause-and-effect, fishbone, Ishikawa diagram	Displays possible causes for a problem by sorting ideas into categories			X	X	X	X		X	X	X	Ad Lib	NA	NA	NA
Chord diagram	Relationships shown on a circle				X	X			X		X	Circular	NA	NA	O
Concept map	Depicts relationships between concepts				X	X			X		X	Ad Lib	NA	NA	NA
Context diagram	Block diagram showing relationships between attributes of an entity				X	X			X		X	Ad Lib	NA	NA	NA
Euler diagram	Venn diagram showing only relevant relationships			X	X	X			X		X	Ad Lib	NA	NA	NA
Flow chart	Drawing linking actions				X	X			X		X	Ad Lib	NA	NA	NA
Gantt chart	Bar chart for times and durations				X	X	X		X		X	Rectangle	C	N	NA
Iceberg diagram	Depictions of levels of a concept sorted from overt to hidden				X	X			X	X	X	Ad Lib	NA	N	NA
Mind map	Diagram used to visually organize information into a hierarchy, showing relationships among pieces of the whole				X	X			X		X	Ad Lib	NA	NA	NA
Network diagram	Chart showing connections between objects or events				X	X			X		X	Ad Lib	NA	NA	NA
Organization chart	Chart showing hierarchical arrangement of a group				X	X			X		X	Ad Lib	NA	NA	NA
Pick chart	Six Sigma window diagram for decisionmaking					X			X	X	X	Rectangle	NA	NA	NA
Sankey diagram	Chart showing information quantity and flow			X	X	X			X	X	X	Rectangle	C	C	NA
Spider diagram	Points on a venn or Euler diagram linking attributes				X	X			X	X	X	Ad Lib	NA	NA	NA
Sunburst diagram	Circular organization chart				X	X			X	X	X	Circular	NA	NA	NA
Timeline	Line showing event times or dates				X	X	X		X	X	X	Line	C	NA	NA
Tree diagram	Organization charts of groupings				X	X			X	X	X	Ad Lib	NA	NA	NA
Venn diagram	Intersecting shapes representing ideas			X	X	X			X		X	Ad Lib	NA	NA	NA
Window diagram	Matrix diagram for text					X			X	X	X	Rectangle	N	N	NA

Layout: Line=one axis; Rectangle=2 axes at 90°; Triangle=three axes at 60°; Cube=four axes at 90°; Circular=Multiple axes arranged radially; Ad Lib=No set layout.

Axis Scales: C=continuous ratio & interval scales; O=ordinal scales; N=nominal grouping scales; NA=no scale. Horizontal & vertical axes can be switched in some graphs.

FLOW CHARTS

Flow charts show connections between concepts or events, usually involving processes in which the order of the connections is important. Consequently, they are directional unlike some concept maps. Flow charts are also unique in their layout and content so they are usually drawn manually.

There are many applications for creating flow charts, most of which involve using various shapes, including rectangles, diamonds, circles, ovals, triangles, and arrows. The shapes are placed in an ordered path, connected by arrows, and labeled with pertinent information.

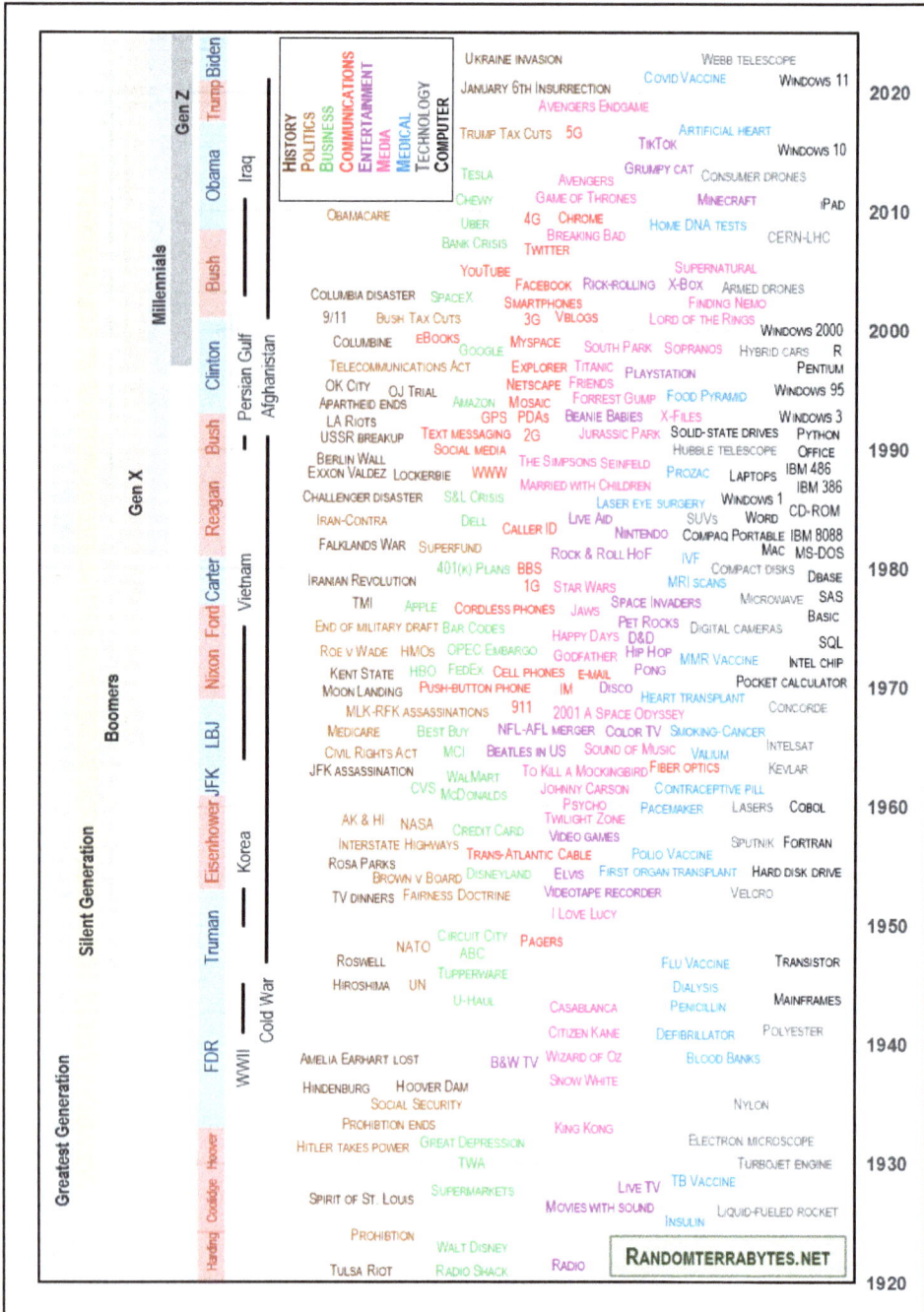

Figure 38. Context diagram showing different life experiences of sociological generations.

Graphics.

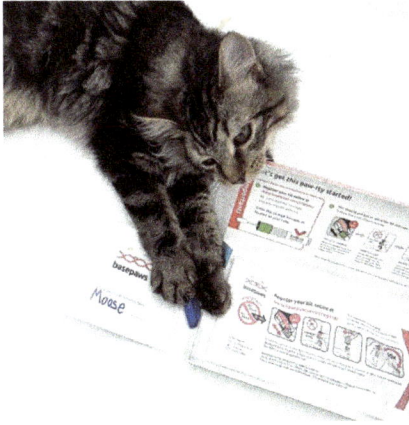

Follow the directions in the flow chart.

Flow charts are used to describe, for example, how work activities proceed in manufacturing and other businesses, how organizations or individuals in an organization interact, and how data moves in an information system. There is an example of a flow chart in Chapter 8 (questions for selecting a statistical method).

SUNBURST DIAGRAMS

A *sunburst diagram* is an organization chart presented in a radial format. Each ring of the chart represents a different organizational level. The center of the circle represents either the top or the root of the hierarchy and the levels progress outwards. The sizes of the levels can be made proportional to data values so the chart can function like a pie chart.

Figure 39 is a sunburst diagram that shows the organization structure of the National Football League. Sunburst diagrams are also known as *ring charts*, *multi-level pie charts*, and *radial treemaps*.

ICEBERG DIAGRAMS

Iceberg diagrams explore the idea that there are some things about a concept that are overt, well understood, believable, or known to most people, there are other things about a concept that are hidden, misunderstood, unconvincing, or unknown, and there are many things in between. The graphic uses an iceberg as a visual metaphor because 90% of a floating iceberg extends below the water's surface.

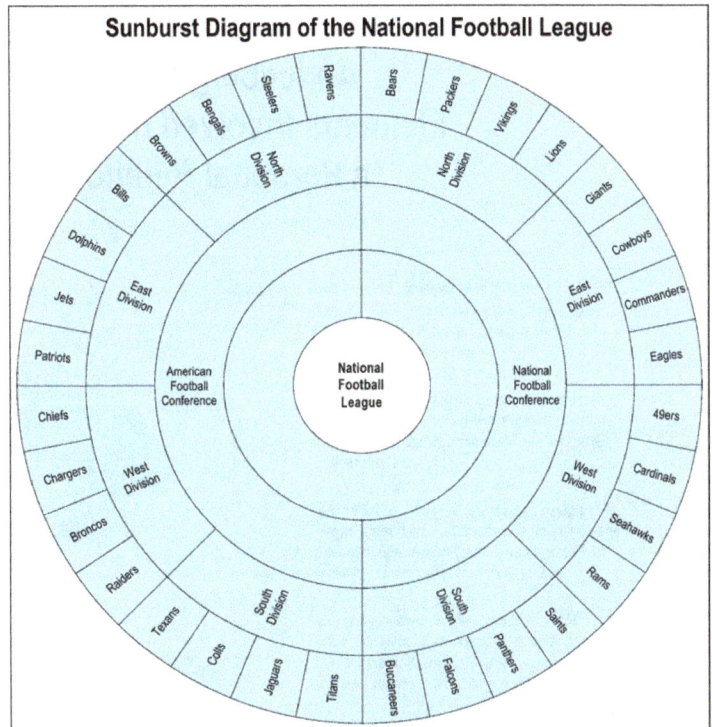

Figure 39. Sunburst diagram of the organization of the NFL.

Graphics.

To create an iceberg diagram, you don't necessarily need data. Nothing gets "plotted" in the traditional sense. You start with a template of an iceberg (or draw one) and add text to the levels of the iceberg to represent what's overt versus what is hidden.

The information behind the graphic could represent anything from the academic consensus on a topic to just informal opinions. As someone who is looking at an iceberg diagram, you have to be aware of the legitimacy of the information behind it.

I'm hiding from the iceberg.

Figure 40 is an iceberg diagram of how human personal relationships develop. It's not based on specific scientific research, but rather a collection of information from popular psychology websites.

The concept is that at the uppermost level of the iceberg someone might take notice of another person from afar based exclusively on physical characteristics. Then at the

An Iceberg Diagram of Elements Involved in Establishing a Personal Relationship

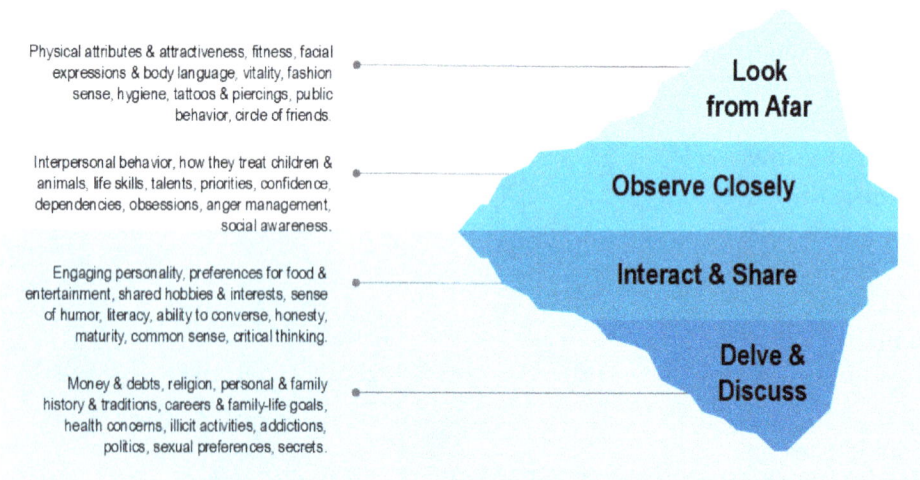

Physical attributes & attractiveness, fitness, facial expressions & body language, vitality, fashion sense, hygiene, tattoos & piercings, public behavior, circle of friends.

Look from Afar

Interpersonal behavior, how they treat children & animals, life skills, talents, priorities, confidence, dependencies, obsessions, anger management, social awareness.

Observe Closely

Engaging personality, preferences for food & entertainment, shared hobbies & interests, sense of humor, literacy, ability to converse, honesty, maturity, common sense, critical thinking.

Interact & Share

Money & debts, religion, personal & family history & traditions, careers & family-life goals, health concerns, illicit activities, addictions, politics, sexual preferences, secrets.

Delve & Discuss

Figure 40. Iceberg diagram of how human personal relationships develop.

Graphics.

next level, they observe additional characteristics involving observations of less tangible characteristics like behaviors. The two individuals might then interact and discover more subtle characteristics about each other in the third level. Finally, they engage in deep discussions of sensitive topics at the bottom of the iceberg.

These levels are not necessarily time dependent. For example, a deep discussion about religion might occur on a first date. If you don't touch all the levels in establishing a relationship, it may not be as strong as you might believe.

CAUSE-AND-EFFECT DIAGRAMS

Cause-and-effect diagrams are used to present ideas for what conditions or events might be responsible for an observed effect. Unlike *iceberg diagrams*, cause-and-effect diagrams are usually based on technical research, or at least, informed opinions. They are commonly used in quality control activities, especially in the statistical field called *Six Sigma*. Cause-and-effect diagrams are also

We were hoping for tuna.

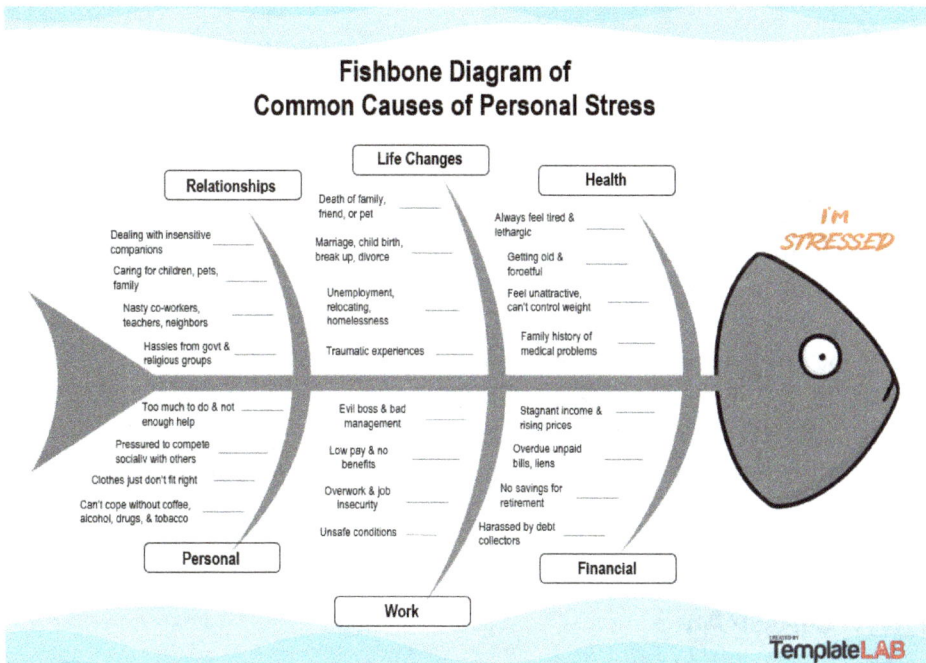

Figure 41. Fishbone diagram of common life experiences that cause stress.

Graphics.

called *fishbone diagrams* or *Ishikawa diagrams* (after their creator Kaoru Ishikawa, 1915-1989, a Japanese professor of engineering and quality control expert).

Cause-and-effect diagrams are usually drawn from templates or by hand. Traditionally, the graphic looks like a line with angled lines attached on either side of the line and pointing in the same direction. More creative depictions draw the graphic like a fish skeleton, hence the name fishbone diagram.

As with an iceberg diagram, creating a cause-and-effect diagram consists of adding text to an appropriate template. With a cause-and-effect diagram, the template would consist of a drawing consisting of lines or a fish.

Figure 41 is a fishbone diagram of common life experiences that combine to cause an individual to be stressed. It's not based on specific scientific research, but rather a collection of information from popular psychology websites.

VENN DIAGRAMS

Venn diagrams have been around for more than a century and precursors existed for centuries before that. They are a well-established concept in mathematics and are even popular on social media as a component of memes. They are meant for visualizing logical relationships but are also used to present mixtures, concepts, group membership, and of course, social memes. Venn diagrams are named after their creator John Venn (1834-1923), an English logician and philosopher.

Venn diagrams are usually drawn from templates or hand drawn. They traditionally use circles but any overlapping or contiguous shapes are possible.

Figure 42 is a Venn diagram showing how the basic ingredients of candy bars—chocolate, caramel, nougat, wafer, fruit, and nuts—are combined in popular consumer products. The figure doesn't show any quantification of the mixtures, only the concepts of the combinations. More importantly, the figure doesn't use circles or rely on overlap to express combinations. It's a completely unique version of a traditional concept.

We found the Snickers.

What's your favorite candy bar?

WINDOW DIAGRAMS

Window diagrams are a type of *matrix plot*, which are just grids for organizing information. The cells of a matrix plot can contain data, tables, graphs, or text. Window diagrams usually contain only text.

 Graphics.

VENN DIAGRAM OF CANDY BARS AND PIECES

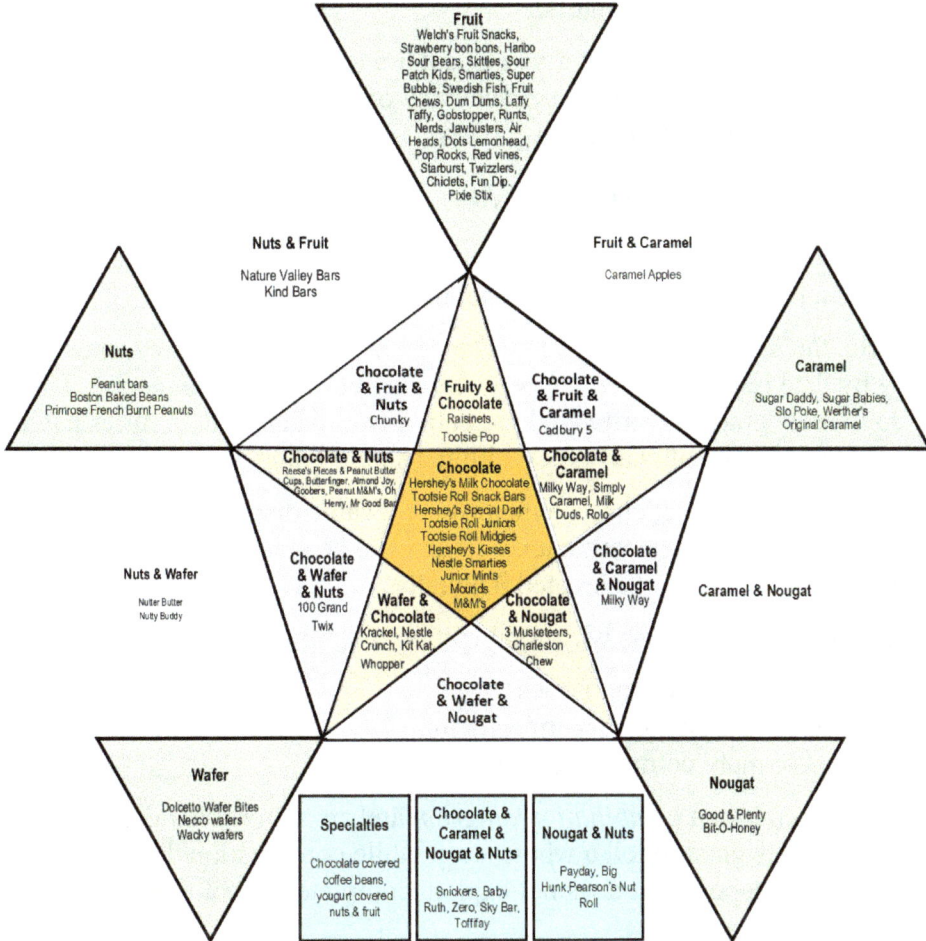

Fruit
Welch's Fruit Snacks, Strawberry bon bons, Haribo Sour Bears, Skittles, Sour Patch Kids, Smarties, Super Bubble, Swedish Fish, Fruit Chews, Dum Dums, Laffy Taffy, Gobstopper, Runts, Nerds, Jawbusters, Air Heads, Dots Lemonhead, Pop Rocks, Red vines, Starburst, Twizzlers, Chiclets, Fun Dip, Pixie Stix

Nuts & Fruit
Nature Valley Bars
Kind Bars

Fruit & Caramel
Caramel Apples

Nuts
Peanut bars
Boston Baked Beans
Primrose French Burnt Peanuts

Chocolate & Fruit & Nuts
Chunky

Fruity & Chocolate
Raisinets,
Tootsie Pop

Chocolate & Fruit & Caramel
Cadbury 5

Caramel
Sugar Daddy, Sugar Babies,
Slo Poke, Werther's
Original Caramel

Chocolate & Nuts
Reese's Pieces & Peanut Butter Cups, Butterfinger, Almond Joy, Goobers, Peanut M&M's, Oh Henry, Mr Good Bar

Chocolate
Hershey's Milk Chocolate
Tootsie Roll Snack Bars
Hershey's Special Dark
Tootsie Roll Juniors
Tootsie Roll Midgies
Hershey's Kisses
Nestle Smarties
Junior Mints
Mounds
M&M's

Chocolate & Caramel
Milky Way, Simply
Caramel, Milk
Duds, Rolo

Nuts & Wafer
Nutter Butter
Nutty Buddy

Chocolate & Wafer & Nuts
100 Grand
Twix

Wafer & Chocolate
Krackel, Nestle
Crunch, Kit Kat,
Whopper

Chocolate & Nougat
3 Musketeers,
Charleston
Chew

Chocolate & Caramel & Nougat
Milky Way

Caramel & Nougat

Chocolate & Wafer & Nougat

Wafer
Dolcetto Wafer Bites
Necco wafers
Wacky wafers

Nougat
Good & Plenty
Bit-O-Honey

Specialties
Chocolate covered
coffee beans,
yogurt covered
nuts & fruit

Chocolate & Caramel & Nougat & Nuts
Snickers, Baby
Ruth, Zero, Sky Bar,
Tofffay

Nougat & Nuts
Payday, Big
Hunk,Pearson's Nut
Roll

Figure 42. Venn diagram showing mixtures of ingredients in candy bars.

Simple window diagrams consist of two criteria, or dimensions, defined by rows and columns. Each dimension usually has two categories, or levels, resulting in four cells, or *panes*. The beauty of a window is the way it can organize complex information into simple binary categories that most everybody can understand. As a consequence, windows are used in many ways to analyze data and other information, and present results to general audiences.

Graphics.

Most window diagrams are either *information windows* (e.g., *Johari windows*, *Eisenhower windows*, *Rumsfeld windows*, and *boundary windows*) or *statistical windows* (e.g., *definition windows*, *variance windows*, *performance windows*, *pick charts*, and windows on scatter plots).

There are examples of window charts in Chapter 5 (precision and accuracy in political polls) and Chapter 8 (size and accuracy of models). Chapter 24 in **Stats with Cats** is also dedicated to window charts.

6. COMBINATION DIAGRAMS.

Combination diagrams, also called *combo charts*, are simply two or more charts, or chart elements, placed together in the same graphic.

For example, the *histograms* shown in Chapter 3 combine a *bar chart*, showing frequency by data bin, with a *scatter plot*, showing cumulative frequency. This is easy to do in most graphing software by creating two different data series having different data and different chart types within a single graph.

Sometimes, plotting distinct chart elements on the same diagram can be impractical. These situations require the combination diagram to be created manually in the same way infographics all have to be created manually.

There are at least three reasons why combination diagrams are used:

- To accommodate scaling issues
- To provide complementary information
- To provide more complete information.

The difference between a *combination diagram* and an *infographic* is that infographics are designed to tell a whole story while combination diagrams are still a visualization that supports but does not explain completely a point being made.

SCALING ISSUES

Sometimes graphs need to use two or more different scales. Usually, this can be accommodated in the software used to create the graph. When it can't, there are two options. One is to create separate standalone plots, the other is to add a small separate graph placed within the plot area of the main graph, called an *inset plot.*

Figure 43 is an example of a *combination diagram* used to accommodate two different scales. It shows the number of reports of Unidentified Flying Objects (UFOs) in the U.S. from 1910 to 2013.

In the diagram, there is an inset plot placed within the main plot. This inset plot had to be placed on the main plot manually. Both the inset plot and the main plot use the same data, only their scaling is different.

Graphics.

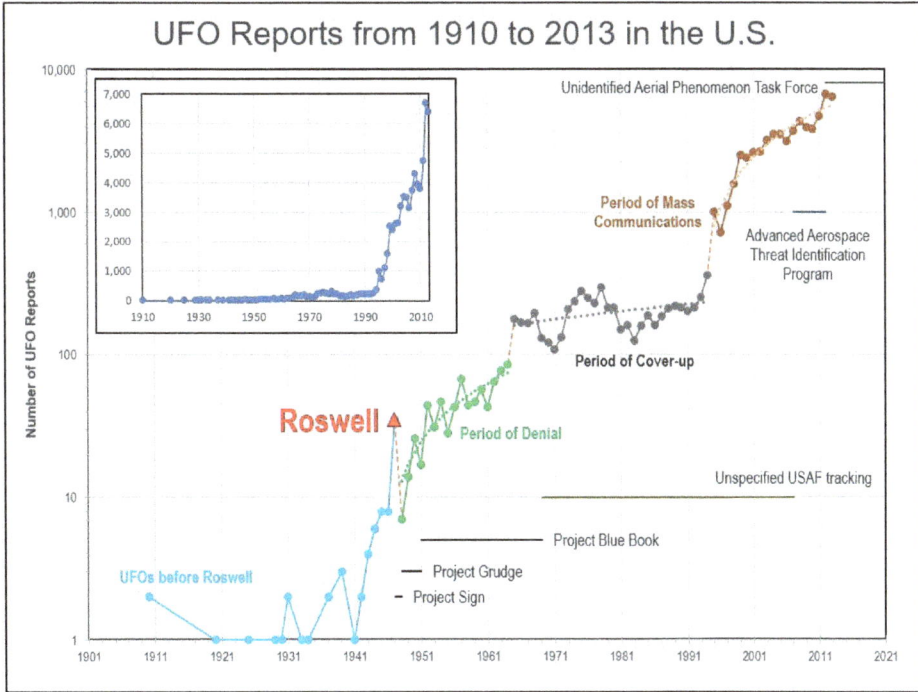

Figure 43. Combination diagram showing reports of UFOs in the U.S. using two different scales.

In both graphs, the horizontal axes represent the year on an ordinal scale. The inset graph shows the number of UFO reports on the vertical axis using a *linear scale*. The main graph shows the number of reports on the vertical axis on a continuous, *logarithmic scale*.

I swear, Bella, I just saw a UFO and the pilot was a cat just like us.

A logarithmic (or log) scale is used in graphs to display data that have a large range spanning orders of magnitude. The scale is nonlinear with each major division representing an exponential increase or decrease. Earthquake and hurricane magnitudes are measured on logarithmic scales.

The data points in the inset graph are connected by solid straight lines of the same color. Only two trends are apparent—a horizontal trend from 1910 to 1992 and a rapidly increasing trend from 1993 to 2013, perhaps attributable to the spread of the internet. In contrast, the data points in the main graph are connected by solid straight lines and colored to emphasize time periods having visually

Graphics.

different trends in the data. These trends may be attributable to different government programs for investigating UFO reports.

Patterns in UFO reporting are highlighted by the scales used for the two vertical axes. The drastic increase in the number of reports since the 1990s can only be appreciated when viewed on the linear scale of the inset plot. However, changes in the patterns of UFO reporting are apparent only when viewed on a logarithmic scale of the main plot.

This combination chart shows an example of a data set having *hidden trends* (see Chapter 8). The axis scaling makes the two graphs of the same data appear remarkably different. The combination of the two charts is an example of how a chart's *Framework* contributes to achieving its *aim*.

COMPLEMENTARY INFORMATION

Combination diagrams can also be used to convey more relevant information. The purpose is to augment the original data with complementary information in order to put the data into a better context.

I'm the government oversight auditor. I see everything. I am Ghost.

Figure 44 is a combination diagram exploring the size of the Federal workforce. Line plots on the top of the figure depict the sizes of the U.S. population and the Federal workforce, and the ratio of the two (i.e., the average number of citizens served per civil servant). Area charts on the bottom of the figure summarize which political party held control of the two Houses of Congress. Major tick marks on the horizontal scale represent decades. Minor tick marks on the horizontal scale represent presidential election years.

The diagram indicates that while the size of the Federal workforce has remained stable since the 1970s, the growth of the population has resulted in an increase in the average number of citizens served by the government employees. The political divisions appear to also have some influence over the employment trends.

The size of the workforce increased from 1960 to 1990 when Democrats mostly controlled Congress and Republicans mostly controlled the Executive Branch. Then, the workforce contracted substantially under the **National Partnership for Reinventing Government** in the 1990s. During this period, the Presidency was held by a Democrat and control of Congress alternated.

 Graphics.

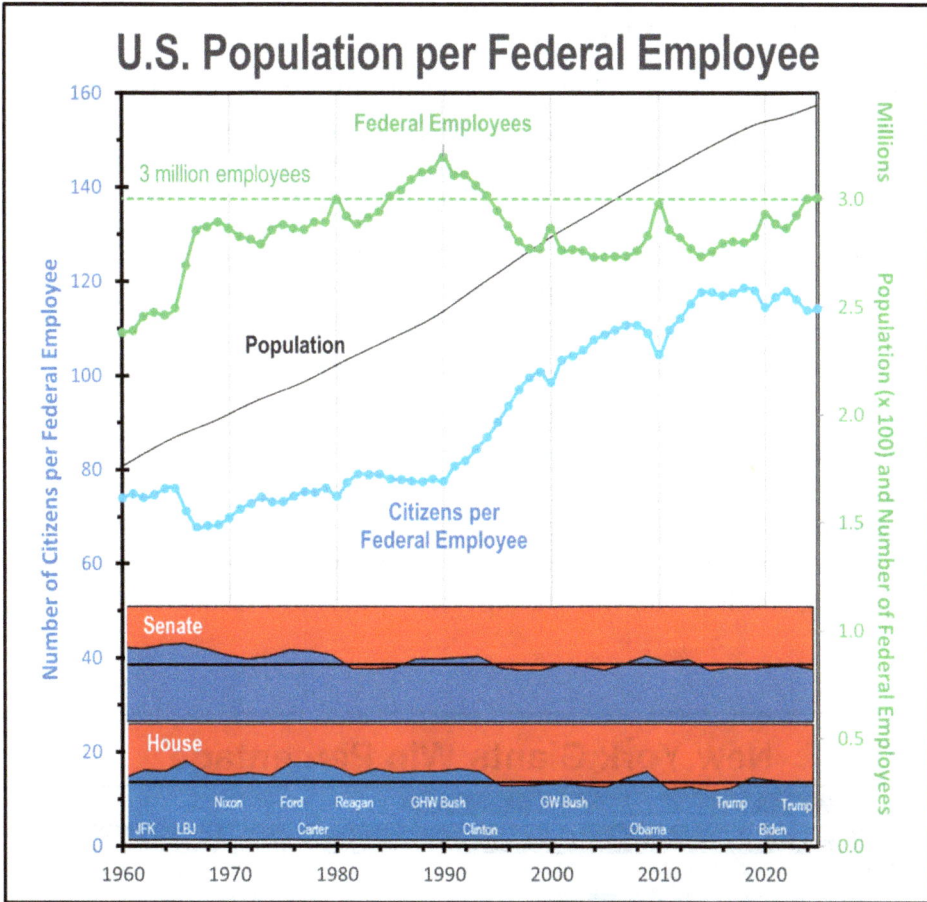

**Figure 44. Combination diagram of the
size of the Federal workforce.**

Following a decade of stability, the workforce has been increasing slightly since 2015 under both Republican and Democratic leadership. (Note: the number of Federal employees increases temporarily once every decade to conduct the national census. Those are the small spikes in the **green** line and dips in the **blue** line.)

While all of these changes in the Federal workforce were taking place, the U.S. population has grown steadily. The result has been that the ratio of the population to the workforce has increased substantially. That finding would be difficult for some people to understand if only the **blue** line representing the number of citizens per civil servant were shown.

Compare this chart to the *waterfall chart* of the same time period (1960-2024) presented earlier in this chapter.

Graphics.

COMPLETE INFORMATION

Some combination diagrams include an abundance of information. They provide viewers with enough diverse information that they can construct a whole data story in their imaginations. But, they don't actually tell the story. That's what *infographics* do.

Figure 45 is a combination diagram involving the non-player side of the NFL's New York Giants. The Giants are the fourth oldest franchise having been established in 1925. The diagram combines four bar charts and two line plots highlighting the ownership, management, and coaching of the franchise from 1980 to 2022.

During that forty-two year period, the team had four owners, five General Managers, and ten head coaches. They finished first in the NFC East Division eight times, played in five Super Bowls, and won four. But, they had their ups-and-downs. They won more games than they lost 20 out of 44 seasons but never maintained that winning momentum for more than three years at a time. Ultimately, it's the players who win games, coaches who win seasons, and owners and managers who build dynasties.

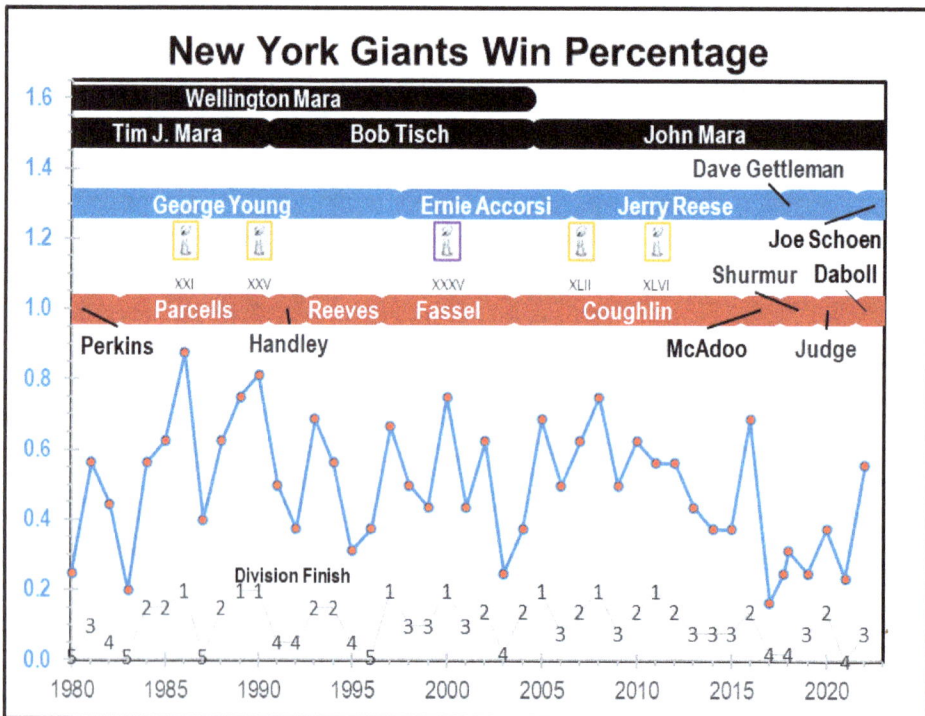

Figure 45. Combination diagram showing the management and record of the NFL's New York Giants.

Graphics.

This is an example of how a combination chart can tell a data story in a glance. It is an example of what David McCandless calls *knowledge compression*.

Data stories can be told in two ways, as a sequence of simple graphics showing just one important point or as a complex combination diagram that includes all the relevant data in one graph.

A series of simple graphs is like a comic strip. Each graph is like a panel of a strip that conveys only a part of the story. Several are needed, in the correct sequence, usually with explanations.

Combination diagrams are like a painting. Everything is presented at the same time. Viewers unravel the story just by looking and thinking.

Each statistician has their own preference for how to tell a data story. They choose what they believe is the right graphic for their goal.

THE RIGHT CHART FOR THE JOB

There are a virtually uncountable number of graphs, subspecies of the basic graphs, variations and extensions of these graphs, and combinations of graphs. For now, focus on the basic types of graphs. That will provide you with a good *Foundation*.

Selecting the appropriate type of graph depends in part on whether the variables to be plotted are measured on ordinal or continuous scales or both. Nominal scales don't enter into selecting a type of plot even though they are used often in graphing to show groups of data.

Table 23 provides an overview of the relationships between variable scales and the most basic types of graphs they can be used with. From there, more complex types of charts can be explored.

Table 23. Graphs that can be used with different combinations of scales.

First Variable Scale	Second Variable Scale	Appropriate Graphs
Continuous	Ordinal: equal intervals	Line; scatter if the ordinal variable has many levels
Continuous	Ordinal: unequal intervals	Bars
Ordinal: equal intervals	Ordinal: equal intervals	Line; scatter if the ordinal variables have many levels
Ordinal: equal intervals	Ordinal: unequal intervals	Bars
Ordinal: unequal intervals	Ordinal: unequal intervals	Bars

THE CHAOS IN CHARTS

In addition to the basic types of charts for properties, mixtures, distributions, relationships, and concepts, there are many variations of the graphs:

Graphics.

- Extensions of basic graphs (e.g., *side-by-side* and *stacked-bar charts*).
- Alternatives for basic graphs (e.g., *area graphs* based on *bar charts*).
- Enhancements of basic graphs (e.g., *box plots* with box width showing frequency of data values).
- Combinations of basic graphs (e.g., *histograms* with *line chart* of total frequency.

There are graphs with confusing nomenclature:

- Graphs that look different but go by the same or similar names. For example, *block diagrams* can refer to a graphic for organizing components using geometric shapes (like a flow diagram) or to a geologic drawing consisting of a map and two cross-sections. *Spider diagrams* are extensions of Venn or Euler diagrams while *spider charts* are another name for *radar charts*.
- Graphs that go by a variety of different names. For example, a *sunburst-chart* is also called a *ring-chart* and a *radial-treemap*. A *radar chart* is also called a *web chart*, *spider chart*, *polar chart*, and *Kiviat chart*.

There are graphs with uncommon formatting:

- Graphs that don't use right-angle axes (e.g., *ternary diagrams*).
- Graphs that combine graphs (e.g., *motion charts, matrix diagrams, combination charts*).
- Graphs that use color instead of data scales (e.g., *heatmaps*).
- Graphs for one observation of many variables (e.g., *radar plots*).

There are graphs for specific applications:

- Graphs that show stock market data (e.g., *candlestick charts, open-high-low-close charts*).
- Charts that show time durations for tasks (e.g., *Gantt charts*).
- Charts that show temporal variance (e.g., *periodograms* and *autocorrelation plots* in time-series analysis).
- Charts that show spatial variance (e.g., *variograms* in geostatistics).
- Graphs that show word frequency (e.g., *word clouds*).
- Graphs for cyclic data scales (e.g., *rose diagrams*).

Time-series charts are especially important in graphing because they are the basis for many statistical relationships. Time is always measured on at least an ordinal scale, but the intervals between time periods are not always equal.

You can call me by any name you want as long as you take care of me.

Graphics.

I wonder what geologic time these rocks came from. It was before I was born.

Geologic time, for example, has more to do with rock layers and fossils than with age even though the unit of the time scale is years. Real-time can be measured in years, months, weeks, days, hours, minutes, and seconds. As long as you don't mix the units, the intervals will be equal.

Dates in spreadsheets like Microsoft Excel actually represent the number of days after an arbitrary starting date, usually set as January 1, 1900. Thus, spreadsheets treat time as an interval scale. What appears as a date is actually a number that has been formatted to look like a date. Try reformatting a date as a number and you'll see. That's why you can find the number of days between two dates by subtracting one date from the other. You can use a date in any spreadsheet operation or graph because they are really numbers.

Here are two other things you should be aware of about time scales in spreadsheets.

First, if you enter a date before the starting date used by the software (e.g., January 1, 1900), the spreadsheet will store it as text rather than a number. This will cause calculations involving dates before the software's assigned starting date to result in a #VALUE error.

Second, the year 1900 has an extra day. When Lotus 1-2-3 was first released in 1983, the program assumed that 1900 was a leap year, which it wasn't. When Microsoft Excel was first released two and a half years later, Microsoft also assumed that 1900 was a leap year, for compatibility. The rest of the leap years after 1900 are correct. But, calculations using data from 1900 may encounter discrepancies. Fun times.

Don't worry if you run into a kind of chart that you've never heard of before, just look it up on the internet. It's probably something derived from a simpler basic graph. Nobody knows every chart, even professional statisticians.

If you need to create a chart for your own data, you will be limited mostly by the capabilities of the software you have. Given the number of free applications for graphing that are available on the internet, this isn't much of a limitation.

WHAT TO LOOK FOR IN GRAPHS

Before you become too excited about a statistical graph, there are two things to consider—is the dataset sound and is the graph appropriate for the data. Then you can concentrate on what the graph means.

Graphics.

DATA LEGITIMACY

Things to look for in assessing the quality of a dataset are detailed in Chapter 3.

Is the source of the data reliable? Is it a primary or secondary source? Government and academic organizations are the best. Sources that provide data without adequate metadata are the worst.

If the aim of the analysis involves probabilities, the sample should be representative of the population. The number of observations should be adequate for the resolution needed, particularly if groups are involved. Will the data be used to analyze snapshots, changes, or general patterns and trends? The metrics being graphed should be appropriate for the aim of the graph. The scales of the variables should be appropriate. Metadata should be provided.

The data should have been scrubbed of errors and other data issues, like: invalid data; incorrectly-recorded data; incorrectly-coded data; data quality information; missing data; extraneous data; dirty data; useless data; invalid fields; out-of-spec data; out-of-bounds data; messy data; corrupted data; and mismatched data.

I'm looking to get a snapshot of my data.

GRAPH LEGITIMACY

Assess the appropriateness of the graph in terms of its *foundation* (aim and focus), *framework* (axes and scaling), and *facade* (embellishments).

Is the graph appropriate for the intended aim and focus? What is the graph supposed to show—properties, mixtures, distributions, relationships, or concepts? This is fundamental, and hopefully, hard to get wrong.

Are the axes appropriately scaled for the variables? This is where a graph is most likely to have issues. It's what critics talk about the most.

Are the labeling and other graph embellishments helpful or intrusive? Formatting draws a lot of criticism but it is seldom a fatal flaw, more like an annoyance.

GRAPH INTERPRETATION

Interpretation of a graph will depend on its type and intended purpose.

For *property graphs* (e.g., *bar charts*, *area charts*, *line charts*, *candlestick charts*, *control charts*, *means plots*, *deviation plots*, *spread plots*, *matrix plots*, *maps*, *block diagrams*, and *rose diagrams*), **look for the unexpected**. Are the *central tendency* and *dispersion* what you expected? Where are any big deviations?

Graphics.

For *mixture graphs* (e.g., *pie charts*, *bar charts*, *proportional area charts*, *word clouds*), **look for imbalance**. If you have some segments that are very large and others very small, there may be common and unique themes in the mix to explore. Maybe the unique segments can be combined. This will be useful information if you do break out subgroups later.

I'm balanced.

For *distribution graphs* (e.g., *box plots*, *histograms*, *dot plots*, *stem-leaf diagrams*, *Q-Q plots*, *rose diagrams*, and *probability plots*), **look for symmetry**. That will separate many theoretical distributions, say a symmetrical Normal distribution from an asymmetrical lognormal distribution. This will be useful information if you consider any statistical testing later.

For *relationship graphs* (e.g., *icon plots*, *2D scatter plots*, *contour plots*, *bubble plots*, *3D scatter plots*, *surface plots*, and *multivariable plots*), **look for linearity**. You might find linear or curvilinear *trends*, repeating *cycles*, one-time *shifts*, continuing *steps*, periodic *shocks*, or just random points (Chapter 8). This is the prelude for looking for more detailed patterns and trends.

For *concept diagrams* (e.g., *brainstorm diagrams*, *cause-and-effect diagrams*, *flow charts*, *iceberg diagrams*, *Venn diagrams*, and *window diagrams*), **look for confusion**. Are the linkages between elements logical, complete, and consistent? Do they make intuitive sense? Are linkages missing?

For *combination diagrams* (based on scaling issues, complementary information, or complete information), **look for storytelling**. Are the scales useful and unambiguous? Is the additional information provided by the primary and secondary graphs consistent and, most of all, useful?

THE NEXT STEP IN GRAPHING

David McCandless has said that "**visualizing information is a form of knowledge compression**." Statistical graphics make data more engaging and easy to understand because we process visual information faster than text.

Sometimes you'll see graphs that illustrate attributes of a single sample or variable. More often, a graph will involve many samples and variables. Occasionally, a graph will contain all the information needed to tell the story of an entire analysis.

Graphics.

There are two approaches to telling a story about data with statistical graphics. The *stepwise approach* involves presenting a series of simple graphics, each of which illustrate a single finding. The creator of the graphics steps the audience through each graph to lead them to the conclusion.

Tell me a story.

The *all-in-one approach* involves combination graphs and *infographics* that show the audience the *big picture.* The creator of the graphic points out to the audience all the details and interconnections in the graph to lead them to the conclusion.

The *stepwise approach* is like a comic strip with each graph being like a panel in the strip. One panel leads to the next, but taken singly, each is inadequate to understand the whole story. The audience becomes engaged by following each panel.

The *all-in-one approach* is like an intricately-detailed Renaissance painting. The audience becomes engaged by spending time understanding the meanings in all the details.

Graphs are tools. If you only have a hammer and a screwdriver in your toolbox, you won't be able to accomplish very much. You'll want to be able to pull out 45-degree, external, snap-ring pliers when you need them.

If you're doing your own analysis, don't worry about the attractiveness of your graphs in the beginning. Produce them honestly so that you can understand your data. When you decide which ones you need to tell your data's story, you can focus

There are more kinds of charts than there are kinds of kittens.

 Graphics.

on making them appropriate for your audience. That's where the *art of the chart* comes in.

Don't be discouraged if you're surprised by how much more there is about graphing than you heard about in high school. There's even more but you probably won't need to worry about it … at least until you decide to become a statistician.

We're ready to start graphing data.

 Graphics.

What's your favorite kind of chart?
We should do a survey about it.

Graphics.

CHAPTER 5. SURVEYS.
THEY'RE EVERYWHERE
EXCEPT STATS 101.

INTRODUCTION

Not a day goes by that we all don't hear about or get invited to participate in some survey about our opinions or preferences. Everybody wants to know what we're thinking—governments, businesses, the media, and each other.

Surveys are really the only way to measure opinions and preferences quantitatively. But surveys are annoying to people who don't want to take them even though those same people want to know the results and complain when they aren't asked what they think.

People believe that conducting a statistical survey is easy and that they know all they need to know about surveys. After all, it's rare for statistical surveying to be taught in high school or even in Stats 101. How difficult could it be? Isn't it just intuitive? Plus, there are free apps everywhere on the internet. What's the catch?

The polls say I have this election in the bag.

The polls say I have this election in the bag.

So why do political pollsters have such a hard time predicting election results? Is conducting a survey more challenging than people assume? And how can you tell a bad survey from a good survey?

That's what this Chapter is about.

SURVEYS ARE EVERYWHERE ... ALMOST

People have been trying for centuries to find ways of gauging public opinion. Before scientific polling, decision makers had to resort to collecting anecdotes, asking experts or informants, or just guessing. King George III didn't think his harsh treatment of the Colonies would spur a revolution. Bad guess.

Government leaders have long relied on traditional public opinions to make policy. But, opinions change over time, sometimes suddenly. American isolationism ended after the attack on Pearl Harbor. Unwanted authoritarian surveillance of the public,

Surveys.

made infamous by Orwell's Big Brother character in his novel **Nineteen Eighty Four**, became acceptable after 9/11/2001.

Statistical surveys were the answer to unreliable, informal methods of assessing group opinions. Surveys have a basis in science, a long history of usage, and most importantly, the ability to quantify results and uncertainties. They are the only way to measure opinions quantitatively.

Surveys aren't just about opinions, for the most part, they are opinions. Different surveys on the same topic can have different results based on who conducts the survey, and when and how they conduct it. Unlike many measurements, survey questions have no reference point or benchmark that they can be compared to. Furthermore, the results of most surveys can't be tested for accuracy, the two exceptions being censuses of entire statistical populations and poll results for completed elections.

The National Census, required in the U.S. Constitution, was first conducted in 1790. Straw polls were popular even then for discerning opinions of groups. Within a generation, newspapers were conducting polls of their readers.

No matter what career you go into, you will participate in, hear about, and experience the results of surveys every day of your life. Governments conduct thousands of surveys every year to collect information on their operations and the statuses of businesses and the citizenry. Companies, small and huge, conduct surveys about their products, their markets, and their services in order to make critical business decisions. They survey everything from the satisfaction of customers to the morale of employees. Universities and independent research organizations conduct a variety of surveys on society. Schools survey students about their teachers and the lunch menu.

I hope they serve tuna muffins for lunch.

But the one place that surveys aren't is in introductory statistics textbooks. Statistical surveying isn't commonly taught in Stats 101 either. The topic is reserved for upper-level courses, usually as an option. You can even get a statistics degree without taking a course on surveys.

The reason that statistical surveying is absent from many educational programs is not because polling is easy, it's just the opposite. Polling is the most challenging analysis in statistics. It is expensive, difficult to implement, and susceptible to the influence of all kinds of unexpected conditions and events.

Surveys.

SURVEYS, POLLS, AND CENSUSES

There are several terms that are used for statistical surveys.

A *census* involves questioning every member (or at least most members) of a group of interest, that is, the *population*. The U.S. conducts a *census* of all Americans once every decade as required by the Constitution.

A *poll* was formerly used to refer to a simple, one-question, informal interview that was often conducted in person. *Straw polls,* used to quickly assess the preference of an existing assembly on a specific question, were common as early as the 1700s.

In contrast to a poll, a *survey* was a formal, elaborate, and more comprehensive data-gathering effort that was conducted with as much statistical rigor as possible. In contrast to a census, survey participants were sampled from a population rather than having the entire population participate.

Today, the terms *poll* and *survey* are used synonymously. Surveys are sometimes referred to as *statistical surveys* to distinguish them from *land surveys*.

Figure 46 shows the number of Google searches for the terms *census*, *poll*, and *survey* from 2004 to 2024. The term *census* is most closely associated with the National Census, which is conducted every ten years in years ending in '00. Google searches for the term peak in those years. The term *poll* is most closely associated with elections, especially presidential elections. Google searches for the term peak in

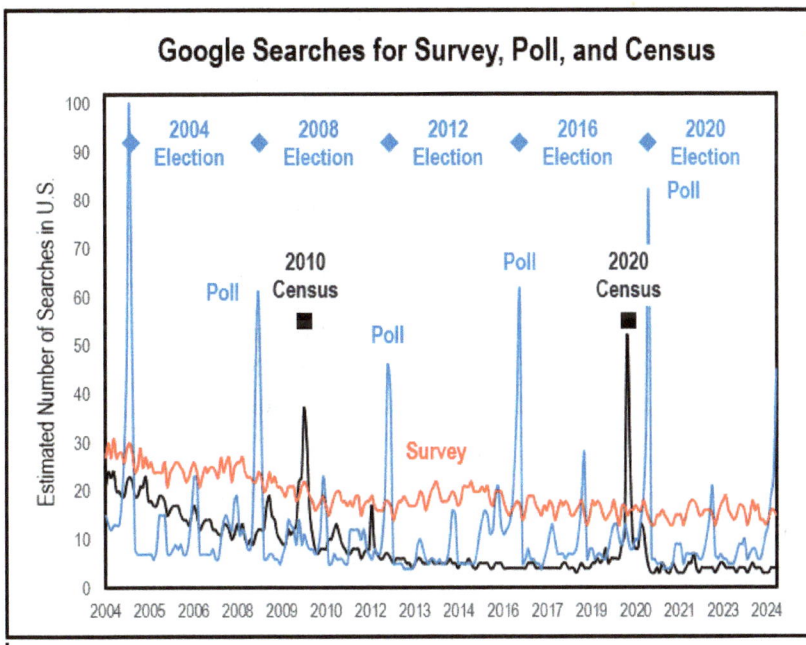

Figure 46. Google searches for census, poll, and survey from 2004 to 2024.

Surveys.

those years. The term *survey* is used generally, mainly for discussions involving statistics, and doesn't peak at any particular time.

Surveys, polls, and censuses are *descriptive*; they don't *predict*. They gauge opinions at a single point in time, a *snapshot*. Imagine taking a photograph of a scenic landscape. The photograph shows the landscape at the moment it was taken, which will look different at other times because of seasons, weather, time-of-day, and occupants in the scene. To use poll data to make a prediction, you need a model (Chapter 8).

WHERE POLLS COME FROM

Polls are used wherever opinions have to be measured and evaluated. They are used extensively in business, marketing, educational and sociological research, government, and especially, politics.

Polling for presidential elections began almost as soon as there were competitive elections after George Washington left office. Everybody wanted to know how the election, the ultimate poll, would turn out. In fact, voting is often called *going to the polls*.

Political polls are like interim votes. Politicians use them to decide which policies to support and how to allocate campaign resources. They use them throughout a campaign, not just at the end when races have to make their final promises.

By the 1940s, polling had become a business, but it was labor intensive and there were only a few national pollsters. What really made polling so pervasive in our society was the development of computers, then personal computers, and then, the internet. Political polls expanded rapidly during the 1990s and after, as shown in Table 24.

Surveys and voting go paw-in-paw.

After 2000, polls and polling organizations were everywhere. Simple polls grew in size and complexity. The terms *survey* and *poll* began to be used interchangeably. Political *polls*, for example, evolved into long, complex, and meticulously-designed *surveys*.

Polling wasn't limited to big organizations. Specialty firms entered the profession focusing on smaller clients, like advocacy groups and political campaigns. Polling increased tenfold from the 1970 to the 1990s. Today, there are too many pollsters to count.

Surveys.

Polls on social media are for entertainment. Serious surveys of opinions and preferences are quite different. There is a lot that goes into creating a scientifically valid survey. Scores of textbooks have been written on the topic. Furthermore, the state-of-the-art is constantly changing as technology advances and more research on the psychology of survey response is conducted.

Table 24. Number of U.S. Presidential polls from 1969 to 2001.

President		Weeks of Polls	Number of Polls	Polls per Month	Number of Polling Groups	Most Nationwide Presidential Polls Conducted
G. W. Bush	Second Term *	116	408	14.0	33	Gallup/CNN/USA (14%) Fox/OpinDynamics (10%)
	First Term	208	782	15.0	41	Gallup/CNN/USA (18%) Fox/OpinDynamics (10%)
Clinton	Second Term	207	506	9.8	27	Gallup/CNN/USA (20%) CBS/NYT (10%)
	First Term	209	331	6.3	5	Gallup/CNN/USA (33%) CBS/NYT (22%)
G. H. W. Bush		207	135	2.6	1	Gallup (100%)
Reagan	Second Term	205	54	1.0	1	Gallup (100%)
	First Term	207	83	1.6	1	Gallup (100%)
Carter		202	92	1.8	1	Gallup (100%)
Ford		122	37	1.2	1	Gallup (100%)
Nixon *		289	97	1.3	1	Gallup (100%)

REASONS FOR POLLING

Surveys in business, research, and government, aim to discover information about opinions that people have. The information sought is different for every survey, but it always involves opinions. Survey sponsors want accurate information that they will use directly to make policy or business decisions. Typically, they won't release information from the survey to the public.

Political surveys are different. They are used for both constructive and nefarious purposes. Most surveys are legitimate, like nonpolitical surveys, but some have an extra dimension in which sponsors really only want information that aligns with their

Surveys.

ideology. Information from unbiased political surveys may or may not be released to the public. Information from biased political surveys is always released to the public, usually filtered through news outlets and social media. That's their whole point.

To be a discerning consumer of political polls, you have to understand how polls are created, how they might be biased, and perhaps most importantly, how their presentation can be misleading.

I'll tell you my policy preferences but it's too early to decide on a candidate.

Political polls usually focus on one of two themes—policies and priorities or parties and people. Polling opinions about government policies and societal priorities is always relevant. The information and people's opinions about it are always changing. Polling opinions about political parties and their candidates (called *horse-race polls*) should only be relevant to campaign organizations. Unfortunately, horse-race polls make for attention-grabbing stories in the media so they are conducted virtually nonstop, even a year or more before an election. Reporting them to the public does more harm than good.

Political polls between elections should focus on policies and priorities rather than political parties and their candidates. They should aim to find out **what** the electorate wants so voters can decide **who** is best able to provide it. Media outlets should champion this concept. This hasn't been the trend, however.

Political polls are usually conducted by candidates running for office, their political opponents, news sources, and researchers. Candidates conduct polls for a number of reasons:

- To gauge voter demographics to allocate campaign resources
- To gauge support for policies
- To test alternative messaging
- To solicit donations.

Candidate polls are typically short and focused. They are usually not statistically rigorous. For example, they might poll voters in their electoral district without any concern for matching a population demographic (called an *area sample*). The results of the polls won't be released to the media as they are irrelevant to most voters.

Besides being annoying, requests for donations reveal a lot about who is conducting the poll and what it'll be used for. Donation requests are a clear signal that the polling is being done by a candidate for their own purposes, for example, to assess policy priorities. The request for donation is just a convenient add-on. If all the candidate

Surveys.

wanted to do was fundraise, they wouldn't have bothered programming a poll into their message.

Some candidates use polls for improper and unethical purposes. For example, a candidate might try to disseminate false or misleading information to survey participants. This is called *push polling*. A candidate might use polls to try to influence media stories and the electorate with biased results. They might also try to confuse poll aggregators by producing multiple biased polls right before an election. That's why it's always important to know who funded and who conducted the poll.

This will come as a surprise.

Polls conducted by or for the media could be legitimate or not. Their goal is to sell content to their audiences. The legitimacy of poll results isn't always their primary concern. Again, it's important to know who funded and who conducted the poll.

THE MOST DIFFICULT ANALYSIS

It should come as no surprise that most people think conducting a survey is easy. And why not? There's plenty of free software available. Google has a site for creating polls. Social media sites and blogging sites provide capabilities for conducting polls. There are also quite a few free online survey tools. The math for analyzing the responses is simple too, just counts and percentages. So why wouldn't people believe that just anybody could conduct a survey?

What makes surveys so difficult to conduct correctly enough to obtain legitimate results isn't the math or the software, it's the nature of the *phenomena* and the *populations*. Surveys aim to measure the intangible.

Surveys are different from most statistical analyses in STEM, business, or most other fields of inquiry for four reasons:

1. Some phenomena are elusive. Most statistical analyses study phenomena that are well established. Surveys explore phenomena that are new or unique. There are often no theoretical bases to establish an expectation; the survey is meant to do that. Furthermore, the phenomenon can change. A product can be modified or events can change the perception of a politician, for instance. Surveys have to characterize phenomena that can evolve unexpectedly.

2. Phenomena are characterized by unproven metrics. Most statistical analyses use well defined metrics to characterize the phenomenon of interest (Chapter 3). Whether they use electronic instrumentation, laboratory procedures, or routinely

 Surveys.

collected performance indicators in business or sports, the metrics have all been tested to ensure they are *repeatable* and *reproducible* for characterizing a phenomenon. They are also checked to ensure they are employed consistently.

Surveys, for the most part, can't do that. Every question is a new, untested metric, presented in a unique context. The way a survey question is worded or presented in a survey can produce unintended interpretations. Surveys have to grapple with uncontrollable *measurement variability*.

3. Representative samples of the population are hard to capture. In most statistical analyses, experimenters control either the selection of the samples or the process by which the samples are selected. This is not entirely true with surveys because of three issues.

Wow, surveys are a lot more complicated than social media makes them out to be.

First, it may be impossible to identify the individuals who are in the population. For example, while it may be easy to identify who purchased a product to survey their satisfaction with the product, it is difficult to identify who is likely to vote in an election. Increased voter turnout and voter suppression, especially when focused on key demographics, are critical unknowns.

Second, surveys aren't like statistical analyses in STEM and business in that the subjects of surveys are all humans who are free to do whatever they choose. They decline invitations to participate, they don't answer every question, and they quit the survey before completing it. An opinion survey may need to invite a thousand members of a population to get one legitimate survey response.

Third, how survey participants are contacted has evolved with innovations in technology. Surveys were all conducted face-to-face until the telephone was invented, and then the mobile phone, and then the internet. All of those methods are still used in combinations to ensure all members of the population have an equal chance of being surveyed. Sampling for a survey is much more complicated than for most types of research.

4. Data aren't always consistent and valid. Data collected for a statistical analysis using a calibrated instrument or established process can be tested to ensure they are valid. The same isn't true of human opinions.

An individual's opinion can change, even from day to day. They may answer questions involving topics they have little knowledge of. And worse of all, they may provide insincere responses. For example, in *open polls* in which participants are

Surveys.

Are you questioning my opinions?

self-selected by volunteering, *freepers* may provide multiple responses to make a certain opinion appear to be more popular.

It's not always possible to identify which responses, if any, are questionable. Surveys have to compensate for unreliable responses without knowing which ones they might be. That's a challenge for even professional pollsters.

Surveys aren't just about asking questions; they are a lot more complicated than most people imagine. They are easy to do for entertainment but hard to do scientifically. In fact, statistical surveying may be the most difficult analysis in statistics. And that's why polling typically isn't included in introductory textbooks or taught in Stats 101.

CRITICAL ELEMENTS OF SURVEYS

There is a lot to know about surveys that can affect their legitimacy. Here are more than a few things you should be aware of even if you don't know the details.

TYPES OF SURVEYS

There are three common types of survey based on how individuals are invited to participate—*probability*, *area*, and *open*.

Probability surveys, also called *controlled surveys*, involve inviting participants to reflect the characteristics of a statistical *population*. The characteristics are usually sociological demographics, like sex, age, race, income, education, and political party.

To conduct a probability-based survey, information has to be available on the population demographics and participants have to be selected on that basis. Then, statistical inferences can be made back to that population from the results of the survey.

Probability surveys are difficult and expensive to conduct, leaving them to the domain of professional pollsters. National political polls reported in the media are usually probability surveys.

Area surveys involve inviting participants from a physical or alternately-defined area and assuming that the participants reflect the characteristics of the area. Examples of *areas* include buildings, communities, users of a business's product or service, formal associations, and even informal groups.

The challenge in area surveying is to distribute invitations either randomly or to every individual in the area. Statistical descriptions of the area can be calculated but

 Surveys.

inference can't be made unless the sample can be shown to be representative of some real population.

Area surveys are challenging to conduct but not as difficult as probability surveys. They are conducted by a variety of pollsters from professionals to trained amateurs.

Another example of the use of area survey is **Family Feud**, a TV game show in which two families try to name the most popular responses to questions that were asked of a studio audience. The questions were part of an area survey of the studio audience.

Most area surveys can't be used to make inferences to a population but they can still be a lot of fun.

Open surveys involve allowing anyone who finds the survey to participate. Surveys that appear on social media are invariably open surveys. These surveys are often produced by untrained individuals and are of low quality. No statistical inferences can be made from open surveys and even simple descriptions are dubious.

SURVEY SOURCES

As you might expect, the source of a survey—both the pollster and the funding organization—is important. This is especially true for political surveys. If you want to assess the legitimacy of a survey, that's where you should start.

Surveys, especially political "who-do-you-plan-to-vote-for" (called *horse-race*) polls, have evolved into broad instruments to explore all kinds of preferences and opinions. You can blame the evolution of computers, the internet, and personal communications for that. There's also the profit motive.

Before 1988 there were on average only one or two presidential approval polls conducted per month. Within a decade, that number had increased to more than a dozen. Similar growth took place in business, marketing, research, and other fields.

By 2022, there were 8,235 active companies in the **Marketing Research and Public Opinion Polling** (NAICS Code 541910) sector employing 101,060 people and having revenues of over $20 billion. Many of the pollsters are small to midsize companies, but there are more than a few big, international companies, like Gallup.

Every pollster conducts surveys in their own ways, and as you will see, there are a lot of ways to conduct a survey. Most survey firms are very good; they wouldn't stay in business if they weren't. Nevertheless, there are some that are better than others. There are also some that are not always as statistically rigorous as they might be.

Surveys.

Polling is big business.

In 2021, fivethirtyeight.com graded 93% of 494 political pollsters with a grade of B or better; 2% failed. Of the pollsters, two-fifths leaned Republican and three-fifths leaned Democratic. Notable Republican-leaning pollsters included: Rasmussen; Zogby; Mason-Dixon; and Harris. Notable Democratic-leaning pollsters included: Public Policy Polling; YouGov; University of New Hampshire; and Monmouth University.

The organizations that fund or conduct a survey are always worth knowing. Most will be reputable and will be disclosed where the survey results are provided. Once the key players are identified, it is prudent to search for additional information on them to understand their backgrounds. If they are not disclosed anywhere, that could be a red flag.

TOPICS

The topics of a survey are simply what the pollster (actually, the organization funding the survey) wants to know. Good surveys define what they mean by the topics they are investigating. They do not push biases or misinformation. They account for the relevance, changeability, and controversiality of the topic in the ways they organize the survey and ask the questions. This is especially important in political surveys.

Some survey topics tend to not be controversial, such as those conducted for business reasons or research in psychology and sociology. Political topics are at the other end of the spectrum; every topic is likely to be highly controversial. As a consequence, it is sometimes difficult to view a survey without preconceived notions. That's a reason why some people mistrust polls.

METHODS

Pollsters use a variety of methods because no one procedure would be suitable for every type of survey and polling organization. Most of the differences involve the ways that survey participants are invited and interviewed, that is, going from a *population* to a *sample* using a *frame* in a probability survey to generate responses. Other methods important in a survey are the number of samples (called *sample size*), the types of questions used, and how the responses are processed and analyzed.

Pollsters control for demographics of the population in probability surveys but usually not the survey methodologies, such as method of invitation or interview. Most polls do use multiple methods and some test whether alternative survey

 Surveys.

methodologies affect results. These tests are usually used for quality control and are not reported.

Information about how a poll was conducted is provided in the survey documentation, not in the media stories about it. You have to search the pollster's website for the documentation on each poll.

This sample doesn't look like it would be representative of all kittens.

POPULATION

The *population* for a survey is the group to which a pollster wants to extrapolate their findings, just like in any statistical analysis. Surveys involving human populations usually target groups with specific characteristics. For example, businesses might survey their customers, or more specifically, customers who bought a certain product or who have been customers for a certain duration. Surveys of political candidates might survey registered voters, or more challenging, *likely voters*.

Election surveys that look at *likely voters* rely on models and predictions about who will turn out to vote. That's a challenging assessment. Even the best models may introduce errors.

Unlike *natural error*, which can be estimated from the *sample size*, these other sorts of error arise largely from having to depart from simple random sampling. These errors, called *design effects*, are much more difficult to quantify and may be larger than the natural error. They are not always reported but they are present nonetheless.

Elections aim to be a *census* of voters. Unfortunately, a third or so of eligible voters don't vote, hence the need to consider *likely voters* in surveys. The demographics of actual voters don't always match the population of registered voters, not to mention the population of eligible voters. That's why turnout and voter suppression are important issues in every election and every election poll. Those are important factors that pollsters have no control over. Even a post-election analysis can't quantify the effects of voter suppression.

SAMPLE

The survey *sample* is the individuals to be interviewed. The *sample* is usually selected from the *population* by some type of *probability sampling*. Usually, stratified-random sampling is used to ensure all the relevant population demographics are adequately represented. The quality of the

To improve turnout, help your friends get to the polls.

Surveys.

samples, that is, how representative they are of the population, establishes the survey's *accuracy*.

Sample size is simply the number of individuals who respond to the survey. More individuals need to be invited to participate in a survey than the number of samples desired because some individuals, often many individuals, will decline to participate. Sample size (and a few other survey characteristics) determines the *precision* of the results.

One of the first things critics of polls cite is how few subjects are interviewed. This criticism is always unwarranted because of the *Central Limit Theorem* (Chapter 3). The real challenge in survey design is to select a large enough sample size to provide adequate *precision* yet not too many samples that would make the cost of conducting the survey unreasonable. Furthermore, bigger sample sizes aren't helpful if the additional samples aren't representative of the population.

Another factor that has to be considered in selecting a sample size is how many categories of the sample will be analyzed. Whenever a category of respondents is broken out of the full sample, there are fewer responses to analyze in the category. As a consequence, the precision of that part of the analysis will be lower than for the whole survey. For example, a survey with 1,000 respondents would have a margin-of-error (MoE) of 3% whereas a category of that survey having 100 respondents would have a MoE of 10%.

FRAME

Deciding how to collect samples of the population can be the most difficult step in conducting a survey. That's the purpose of the *frame*, that is, how to collect the samples randomly. The frame boils down to a list of subjects that represent the population. While information about survey populations is available for the most part, information sources for frames are more diverse and not centralized.

Assembling a frame from a population of individuals might involve searching telephone directories, voters lists, tax records, membership lists of public organizations, and so on. Sometimes framing involves developing procedures to randomly dial phone numbers or posting invitations on the internet.

There are different ways to create a frame for a survey.

Online invitations may be extended by ads through text or email, on social media or search engines, or even on websites offering rewards in exchange for survey participation. These sources don't typically provide demographics for the

Surveys.

individuals, so questions have to be asked in the survey to categorize the respondents.

Many firms create (or purchase) *panels* as sources for their survey frames. Panels are databases developed by contacting individuals for the sole purpose of establishing their demographic characteristics so they can be invited to participate in future surveys. People who want to express their opinions through surveys can join panels online.

Getting the *population*, *sample*, and *frame* right are the most fundamental aspects of a survey that can go wrong. Professional statisticians agonize over it. When something does go wrong, it's the first place they look. Nevertheless, sometimes definitely identifying problems in surveys is nearly impossible.

MARGINS OF ERROR

As with all data, survey data is made imprecise by uncontrolled variance—*natural, sampling, measurement,* and *environmental* (Chapter 3). That's why you'll see survey results cite a *margin-of-error* (MoE) of plus-or-minus some number.

MOE AND SAMPLE SIZE

For a simple random sample from a surveyed population, the MoE is equal to the square root of a *Distribution Factor* times a *Choice Factor,* divided by a *Sample Size Factor* times a *Population Correction*:

$$\text{Margin of Error in a Survey} = \sqrt{\frac{\text{Distribution Factor} * \text{Choice Factor}}{\text{Sample Size Factor} * \text{Population Correction}}}$$

Distribution Factor is the square of the *two-sided t-value* based on the number of survey respondents and the desired *confidence level*. The greater the confidence the larger the t-value and the wider the MoE. The most commonly used confidence level is 95%, which uses a multiplier of 1.96.

I choose to be in this cup.

Choice Factor is the percentage for a poll's highest response (**p**) times the percentage for the second highest response (**q** or **1-p** for responses having only two choices). MoE is greatest when **p** and **q** are close in value, within ten percentage points. That's why the largest MoE will always be near a 50%-50% split in responses.

Sample Size Factor is the number of people surveyed. The more people you survey, the smaller the MoE.

Surveys.

Population Correction is an adjustment made to account for how much of a population is being sampled. Bigger populations have smaller MoEs (Figure 47).

The MoE formula for a poll can be simplified by making two assumptions.

1. Unless you know differently, assume the population size is large compared to the sample size. If true, you can ignore the Population Correction.
2. Unless you expect a different result, assume the percentages for respondent choices will be about 50%-50%. This will provide the maximum estimate for the MoEs, ignoring other factors.

These assumptions reduce the equation for the MoE to $1/\sqrt{n}$. What could be simpler? This is fortuitous because pollsters can select their sample sizes but they have no control over population size or the percentages of the responses.

Figure 48 illustrates the relationship between the number of responses and the MoE. The MoE gets smaller with an increase in the number of respondents. However, the decrease in the error becomes smaller as the number of responses increases.

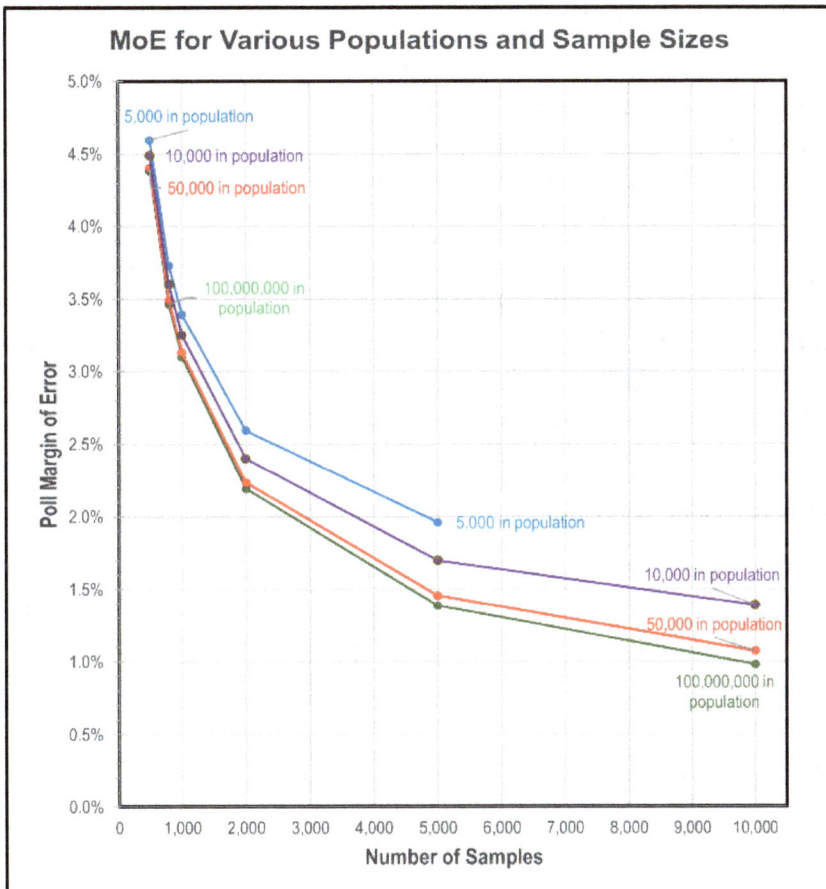

Figure 47. MoEs are smaller for large sample sizes from large populations.

Surveys.

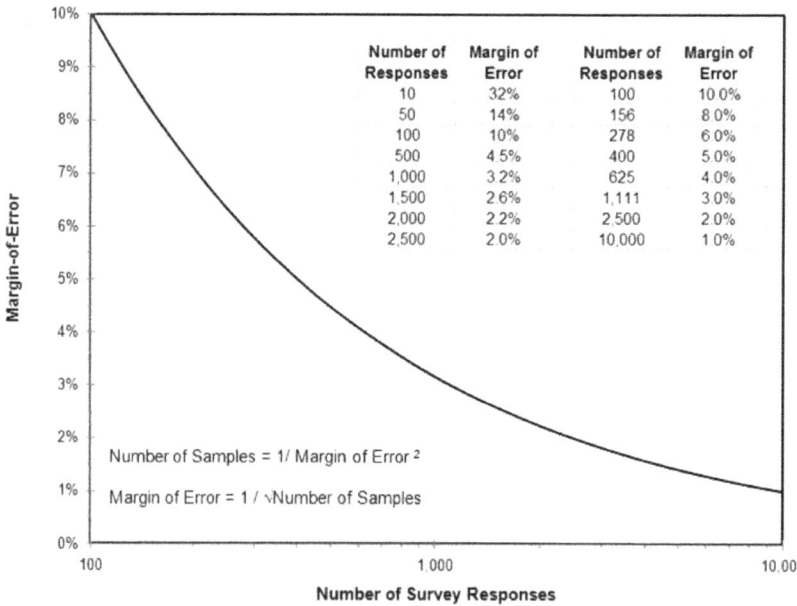

Number of Responses	Margin of Error	Number of Responses	Margin of Error
10	32%	100	10.0%
50	14%	156	8.0%
100	10%	278	6.0%
500	4.5%	400	5.0%
1,000	3.2%	625	4.0%
1,500	2.6%	1,111	3.0%
2,000	2.2%	2,500	2.0%
2,500	2.0%	10,000	1.0%

Number of Samples = 1/ Margin of Error ^{2}

Margin of Error = 1 / √Number of Samples

Figure 48. MoE depends primarily on sample size.

Most pollsters don't use more than about 1,500 responses simply because the cost of obtaining more responses isn't worth the small reduction in the MoE. Most nationwide political polls, for example, use 500 to 1,500 individuals to achieve MoEs between 4.5% and 2.6%. Using more than 1,500 individuals is expensive and doesn't increase precision much (as shown in Figure 48).

Sometimes, sample sizes are increased so that groups within the entire sample can be analyzed with acceptable precision. For example, the U.S. Bureau of Labor Statistics surveys about 110,000 people every month because they have to break down their statistics by several demographics. Their overall poll MoE is less than half of a percentage point but the additional samples are really collected to provide acceptable MoEs for groups within the surveyed population.

MOE AND RESPONSES

The poll MoE represents the natural variation inherent in surveys of a population. However, to compare two percentages of responses from the same question on a survey, a different formula is used based on **p**, **q**, and the sample size. This MoE is equal to the t-value for 95% confidence times the square root of the product of the percentages divided by the sample size. Whew!

Figure 49 shows the MoEs for differences in response percentages for a question at eight sample sizes. The largest MoEs occur near the 50%-50% split of responses, a big problem in close elections, and for smaller sample sizes.

Finally, there is another formula for MoE to compare percentages for the same response in different polls, conducted by different pollsters or at different times. This formula is equal to the t-value for 95% confidence times the square root of the

Surveys.

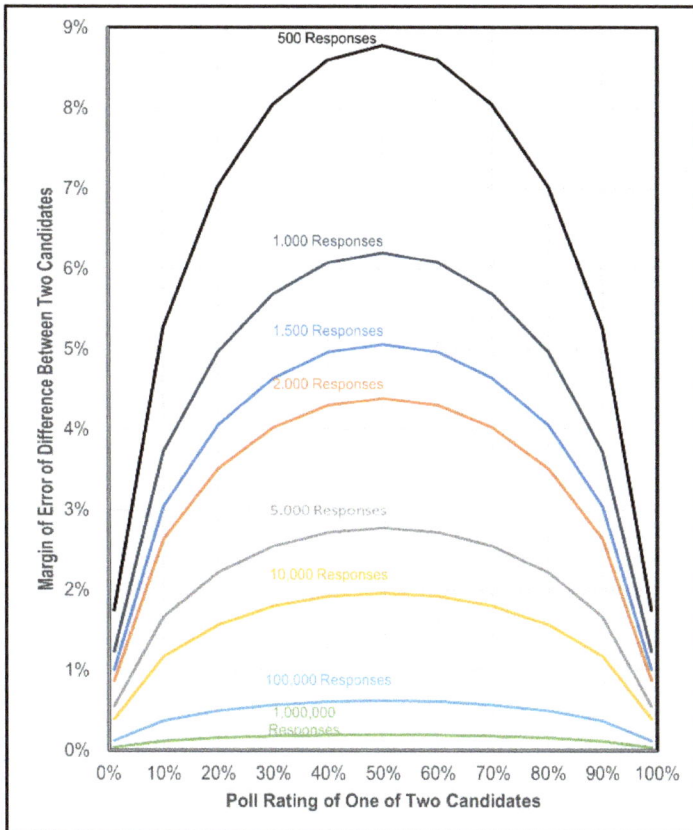

Figure 49. MoE for the difference between response options depends on the size of the difference and the number of samples.

product of **p** and **q** from the first poll divided by the sample size of the first poll, plus the product of **p** and **q** from the second poll divided by the sample size of the second poll. Yikes!

Just remember that the MoE isn't calculated in the same way for every type of comparison of poll results.

The t-values included in formulas for MoE are typically based on having 95% confidence that the actual percentages for survey responses (i.e., **p**) in the population will fall into the interval between **p-MoE** and **p+MoE**. This means that, on average, five times out of one hundred the true percentages will be outside of the limits.

MoE AND CONFIDENCE

What if you wanted to consider different error rates? Formulas for estimating MoE have a term for a *t-value* (or a *z-value*) so you could explore different confidence levels. You could consider accepting one error in every ten (90% confidence) by using a t-value of 1.64 (instead of 1.96 for 95% confidence), which would decrease

Surveys.

If the difference in survey results aren't outside the margin-of-error, ignore them.

the calculated MoE by about a third. You could accept one error in every hundred (99% confidence) by using a t-value of 2.58, which would increase the calculated MoE by about a third. You could even accept only one error in every thousand (99.9% confidence) by using a t-value of 3.30, which would increase the calculated MoE by about a two-thirds.

The real point to remember about MoE formulas and estimates of percentages in horse-race polls is that comparing response percentages isn't as simple as just looking at the difference between two results. The MoE, calculated in the correct way for the comparison, has to be considered. Don't get too excited if a candidate's poll numbers shift a few percentage points between polls.

DESIGN EFFECTS

MoEs based on the numbers of responses characterize the *natural variability* in a survey assuming the samples were selected by *simple random sampling*. Typically, though, samples are selected to match a population demographic using *stratified*, *cluster*, or *respondent-driven sampling*. This introduces variability that has to be accounted for as part of survey elements called *design effects*.

Statisticians use a variety of equations, depending on the situation, to estimate an *effective sample size* that would define the MoE if simple random sampling had been used. The effective sample size is always smaller than the actual sample size resulting in a larger MoE when there are design effects.

Three common design elements that affect effective sample size are:

- 🐾 *Disproportional sampling*, when a pollster purposefully designs their sample to oversample or undersample specific populations, demographics, or clusters to account for anticipated changes, such as increased turnout.
- 🐾 *Non-coverage*, when a frame doesn't include all the individuals in the population. This element can be very difficult to measure and adjust for.
- 🐾 *Non-response*, when invitees do not participate in the survey as they were expected to. These reasons may be related to answering the entire survey or just specific questions.

For example, a pollster might *oversample,* collect additional responses from a certain demographic so that there are enough responses to analyze, when there are not enough responses because the demographic is small or invitees choose not to participate. Voters registered as members of third-parties, for instance, may have to be oversampled so the demographic can be analyzed.

While oversampling works well to augment small groups so they can be analyzed, the full sample has to be adjusted to compensate for the oversampling before it can be analyzed. That's where *weighting* comes in (see Chapter 3).

In order to make their results more representative, pollsters sometimes have to *weight* their data so that it matches the population demographics. Without adjustment, the results would over-represent people who are easier to reach and under-represent those who are harder to interview.

Weighting involves assigning different levels of importance to responses depending on whether they under-represent or over-represent their demographic. This is done using *weights*, which are just values derived by some uniform procedure to correct non-representativeness.

Often, the weights are derived from the percentages of the demographic in the sample versus in the population. Sometimes, more complex weighting algorithms involving several demographics are used. This makes *weighting* a design choice that has to be accounted for in estimating the survey MoE.

While weighting is necessary for ensuring the *accuracy* of poll results, it also has the effect of decreasing the *precision*, that is, making the MoE larger.

Weighting—damned if you do and damned if you don't.

Don't weight the data, you'll bias the results.
I have to weight the data so my sample will match the target demographic.

INVITATION AND PARTICIPATION

There are more than a dozen ways pollsters use to invite individuals to take a survey, including: in-person; mail; live telephone (wireless phone and landline); text; email and other internet-based forms of communication; and panels. Because each method has its own demographic, regulatory, technological, and business limitations, pollsters are forced to maintain multiple capabilities. This of course increases the costs of conducting surveys.

Probability sampling requires that every member of the population to be surveyed has an equal opportunity of being selected. But, that doesn't mean that you will be invited to take a survey.

The probability of being invited at random to take a nationwide political poll is about the same as winning a nationwide lottery, less than 1 in 250 million. The probability is less if you don't respond to emails or answer telephone calls or texts from an unknown number. The probability is greater if you're in a known demographic of interest because pollsters will keep trying to contact you. The probability is also greater if you're a member of a *panel*.

TELEPHONES

Telephones are probably the most common method of inviting people to participate in a survey simply because not everyone is equally capable with internet-based forms of communication.

Landlines were the predominant, if not the only, method of contact by telephone before the 1990s. Caller ID (1988) & regulations under the Telephone Consumer Protection Act (1991) ultimately made random dialing for polls an unproductive methodology. The growth of the internet and wireless phone ownership gradually changed how different population demographics could be contacted.

From the late 1990s to the late 2010s, neither landlines nor wireless phones adequately represented the demographics of the U.S. population. By the late 2010s, pollsters concluded that wireless phones did adequately represent the population. Landlines were still useful because they were cost-effective even if they no longer represented mainstream demographics.

There is a myth that landline users are all senior citizens. In reality, only about a tenth of landline users are over the age of 65.

Cell phone users ignore calls from telephone numbers they don't recognize. Landline users actually do the same.

About a third of households in the U.S. had a landline in 2022 but less than 2% ONLY had a landline. Most households had both a landline and a wireless phone. Some households maintained a landline only because it was part of a provider's bundle of telephone, internet, and television. Less than 1% of households had no telephone service at all.

Landlines are more common among households that own rather than rent, especially in nonrural areas where the wired infrastructure already exists. However, since the Federal Communications Commission scrapped regulations that required homes to be provided copper wires for landlines, most landline homes now have VoIP (Voice over Internet Protocol), which usually sends calls through internet connections.

PANELS

Some pollsters maintain (or purchase) databases, called *panels*, of prescreened respondents who are willing to participate in surveys on certain topics.

To apply to a panel, individuals provide demographic information so that pollsters can decide if they meet the needs of the panel. If accepted to the panel, individuals are invited to participate in future surveys based on their demographics. Acceptance to a panel is controlled; panel members don't see every poll. Invitations depend on the demographics a pollster needs. Someone could lie to get on a panel but their responses on a poll would be flagged for review if their responses were too inconsistent with their profile.

Hi, I'm here to join your panel.

Panels are expensive to create and maintain because panelists drop out over time (called *attrition*) or transition to less-needed, over-represented demographics and have to be replaced. Sometimes, panelists are dropped because of *inactivity* or high rates of *participation refusal*, *question skipping*, or *survey drop-out*.

RESPONSE RATES

One challenge in polling is the large decline in *response rates* (the percentage of people who answered a survey compared to the number of people invited) since the introduction of wireless phones and caller ID. This is a potential issue because of *nonresponse bias*, in which people who answer polls hold different opinions from the ones who don't.

Besides people not answering survey invitations, response rates are also affected by participants who begin but do not complete a survey, called *dropouts*. Dropout rates are affected by survey topic, survey length, types of questions, and ease of completion, which can result in *survey fatigue*. Sometimes, incentives are provided in which the amounts depend on whether the respondent belongs to a part of the population that is hard to reach.

Numerous studies have found that national polls with low response-rates come within a few percentage points of high response-rate Federal surveys. Consequently, low response rates are usually corrected through conventional demographic *weighting*. Nevertheless, it is still a problem if weighted low-response-rate polls are even a few percentage points off if the alternative responses are very close.

Surveys.

INTERVIEW METHODS

There are a variety of methods used to provide questions to individuals in a survey, including in-person, live telephone, texts, recorded message (called interactive voice response, IVR, or *robopolls*), surface mail, email, and websites. Some surveys use more than one method in order to test the influence of the interview methods.

> *Survey Avoidance Disorder. The resistance of survey invitees to respond to important surveys as a result of being bombarded by unnecessary and poorly constructed surveys created by amateurs. This is analogous to the creation of superbugs from the overuse of antibiotics.*
>
> C. Kufs, Survey Avoidance Disorder, posted on 7-18-2012 at statswithcats.net.

Each of these interview methods has its own advantages and limitations. For example, wireless phones are more expensive for conducting interviews than landlines but match U.S. demographics better. Some surveys have to use more than one method in order to obtain enough participants. The percentage of national pollsters that use at least three different methods to sample or interview people increased from 2% in 2016 to 17% in 2022.

QUESTIONS

The questions that are included in a survey are often a focus of critics. The construction of survey questions is an arduous process involving eliciting information on a topic so as to not influence the resulting answer. It sounds simple but it's not, as every professional survey designer knows.

The structure of questions shouldn't be vague or leading, employ double negatives, have compound parts, or be excessively long. The choice of individual words is also important to ensure that they do not introduce bias or are offensive or emotion-laden. They should not be misleading, unfamiliar, or have multiple meanings. Words and phrases with ambiguous common meanings—like entitlement, the economy, [group] values, socialism, and woke—must be defined in the survey. Jargon, slang, and abbreviations/acronyms are particularly unacceptable.

Sometimes surveys have to be presented in different languages besides English, so translation is a consideration. Questions also have to be designed to facilitate the analysis and presentation of results.

We're experiencing survey fatigue.

Surveys.

TYPES OF QUESTIONS

Asking a question in plain conversation doesn't require the rigor that is needed for survey questions. In a person-to-person conversation, you can rephrase and follow-up when you don't get an answer that can be used in an analysis. Surveys don't have that flexibility. There may be hundreds or thousands of participants in a survey.

In-person interviews must strictly follow a fixed script so no inadvertent bias is introduced. Written questions must be unambiguous. Response choices have to be constructed so that respondents are guided to categorize their choices into patterns that can be analyzed. There are quite a few ways to do this.

You just have to ask the right questions in the right way in the right order.

OPEN ENDED QUESTIONS

The most flexible type of question is the *open-ended* question, which has no predetermined categories of responses. This type of question allows respondents to provide any information they want, even if the researcher had never considered such a response. As a consequence, open-ended questions are notably difficult to analyze. They are used to a limited extent in business surveys where the extra work of reading all the responses is worth the effort if some tidbit of information can be extracted to direct business decisions. Open-ended questions are almost never used in political polls.

CLOSED ENDED QUESTIONS

Closed-ended questions all have a finite number of choices from which the respondent must select. That makes them easy to analyze. There are many types of closed-ended questions, including the following eight.

1. Dichotomous Questions—either/or questions, usually presented with the choices *yes* or *no*.

Do you believe that climate change is an important issue that should be addressed?

☐ *Yes*
☐ *No*

Dichotomous questions are easy for survey participants to understand. Responses are easy to analyze. Results are easy to present. The drawback of dichotomous questions is that they don't provide any nuances to participant answers.

Surveys.

2. Single-Choice Questions—a vertical or horizontal list of unrelated responses, sometimes presented as a dropdown menu on web surveys. Responses are often presented to respondents in randomized orders.

Which of the following issues is most important to you? Select 1.

☐ *Abortion*
☐ *Climate change*
☐ *Covid*
☐ *Economy*
☐ *Election fraud*
☐ *Immigration*
☐ *Zombies*

Single-choice questions are easy for survey participants to understand. Responses are easy to analyze. Results are easy to present. The drawback of single-choice questions is that they can't always provide all the choices that might be relevant. These are called *constrained questions.* In the example question, for instance, there are a lot more issues that a participant might think are more important than the seven listed. UFO origins, for instance.

3. Multiple-choice Questions—like a single-choice question except that the respondent can select more than one of the responses. This presents a challenge for data presentation because percentages of responses won't sum to 100%.

Which of the following issues are most important to you? Select up to 3 issues.

☐ *Abortion*
☐ *Climate change*
☐ *Covid*
☐ *Economy*
☐ *Election fraud*
☐ *Immigration*
☐ *Zombies*

Multiple-choice questions are somewhat more difficult for survey participants to understand because participants can select more than one response. Survey software can help to validate the responses. Multiple-choice responses are more difficult to analyze because it's almost like having a dichotomous question for each response. Results are more difficult to present clearly because percentages can be misleading. The advantage of multiple-choice questions is that they provide some comparative information about the choices in an efficient way.

4. Ranking Questions—questions in which respondents are supposed to place an order on a list of unrelated items.

Surveys.

Rank the following issues in order of their importance to you, where 1 is the most important and 7 is the least important.

_____ *Abortion*
_____ *Climate change*
_____ *Covid*
_____ *Economy*
_____ *Election fraud*
_____ *Immigration*
_____ *Zombies*

Ranking questions aren't too difficult for survey participants to understand but rank-ordering takes much more thought than just picking a single response. Responses are also much more difficult to analyze and present. The advantage of ranking questions is that they provide more comparative information about the choices than multiple-choice questions.

5. Rating Questions—questions in which respondents are supposed to assign a relative score on unrelated items. The score is on some type of continuous scale. Responses might have to be written in or indicated on a web site with a slider.

On a scale of 1 to 10 where 1 means unimportant, how important an issue do you believe climate change is to you?

_____

Rating questions are relatively easy for survey participants to understand, although anything requiring survey participants to work with numbers presents a risk of failure. Responses are easy to analyze and results are easy to present, though. The drawback of rating questions is that they take participants longer to respond to than Likert-scale questions.

6. Likert-scale Questions—like a single-choice question in which the choices represent an ordered spectrum. An odd number of choices allows respondents to pick a middle-of-the-road position, which some survey designers avoid because it masks preferences to *lean* one way or the other.

How important an issue do you believe climate change is to you. Select 1 choice.

☐ *Very Important*
☐ *Somewhat Important*
☐ *Not Very Important*
☐ *Not Important.*

Likert-scale questions are easy for survey participants to understand. Responses are easy to analyze and present. The drawback of Likert-scale questions is that they are less precise than rating questions.

Surveys.

7. Semantic-differential Questions—like a Likert or rating-scale question in which the choices represent a spectrum of preferences, attitudes, or other characteristics, between two extremes (e.g., agree-disagree, important-unimportant). Semantic-differential questions are thought to be easier for respondents to understand than Likert-scale questions.

How important an issue do you believe climate change is to you.
Select 1 choice.

Very Important ○ ○ ○ ○ ○ ○ ○ ○ ○ ○ Not Important

Semantic-differential questions are easy for survey participants to understand. Responses are easy to analyze once the responses are coded as ratings. Results are easy to present. The drawback of semantic-differential questions is that they are not supported by some survey software.

8. Matrix Questions—Questions that allow two aspects of a topic to be assessed at the same time. Matrix questions are very efficient for providing information but are not available in all survey software.

How important do you believe the following seven issues are to you? Select one of the four columns for each row.				
	Very Important	*Somewhat Important*	*Not Very Important*	*Not Important.*
Abortion				
Climate change				
Covid				
Economy				
Election fraud				
Immigration				
Zombies				

Matrix questions are difficult for some survey participants to understand. Responses are easy to analyze and present because they are like multiple Likert-scale questions.

BRANCHING

When you are having an informal conversation with another person on a topic, you might ask a question, and then a follow-up question to clarify the person's previous answer. That can't be done in a survey where the questions all have to be pre-scripted. Surveys handle this dilemma using a technique called *branching*.

In branching, alternative follow-up questions are asked depending on the respondent's answer to the original question. For example, a dichotomous question might *branch* to one set of questions to clarify why a respondent answered *yes* or to a second set of questions to clarify why a respondent answered *no*. It's like having mini-surveys within a larger survey.

Surveys.

Branching is like having a survey within a survey.

Branching helps to clarify a respondent's answer, though in a somewhat awkward way. Branched questions are usually closed-ended but they don't have to be. Branching makes survey analysis more difficult because respondents end up answering different sets of survey questions depending on the branching, thus introducing missing data. The MoEs are also larger because the number of responses will be smaller for branch questions than for full-survey questions. Still, they are easier to analyze than open-ended questions where the possible responses are infinite.

ISSUES WITH QUESTIONS

It should come as no surprise when a survey question is less effective than it was hoped for by its creators. There are several reasons—reliability of questions, dependability of opinions, and susceptibility to biases.

RELIABILITY OF QUESTIONS

Every question in a survey is a metric, a variable that is used in a statistical analysis. In statistical studies that do not involve surveys, a metric might be measured by instrumentation (e.g., laboratory instruments, field meters), devices (e.g., rulers, scales, sensors), references (e.g., comparison charts), or processes (e.g., business rules). All of these data acquisition methods are meant to be used repeatedly so they are routinely tested and calibrated.

The same is not true with most survey questions. They are created for a specific survey and usually used without verification testing. (Some questions, and even whole surveys, that are developed by professionals and are meant to be used repeatedly, do undergo validation.) Consequently, survey questions should be expected to be less reliable than, for instance, a pH meter or a thermometer.

One prominent cause of question unreliability is *vagueness*. Vague questions usually don't *bias* results in a certain direction but they do add unnecessary *variation* (Chapter 3). Some respondents answer the question as it might have been meant and others will answer the question as they think it might have been meant. The results will be a mishmash of opinions unsuitable for analysis.

For example, surveys have for decades asked how respondents feel about **the economy**. However, what that term "the economy" means to people can vary a great deal. Does it refer to traditional measures of the **Wall-Street economy**, like stock trends and yields on U.S. Treasury bonds? Does it refer to traditional measures of the **Federal government economy,** like unemployment, the size of the deficit, and

 Surveys.

GDP? Does it refer to traditional measures of each individual's **personal economy**, like trends in the prices of things such as housing, fuel, and food? All those measures can result in very inconsistent opinions about **the economy**, such as when the National economy is strong and growing while personal economies are stagnant or weakening. This is an important issue in both data acquisition and interpretation that is not always acknowledged adequately.

I don't have enough money so the economy must be bad.

DEPENDABILITY OF OPINIONS

Physical characteristics can often be measured in more than one way and usually do not change suddenly. Measure your weight at home, at the gym, and at the doctor's office and the values will all be about the same.

In contrast, the characteristics that survey questions measure are intangible. They are opinions or feelings that can't be measured in any other way. They may be fleeting or not deeply held. Furthermore, respondents might not be candid about controversial opinions when talking to an interviewer on the phone, or might answer in ways that present themselves in a favorable light (such as claiming to be registered to vote when they are not). It's hard for surveys to be consistently accurate and precise when they capture only a moment in time and their target is moving.

BAD WORDING

Survey questions shouldn't be vague or confusing, nor have compound parts (called *double-barreled questions*). They should not express opinions, called *leading questions* or lead respondents to answer questions in a certain way, called *loaded questions.* They should not use *absolute wording*, like always or never, and all or none.

The choice of individual words is also important to ensure they do not introduce bias or are offensive or emotion-laden. A poll that asks which person "caved" in a negotiation or which product "fails" most often introduces bias because "cave" and "fail" evoke negative emotions.

Words should not be misleading, unfamiliar, misunderstood, or have multiple meanings. Jargon, slang, abbreviations, and acronyms are particularly taboo. Sometimes surveys have to be presented in different languages besides English, which may lead to further confusion.

POOR RESPONSE CHOICES

The two requirements of good response choices are that they be comprehensive and mutually exclusive. Choices that are *comprehensive* do not leave gaps in the list of

Surveys.

possible responses. Choices that are *mutually-exclusive* don't overlap. One choice shouldn't copy or paraphrase from another choice. For example, choices that provide number ranges shouldn't list the end of one range as the start of the next.

Constrained questions are those in which the number of possible responses to the question is large and complex, requiring the response choices that are provided to be limited. The wording of the question may be fine but it can be misleading when the results are presented as the only choices selected by respondents. This happens with multiple-choice, ranking, and matrix questions.

For example, a survey might ask "what's the most important characteristic for you in buying a new car?" with the only choices being "tinted windows," "moonroof," "heated seats," and "cup holders." The media then reports that "Americans want heated seats; don't care about cup holders." Another example are political polls that ask "what is the most important problem facing the country?" then provide just a few secondary issues as choices.

This is way more than I expected.

Creating survey questions is not as simple as critics think it is.

SUSCEPTIBILITY TO BIASES

Questions cannot be engineered like a laboratory instrument or a business process. They can be devised by professionals and peer-reviewed, but they don't have standards they can be compared against.

All aspects of opinion surveys, from creation to response, are created by fallible humans who are subject to all kinds of errors, biases, and unfortunate choices. As a consequence, many forms of bias that affect surveys can be attributed to survey creators, interviewers, and respondents. Focus groups, pilot studies, and simultaneous use of alternative survey forms are sometimes used for evaluating the effectiveness and biases of individual survey questions and the overall survey. Even so, there is no way to eliminate biases completely because it's impossible to eliminate humans from the process.

Here are a few examples of biases.

RESPONSE BIAS

Response bias is a term for when survey participants respond in an insincere or false manner because of some external trigger. The trigger may be the way a survey question is worded, or asked by an interviewer, or how the survey is presented or conducted. Here are five types of response bias:

Surveys.

- An *acquiescence bias* is when participants answer questions positively from ignorance or apathy, or just to be supportive or optimistic. They may want to please the researcher with an answer they think is wanted.
- A *dissent bias* is when respondents answer survey questions negatively just to provoke or antagonize the researcher or provide a pessimistic viewpoint.
- A *neutral bias* is when participants select a middle-of-the-road response or an "other" response because they have no deeply-held opinion or don't want to stand out. Five-choice Likert-scale questions are susceptible to neutral bias while four-choice Likert-scale questions are not.
- A *conformity bias is* when participants are uncomfortable reporting their honest answer and instead, provide a more morally or socially-accepted response.
- An *extreme bias* is when participants select responses more extreme than their beliefs in order to exaggerate their own belief or slant results to a particular goal. This can be a significant problem for some *hot-button issues*, like immigration, foreign aid, gun control, and individual rights.

NON-RESPONSE BIAS

Non-response bias occurs when an *invitee* declines to participate in a survey, or when *participants* included in a survey don't respond to some questions.

There are many reasons why an invitee might refuse to participate in a survey. The survey might be too long for the time the invitee has available. There may have been problems with the communication of the survey, either because of unclear accents and language differences, or technical issues with telephone and internet connections.

Participants in a survey might not respond to some questions if they find them to be objectionable or unanswerable. A certain question might also not have a response choice that aligns with their opinion. Also, some participants might not want to disclose their demographic information or ignore those questions because they don't understand their essential purpose. This is especially true when the demographic questions are placed at the end of a survey.

Sometimes you have to throw away all the responses of participants who don't answer all the questions.

Demographic questions can present problems if respondents fear that their anonymity might be compromised. That's why *personally identifiable information* (PII, like Social Security numbers) is never requested in a legitimate opinion survey. In-house surveys of employees are sometimes *confidential but not anonymous,* meaning that the statistician analyzing the results knows who the

Surveys.

respondents are but does not reveal that information. This practice allows demographic information to be merged from other sources rather than being asked in the survey.

Usually, demographic questions aren't asked until the end of a survey because participants could drop out after seeing what they might consider to be invasive questions. Unfortunately, that means that the whole survey has to be completed first, so the participant faces the dilemma of providing the information or dropping out, in which case, the results can't be used.

Non-response bias results in missing data. Requiring answers is not a good solution, though, because it may cause frustrated respondents to leave the survey, worsening the *drop-out rate*. It's a dilemma faced by pollsters on every probability survey.

ORDER BIASES

An *order bias* causes a *participant* to respond on the basis of the position of a question or response choice.

Question-order bias is displayed when participants answer a question differently because of how they answered a previous question. For example, a participant might select a response choice for their satisfaction with aspects of a product or service on the basis of their choice for overall satisfaction with the product or service. It is usually better to ask about aspects of a topic before asking about the overall topic. This approach can help a respondent clarify their response to the overall question.

Response-order bias is displayed when a participant selects responses on the basis of the position of the response on the list of choices. For example, a participant may select the same choice for all their responses on a product-satisfaction survey

I hate your poll.

because of its position in the list. *Response-order bias* is why candidates in an election want their name to appear first on the ballot. The best way to compensate for *response-order bias* on a survey is to randomize the order of the responses for every participant. This is often done on ballots so that every candidate has the same opportunity to appear at the top of the ballot.

WHY PEOPLE HATE POLLS

Perhaps as a consequence of do-it-yourself polling, there is no end to truly bad, amateur polls. It's not just on social media, either. Businesses who can't afford to hire a professional pollster produce some awful surveys. Usually, bad polls are just a nuisance. If you know what to look for, they are easy to recognize and ignore.

 Surveys.

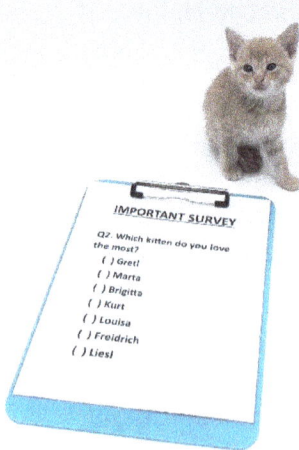

IMPORTANT SURVEY

Q7. Which kitten do you love the most?
( ) Gretl
( ) Marta
( ) Brigitta
( ) Kurt
( ) Louisa
( ) Freidrich
( ) Liesl

*I'm not biased.
I'll reorder the other choices
but my name always
has to come first.*

There are also, however, well-prepared polls that are *meant* to mislead, some overtly and some under the guise of unbiased research. They are really meant to provide support for some desired result. As a consequence, some people have come to believe that information derived from all polls is biased, misleading, or just plain useless.

Like other complex practices such as medicine, statistical polling isn't an exact science and can unexpectedly and unintentionally provide erroneous results. But for the most part, statistical surveying is legitimate and reliable even if the public doesn't understand it.

People are comfortable with polls that confirm their preconceived notions. This is called *confirmation bias*. But, they lambaste polls that don't confirm their beliefs because they don't understand the science behind statistical surveying. This is especially true of political polls and is experienced by both sides of the political spectrum. Nonetheless, surveys are relied on extensively throughout government and business to support their work. And, of course, politicians live and die by poll results.

SIX CRITICISMS OF POLL HATERS

People criticize surveys all the time, especially political polls. Some criticisms are reasonable and valid when the survey is based on flawed methods. Other apparently flawed surveys are just a reflection of the results being different from what the critic believes, their *confirmation bias*.

Critics of surveys probably wouldn't care about most surveys if they didn't have preconceived notions about what the results should be. Mostly, if they do see a poll that doesn't agree with their thinking, they are quick to find fault. Some of their criticisms could have merit, but usually not. They are just red flags that the critic has little understanding of statistics and surveys.

Here are six examples.

1. TOO FEW PARTICIPANTS.

Some critics of polls don't seem to understand that a *sample* of only a few hundred individuals can be extrapolated to a much larger *population*. They either don't understand the science of statistical sampling or they refuse to believe it.

This criticism is common in political polls. The first political poll dates back to the Presidential election of 1824. Probability and statistical inference for other

 Surveys.

Won't you answer just a few questions?

applications is hundreds of years older than that. The science behind extrapolating from a sample representative of a population to the population itself is well established.

For a poll of the total U.S. population of about 330 million, it's not uncommon for pollsters to use 1,000 survey participants to represent the 210 million adults. Political polls typically use samples of fewer than 1,500 to represent 170 million registered voters. An additional challenge is to identify registered voters who are also likely to vote in a time when the effects of voter suppression can't be assessed easily.

Even with a sound survey frame and sample, this criticism would be appropriate if the sample size were very small, say less than 100. This would make the MoE about ±10%, fairly large for comparing preferences for even two alternatives. You can get an *accurate* result with a small sample size but the *precision* of the estimates may be unacceptably large. It's a matter of *resolution*. That's why polls always need to cite their MoEs.

The criticism that a survey, conducted by a professional pollster, has too few participants is easy to assess and is almost always unjustified. The exception to this guidance is when many categories of demographics that have a much smaller number of responses than the whole survey are analyzed. This results in greatly different MoEs than reported for the full survey.

2. THEY DIDN'T ASK ME.

Getting invited to participate in a political poll might have a probability of 1 in 170,000,000. That's akin to winning the lottery. The irony of this criticism is that many of these critics openly acknowledge that they don't respond to invitations to participate in surveys. You can't win if you don't play.

The complaint that they weren't asked for their opinion is really about the frustration they have when the survey results don't match their expectations. The results just mean that the opinion of the critic doesn't match the consensus of the population.

If the survey frame and sample are appropriate, the demographic of the critic is

Nobody asked me.

 Surveys.

already represented. This criticism is always unjustified.

3. ONLY LANDLINE USERS WERE INTERVIEWED.

This criticism has to do with how technology affects the selection of a sample. The issue dates back to the 1930s and 1940s when telephones were first used extensively in polling.

In the 1948 Presidential election, Harry Truman defeated Thomas Dewey despite leading pollsters of the day—Gallup, Roper and Crossley—finding the opposite sentiments throughout the summer before the election. Their polls were based on non-probability *quota sampling*, which allowed interviewers to select participants based on a predetermined number of participants rather than on a frame that would provide random selection. This led to *selection bias* because the dominant method of polling at the time relied on landline telephones, which were more likely to be owned by wealthier, urban, Republican households over lower-income, rural, Democratic households. Consequently, the polls did not fairly represent the electorate.

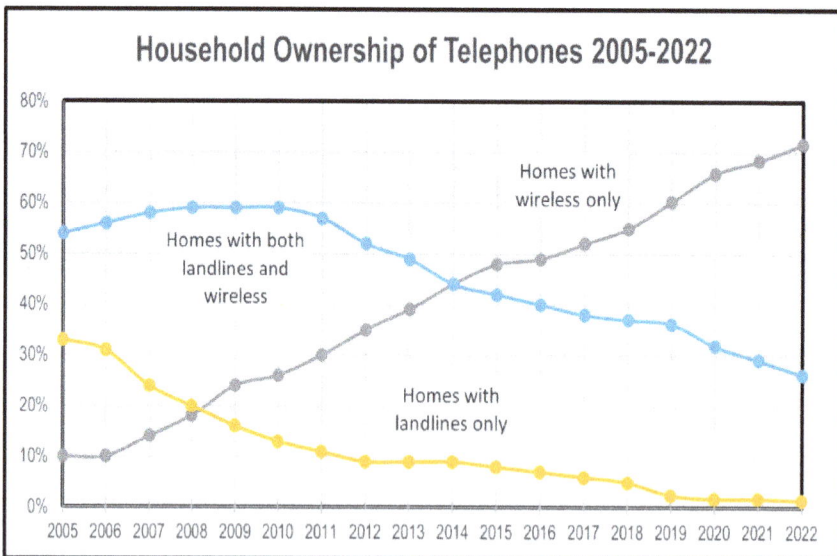

Figure 50. Household ownership of landlines has been decreasing steadily since 2010.

The issue repeated in the late 1990s and 2000s when wireless phones began replacing landlines. For that period, neither mode of telephony could be relied on totally to be representative of the U.S. population.

In 1998, most households had a landline although 36% also had wireless phones. These wireless phones were expensive to own and use. Like in the 1930s, most wireless phone owners were financially stable if not wealthy. Over time, wireless phone technology improved, networks expanded, connection plans became more generous, and consumer costs decreased.

Surveys.

By 2019, over half of households only had wireless phones. Less than 10% still relied solely on landlines. Over 80% of households had both wireless phones and landlines. As a consequence, pollsters determined that wireless phone users were sufficiently representative of the population to be used as a frame. In fact, wireless phone users were more demographically representative of the U.S. than landline users who tended to be older though not as old as some believe. Only about a tenth of landline users are over the age of 65.

Figure 50 shows how the ownership of telephones changed from 2005 to 2022.

Today, using telephone lists exclusively to create frames is an issue that reputable pollsters easily avoid. Most big surveys use several different sources to create frames that are representative of the population. The criticism that only one method of framing a sample is used, especially landlines, is always incorrect.

4. THEY ASKED THE WRONG QUESTIONS.

This criticism isn't about gathering information about the wrong topics, it's thinking that the questions were biased or misleading in some ways. It might even be true that the criticism is made without the critic actually reading the questions because that information is seldom available in news stories. It has to be uncovered in the original survey-analysis report.

This criticism may have merit if the poll didn't clearly define terms or used some other type of misleading wording. Professional statisticians usually ask simple and fair survey questions but may on occasion use vocabulary that is unfamiliar to participants. This can happen if a survey is designed for a certain reading-level or is translated from one language to another.

It's usually not too difficult to tell if a question might be problematic. Still, critics have to actually review the survey questions for their assertions to have any merit.

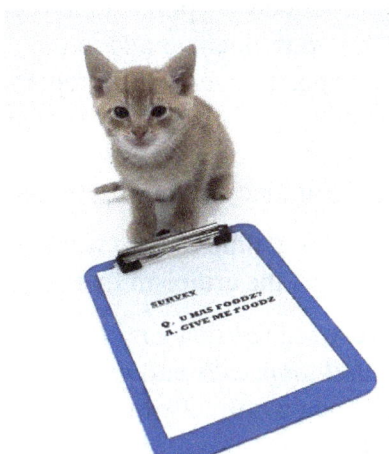

These questions aren't so hard.

5. THE RESULTS WERE PREDETERMINED.

This is a bold criticism that isn't all that difficult to confirm or invalidate.

First and foremost, no professional pollster is likely to commit fraud, regardless of the reward, just because their business and career would be in jeopardy if the allegation were substantiated. Look at the source. If it is any nationally-known pollster who has been in business for a while, the criticism is unlikely to be valid. If it is a non-independent, in-house poll, like a candidate conducting their own poll, skepticism may be warranted.

Surveys.

Nevertheless, the allegation may seem plausible because the process of conducting a nationwide probability survey is so complicated and not well understood by most people. Short of fabricating all the data, there are three ways that a pollster could bias the results of a poll:

- **Questions**. Asking leading questions would be easy but too obvious even to non-professionals. The only thing a reviewer would have to do is find the survey report on the pollster's website, read all the questions, and look for inappropriate wording. What is more subtle is when biased pollsters order a series of questions to put respondents into a frame of mind so that they provide a certain response to a later question.
- **Framing and Sampling**. Biasing the framing and sampling wouldn't be too hard for a pollster to do but the results couldn't be guaranteed, thus making it unlikely. Furthermore, reported demographics of the participants might be a red-flag alerting reviewers to the possibility of a problem. Reviewers would have to find the survey report on the pollster's website, read the pollster's methods (if they're documented in sufficient detail), and look for anything questionable. This would be difficult and time-consuming to prove, and almost impossible for non-professionals.
- **Weighting**. Adjusting results for segments of the participants is done routinely by pollsters to correct for imbalances in the demographics. While always documented in the pollsters records, the weighting procedure is rarely reported even in poll reports. Some pollsters consider their weighting models to be proprietary and don't share them unless obligated to do so for some reason. Reviewers would probably not be able to access let alone judge the legitimacy of a pollster's weighting model.

Consequently, unless there were a thorough professional investigation of a poll, it is unlikely that the criticism that the-results-were-predetermined could be proven.

Two very big red flags in identifying polls that may have been biased by the pollster are *solicitation polls* and *push polls*. Both types of polls begin normally and look legitimate, but they aren't.

A *solicitation poll* begins with a few questions about serious topics then ends with a request for a donation, an email address, a link to sign a petition, or some other way of allowing the pollster to engage the participant further. These may be legitimate attempts by a candidate to collect information from supporters but, because they are area surveys rather than probability surveys, would be inappropriate for much else.

Watch out for solicitation polls and push polls.

 Surveys.

Bogus solicitation polls are easy to spot if there is only a single, leading question on a single topic.

Push-polls are more subtle than solicitation polls because they tend to be longer and more complex, like a legitimate poll would be. Push polls also begin normally, covering a range of topics. Eventually, however, questions will be presented that begin "*Would you be more or less likely to* [do something] *if you knew that …?*" The purpose of the question is to force the participant to acknowledge information the pollster wants to convey.

For example, "Would you be more likely to buy this product if you knew it was rated a best buy?" The implied information doesn't have to be true, and in politics, often isn't. Push polling was used in the 2000 Republican primary in South Carolina when a pollster for George Bush asked "Would you be more likely or less likely to vote for John McCain for president if you knew he had fathered an illegitimate black child?" The information was false but it helped Bush win the primary.

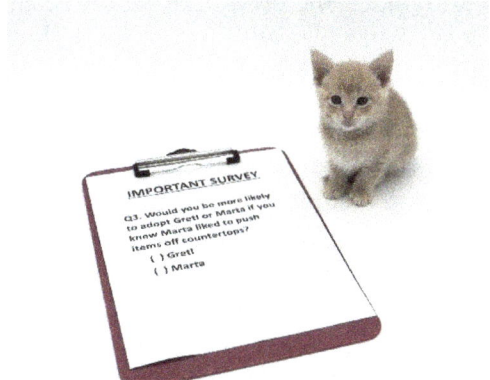

Push polls are really attempts to influence participants.

If the source of a survey is an unknown pollster without a notable track-record, look at the report on the survey methods. These reports can usually be found by conducting an internet search for the poll or the pollster (not the media announcement of the poll). The reports might suggest poor methods but that wouldn't necessarily guarantee a particular set of results. If there were an obvious bias in the methods, like surveying attendees at a gun show about the Second Amendment of the Constitution, it should be apparent. Also consider the source of the funding for the survey; follow the money.

If there is no background report available on the survey methods, this criticism would merit attention. In particular, if the survey results were prepared by a non-professional for a specific political candidate or party, skepticism would be appropriate.

6. THE RESULTS ARE WRONG.

There are many things that can go wrong with a survey. If a survey created and deployed by a professional is wrong, it is usually unintentional and not wrong by much. If a survey is intentionally wrong, it is usually easy to detect either by comparing the results to similar polls conducted at about the same time or by reviewing the survey report (if there is one) for poor polling methods. The first approach is about all that non-statisticians can do.

Surveys.

Criticisms that the results of a survey are wrong are usually suppositions based on *confirmation bias*. The criticism that a poll is invalid is sometimes expressed by saying that the results apply only to the people questioned. This is in complete denial of the long-accepted science of statistical sampling. Again, compare the suspect poll to other polls researching the same topics during the same timeframe. If the results are close, especially within the MoEs, the poll is almost certainly legitimate.

If my candidate doesn't win, it must be because of election fraud.

Figure 51 shows the differences between final election polls and the actual popular votes for elections from 1936 to 2020. For the most part, the polls accurately reflected the popular vote results within the MoEs. (Remember, national polls assess popular votes not electoral votes.) The elections of 1936, 1948, 1952, and 1992 all fell outside the 99% MoE. The elections of 1980 and 2016 fell outside the 95% MoE. The other 42 elections all fell within the 95% MoE. That's not too bad considering everything that impacts polls. (Data from en.wikipedia.org/wiki/Polling_for_United_States_presidential_elections\)

Criticisms based on suspect survey methods are difficult to prove. The only way to determine that a political poll was truly wrong is to wait until after the election and conduct an after-action review. Even elections themselves, which are also instances of surveys, can be tested by *exit polls*.

Exit polls are in-person interviews with voters as they leave their polling place to collect preliminary information on their demographics and who they voted for. In some cases, exit polls may also reveal discrepancies between the poll and the actual vote, which may merit investigation. Exit polls are used as a check for election fraud and voter suppression both in the U.S. and foreign countries.

POLLING DENIALISM

Polling denialism is the notion that statistical surveys, especially in horse-race polls, have all become illegitimate and shouldn't be believed or even participated in. This phenomenon began in the U.S. during the 2016 Presidential election and escalated after that.

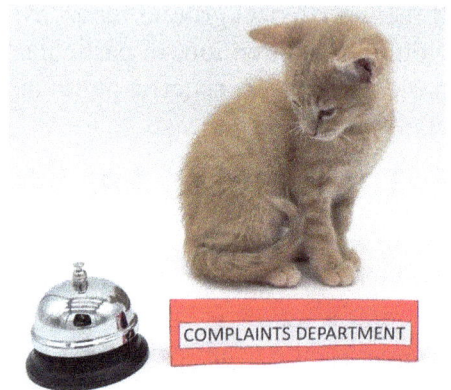

Aggregated polls conducted just before the 2016 election indicated that Hillary Clinton had a 4% (47% to 43%) lead over Donald Trump. Pundits predicted that the difference would mean that Clinton would win the election,

The customer isn't always right.

Surveys.

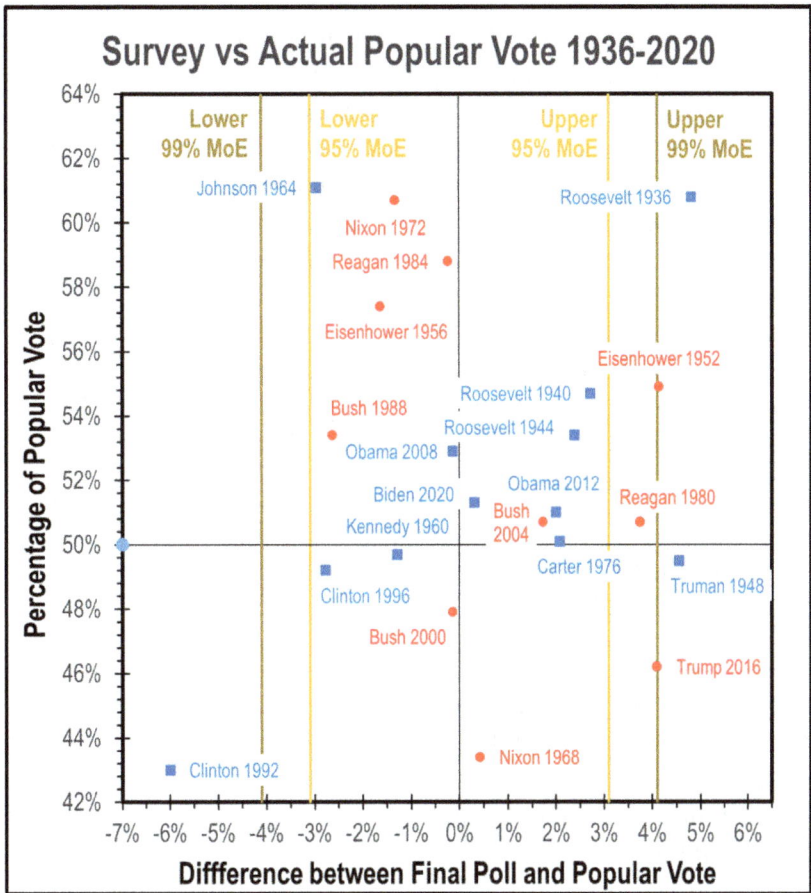

Figure 51. Differences between final election polls and the actual popular votes for elections from 1936 to 2020.

which was amplified in the media. On election day, Clinton did win the popular vote 48.2% to 46.1%, within the MoEs of the polls. However, she lost the Electoral vote 227 (42.2%) to Trump's 304 (56.5%).

The pundits and the media didn't consider the Electoral College in making their predictions. If they had looked at State polls, they might have seen Pennsylvania, Michigan, and Wisconsin, and hence, the Electoral College, being won by Trump. Nevertheless, the media portrayed Trump's win as a failure of polling rather than their interpretation of the polls. As a result, people who believed stories from the media rather than understanding the science of surveying became *polling denialists*.

Polling denialism is now a worldwide phenomenon. It is not the same thing as justified caution or even healthy skepticism. It's refusing to accept the legitimacy of statistical science and is no different than denying any science.

Surveys.

> *Polling itself is blamed for the way that media sources misinterpret polls (often to play up false narratives that an election is close, but also sometimes to pander to an audience, or at times through more innocent errors.) This confusion between polling and interpretation means that polls that may have been accurate are falsely blamed and unjustly distrusted, while media figures who have willfully or cluelessly misinterpreted the polls get away Scot free. Similarly, deniers blame good polls for the results of bad polls, again meaning that the good polls are wrongly suspected while the bad ones go unpunished. In this way, poll denialism enables more bad poll reporting.*
>
> Dr Kevin Bonham, 9-6-2023,
> Australian Polling Denial and Disinformation Register,
> https://kevinbonham.blogspot.com/2023/09/australian-polling-denial-and.html

Pundits warn that some people may be less likely to vote if they are told that a certain candidate is extremely likely to win. This, of course, is a responsibility of media prognosticators who interpret the pollster's data rather than the pollster itself. Many people don't understand the distinction, though, and call for ALL polls to be ignored, hence polling denialism.

The admonition to ignore all polls is misguided. It is better to participate in polls and learn how to assess them. Not participating in polls just means your opinions will be ignored and you'll have to live with what others think. It's like not voting.

Participating in polls is like casting interim votes. You can only vote once but you can answer many polls.

WHAT TO LOOK FOR IN STATISTICAL SURVEYS

No poll is perfect so any time you see poll results that seem unusual or unlikely, you have to first ask yourself whether it is a misleading poll or is legitimate. You have to ask yourself if your judgment about the poll is valid or is it a reflection of your own *confirmation bias*. Moreover, you have to ask yourself what your motivation is for looking into it. Do you want to understand how the curious results were obtained to learn from it or to criticize it?

If you want to criticize a poll, you'll need to be sure you understand how polling does and doesn't work, otherwise you'll just embarrass yourself and spread misinformation. You'll have to investigate in detail what the pollster actually did, well beyond the initial news story that originally caught your attention.

Surveys.

*If you don't answer surveys
how can we understand
what you're thinking?*

You'll have to find a link to the documentation of the original poll. If there is none, you'll have to search the internet for the polling organization, topic, and date. If there is no link to the poll, or if the link is dead or leads to a paywall, the legitimacy of the poll is suspect.

Even if you do find a link, you probably won't be able to find all the information you might need. Like any business, pollsters don't disclose their proprietary processes. You may not be able to answer everything.

Follow these six steps to help you assess the most obvious and easy-to-evaluate criteria to the most detailed and obscure.

STEP 1. OBVIOUS RED FLAGS.

Some polls should be considered laughable at a glance, but not because of the results. They are dubious because of when and where they were conducted or how they are portrayed. Here are a few guidelines.

- Any poll that does not provide readily-available documentation is not worth pursuing. You may have to look for it on the pollster's website, however.
- Any *open-invitation poll* isn't worth critiquing. They are for entertainment only. If they happen to show any result that might make sense, they are still irrelevant.
- Any results from *solicitation polls* or *push polls*, if you can identify them, are irrelevant and not worth pursuing.
- Any *area polls* that are undefined or poorly-bounded can be ignored. For example, ignore product polls that are not restricted to confirmed owners who have used the product for a sufficiently long period or building satisfaction surveys that don't distinguish between full-time occupants and visitors.
- Polls conducted at inappropriate times or locations can be ignored. For example, ignore *horse-race polls* conducted too early before the election. They are only good as media hype. Many voters don't decide on a candidate until right before the election, perhaps just weeks. Polls about policies and priorities are worthwhile anytime.

Don't worry about the sample size of a poll unless the MoE is large compared to the margin between the candidates. Sample size is not as important as the representativeness of the sample which would be too difficult to determine at this point. Wait for Steps 4 and 5.

Surveys.

A single poll may be an anomaly. Look for consistent results from multiple independent pollsters before you decide to investigate further. It could save you a lot of unnecessary research and evaluation time.

STEP 2. SOURCES.

This is where the research begins. You probably have at least one media report, which hopefully, will disclose the pollster who conducted the poll and who funded it. If it doesn't, search the internet for polls conducted on the dates and locations mentioned in the media story. If you can't find anything, you're done. The poll lacks credibility.

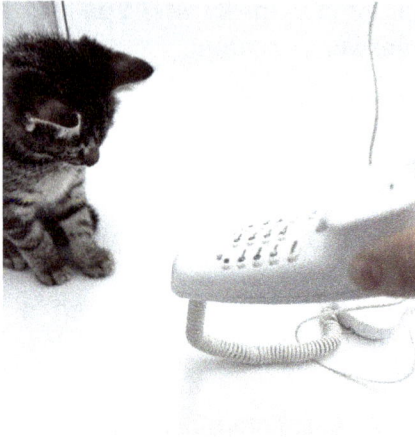

It's for you. It's a pollster.

If you can identify the pollster, you will be able to find the poll documentation on their website. The poll report will identify who funded the survey. From there, you'll want to identify and research who owns or manages the funding organization, the pollster, and the media outlet that reported the poll results.

With this information, you can decide if the players seem to be independent, unbiased, and reputable, at least in your own judgment. You can visit sites that rate pollsters, like fivethirtyeight.com. There are also sites that rate biases on media sources, like allsides.com and mediabiasfactcheck.com. Wikipedia will also have information on the businesses. Any poll conducted by a candidate or a political party is not likely to be totally legitimate.

STEP 3. QUESTIONS.

Questions aren't the next most important survey element to look at, that would be the sample, but survey questions are easier to identify and assess.

The media reports about the survey certainly only mentioned a few of the questions so you'll have to find the pollster's report to see all the questions. Once you do, the evaluation is straightforward.

Were appropriate close-ended question types used? Was there branching used that would reduce the sample size? Was the wording of the questions clear and concise? Were the questions simple and unbiased? Was the sentence structure of the questions understandable? Were any confusing or emotion-laden words used? Did the questions directly address the topics of the survey?

Also consider the order of the questions. Could the order confuse respondents by shifting the focus of the questions (e.g., mixing candidate affiliations or mixing candidates with non-candidates)? Were suspect questions asked very early or very late in long surveys when participants might not be fully attentive?

Check to see how the interviews were conducted, whether autonomously over the internet, by telephone or video meeting, or in-person. It probably won't matter. Sophisticated surveys might use more than one interview method and then compare the results. See if that was done.

Finally, consider the number of questions. Some surveys have many questions, requiring considerable time to complete. Figure that it might take a minute for every two or three questions. Sixty questions could take a respondent a half hour to complete. Many political surveys are much longer.

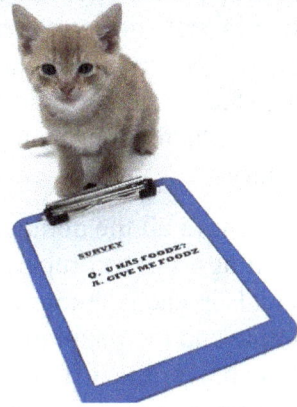

I want to pick both options.

Long surveys are susceptible to respondent dropout or, at least, loss of focus. Check to see if the survey participants were part of a panel. Panelists would be expected to endure more questioning than participants selected by random contact.

STEP 4. DEMOGRAPHICS.

As with the questions, how the samples were selected is more important than the sample demographics, but information about demographics is easier to identify and assess.

If you're evaluating any kind of *probability survey*, you must consider the demographics of the population and the sample. Legitimate surveys will always ask survey questions about the backgrounds of the respondents—sex, age, race, income, education, and so on—unless participants are part of a panel that already provides the information. If the probability poll doesn't consider that information, it's probably not legitimate.

The demographics for the *sample* should be reported in the documentation for the survey, which is usually provided on the pollster's website. The demographics will appear in a table as the number of responses in each category for each demographic.

The demographics for the *population* are harder to find because the survey documentation rarely states them. So first, you have to understand how the pollster defined the population. That information should be in the survey documentation.

Surveys.

Finding the demographics for the population can be a challenge. Surveys involving public groups (e.g., U.S. populations by State and local jurisdictions) are readily available on the internet. Some information might not be entirely available or be up-to-date, like statistics on voter registrations, schools, medical care, law enforcement, and the like. Some information won't be available at all like demographics of **likely voters**, which are determined by models created by the pollster. Information for surveys conducted for businesses usually won't be available either because it is proprietary.

Once you have all the demographic information, do two things. First, compare the percentages of individuals in each demographic category between the sample and the population. They should be close; the bigger the differences the more biased the poll may be. Second, look at the number of samples in the categories of each demographic. This will determine (except for design effects) the MoE for percentages in the categories. For example, a survey of 2,000 responses (MoE of 2.2%) may have a demographic category (*subpopulation*, like "over 65" is a category of the age demographic) having 200 responses (MoE of 7.1%).

How did you weight the poll for demographics?

Differences in response percentages between population and sample of more than a few points may require the pollster to make adjustments. Look in the survey documentation for any mention of *weighting*.

STEP 5. PROGRESSION FROM POPULATION TO FRAME TO SAMPLE.

The pollster's progression from population to frame to sample is probably the most important aspect of a survey. Unfortunately, it is the hardest element to obtain information on and the hardest to evaluate. The reason for this is that the frame and sampling scheme may be complex and may be considered to be proprietary by the pollster.

The process is usually very difficult if not impossible for non-statisticians to assess. It's even difficult for statisticians to understand completely because of the variety of survey elements that might have been used. If documentation isn't available, it will be impossible for anyone to judge. It's not just a matter of polling whoever answers a phone or visits a website. Participants have to be weighted for population demographics and cleared from any potential biases. Every member of the population has to have an equal probability of being selected.

Surveys.

As a rule-of-thumb, if the process is complex and described in detail, it is more likely to have been valid than not. In reality, though, if you get to this step it's probably not worth pursuing unless you're getting paid to do it.

STEP 6. METHODS.

This step is very difficult to complete if you haven't taken courses in statistical surveying. It may not even be possible without insider knowledge of the procedures the pollster actually used. It

This is putting me to sleep.

might also be controversial; what one professional statistician considers to be an effective method another statistician may not.

The point of this step is to consider the validity of criticisms made by others. Just because someone criticizes a pollster's methods, especially on social media, doesn't mean the criticism is legitimate. Things like advanced types of random sampling, such as stratified or cluster sampling, over and under sampling, and the adjustments made to correct for other design effects, are well beyond the expertise of most commenters.

There will always be something that might adversely affect the validity of a poll. Even professional statisticians make mistakes or overlook minor details. However, these flaws may not be fatal and will probably be impossible for most readers to spot.

If you as an average consumer see something in the population, frame, sample, or questions that is dubious, you may have cause to critique. Otherwise, don't expose your ignorance by complaining about not having enough participants or that all the participants were surveyed by landline.

IS POLLING REALLY BROKEN?

It shouldn't come as a shock when a poll lacks the anticipated precision, or worse, accuracy. Star athletes don't win every competition, lawyers lose cases on technicalities, politicians lose elections, A-list actors don't get every part they want, expert stock traders make bad picks, journalists get facts wrong and have to correct or retract their stories. Pollsters sometimes obtain inaccurate or imprecise results through no fault of their own. That is certainly not a reason to allege that polling is broken and no longer useful.

Surveys.

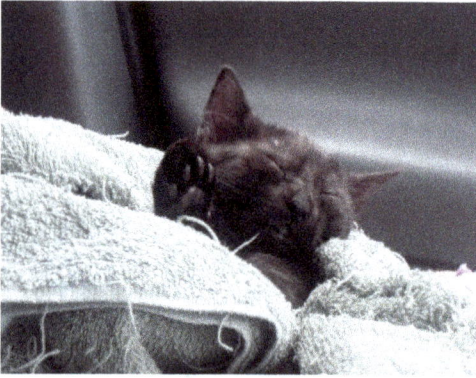

I don't want to hear it.

Over the last eighty years, there have always been discrepancies between polls conducted in the last week before a Presidential election and the election itself. However, the discrepancies are almost always within the MoEs of the surveys. Nevertheless, this can cause a perception of failure because many political contests have been so close.

An important thing to recognize also is that most surveys, including many surveys on politics, are never questioned for no other reason than that the results can't be proven to be wrong. It's only after an election that the legitimacy of the associated horse-race polls are questioned. Hindsight is easy.

Pollsters have to contend with many constantly-changing sources of variability. Of course there's added *natural variability* because populations and candidates change from election to election. There's added *measurement variability* because of new quirks in the polling process. There's added *sampling variability* because potential participants are advised to avoid taking polls and *polling denialism*. There's added *environmental variability* because of all the impactful mainstream media news and social media reactions. But most of all, the fundamental opinions of the population, the targets of the polls, change over time. People just change their minds. Surveys reflect all these things.

Nevertheless, there are three key reasons for why political polls fail to match election results—survey and pollster issues, election issues, and societal issues.

SURVEY AND POLLSTER ISSUES

There are things some pollsters might be able to control successfully or compensate for but they choose not to for a variety of reasons (e.g., scarce demographics). There are things pollsters could control or compensate for but the things either aren't known or understood well enough by the pollster to take action (e.g., turnout or voter suppression). Some things just can't be controlled (e.g., weather).

Of all the things that pollsters have to deal with that can lead to inaccurate and imprecise surveys, two stand out—*demographics* and *professionalism*.

First and foremost is demographics. Pollsters have learned to account for many demographics, like sex, age, race, party, location, and likelihood of voting. But, there are many other population characteristics that may unexpectedly become important in an election, like education, income, family size, gender, and religion.

It's impossible for pollsters to know what demographics might be important in an upcoming election so pollsters use the strategy of *important-before; important-*

 Surveys.

again. That doesn't help when some new population characteristic becomes relevant in an election but isn't recognized until afterwards.

There's a trade-off. The more demographics that have to be accounted for, the more difficult and expensive the survey will be, leading to larger required sample sizes and less frequent polling.

There's a mom and three kittens in this feline household.

Second, pollsters are not required to be licensed or certified, or follow a code-of-ethics. There is no organization that could review consumer complaints. Still, most individual pollsters and survey firms act professionally; they wouldn't stay in business if they didn't. But with over 100,000 individuals working in 8,000 companies, there are bound to be a few black sheep.

ELECTION ISSUES

There are things in elections, either generally present or specific to a particular time, location, or contest, for which pollsters can't control or even make adjustments. For example, voting schedules and rules are different in every State. Some States provide for election-day or even automatic voter registration to encourage voting. Other States limit opportunities for voters to cast their ballots to safeguard against fraud. These differences make it difficult for pollsters to model voter turnout.

Additionally, some races are very close. Close elections require great resolution perhaps even beyond the resolving power of statistical surveying. There's nothing pollsters can do about these issues. Worse, too often critics ignore pollster warnings about MoEs. Pollsters also sometimes neglect to document the error margins attributable to *design effects*.

Finally, there's the issue of the Electoral College used in the U.S.. Nationwide polls reflect the popular vote. Though they are usually accurate at identifying what the entire electorate is thinking, they can fail to identify the winner of a Presidential election because of the Electoral College, which is controlled by States. Horse-race polling in every State is burdensome and expensive. As a consequence, only the most important swing States are polled. The vast majority of Americans who live in uncompetitive States are essentially ignored. Nevertheless, nationwide polling is used as a surrogate for State polling because the results are roughly correlated. This is a predicament not likely to be solved anytime soon.

Figure 52 shows the margins in the popular and electoral votes for the U.S. Presidential elections from 1828 to 2020. They are correlated with a correlation coefficient of 0.75 and an r^2 value of 0.56 (see Chapter 7). Most of the time, the election margins fall within plus-or-minus three standard deviations (*Six Sigma*, one

Surveys.

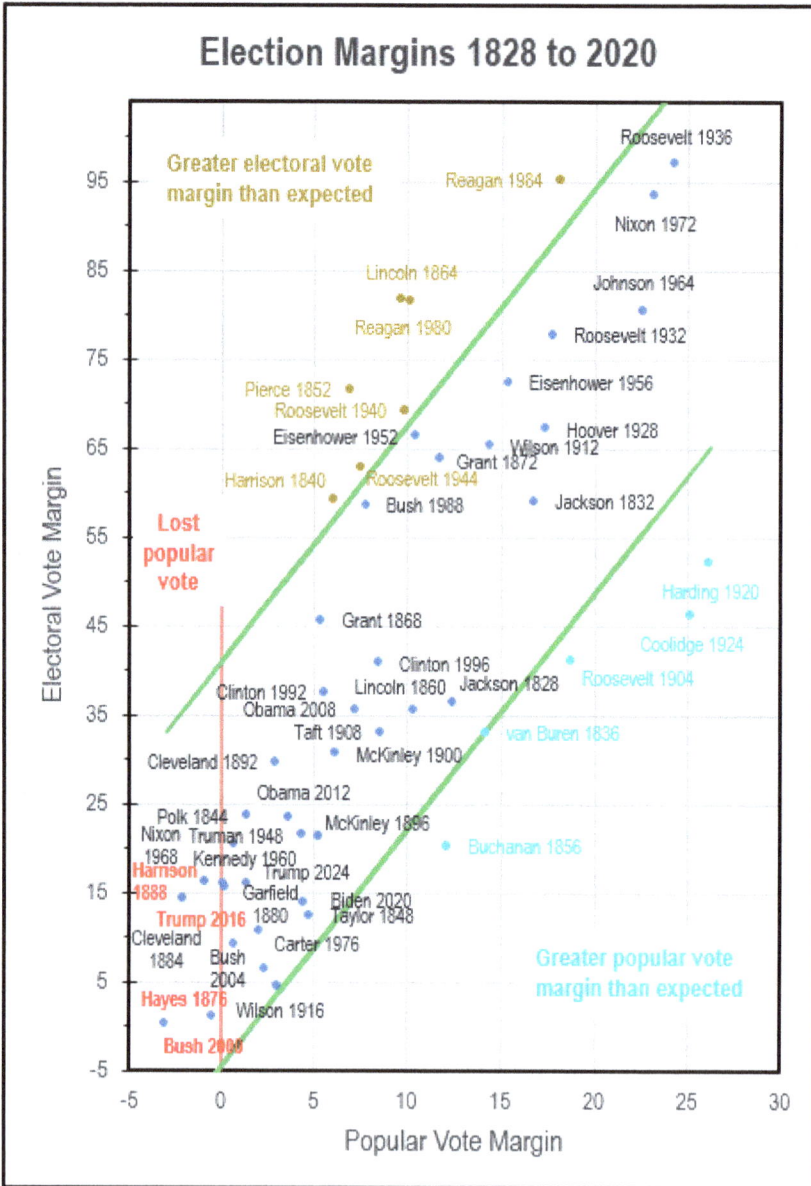

Figure 52. Margins in the popular and electoral votes for U.S. Presidential elections from 1828 to 2020.

in a million). There are, however, quite a few elections that fall outside those limits. Perhaps there should be as much scrutiny on the election process as on the polling.

SOCIETAL ISSUES

There are many influences in society during an election that influence how voters will vote, which pollsters can't control. Examples include:

Surveys.

- Attitudes, especially about hot-button political issues like the economy and individual rights.
- Changes in technologies that affect polling and elections, like telephony and the internet, and the laws and regulations that affect them.
- Statements and endorsements by celebrities, experts in relevant issues, authors, internet influencers, and religious leaders.
- Legal decisions concerning election rules and government operations.
- Mainstream media hyping polls without qualifying limitations and context, which are then amplified in social media.
- Unpredictable voter intimidation, not just by governments but also by partisan organizations and individuals.
- Refusals by individuals to participate in polls because of privacy concerns and poll denialism, leading to non-response bias.
- Weather.

All of these and other factors influence turnout on election day. Over the past two decades, about 40% of eligible voters did not vote, as shown in Figure 53. That's more nonvoters than voters who voted for either party. This pattern is a sad statement about the process of electing the U.S. President. (Data for nonvoter percentages came from www.susqu.edu/live/news/768-research-identifies-four-types-of-nonvoters-and/.)

THE FUTURE OF POLLING

Political polling isn't going to go away. It's the only way to quantitatively evaluate opinions on policies, attitudes, and opinions. Candidates need it, and in reality, the electorate needs it too. But, it should stimulate constructive debates not escalate divisiveness.

You have to have effective tools but sometimes you have to work with what's available.

Polls on the same topic conducted at the same time can produce different results for a variety of reasons, such as unanticipated events or conditions, non-representative samples, poor question design, and of course, natural variability. Some polls might deliberately employ skewed sampling or word questions to try to get a particular result that will influence public opinion. Legitimate professional pollsters don't do this, they have reputations to protect.

Don't pay too much attention to individual polls. Instead, look at aggregations of independent professional polls conducted at about the same time. Also be sure the polls that are included in the aggregation all come from reputable pollsters.

Surveys.

Average Proportions of Voters 2000-2020

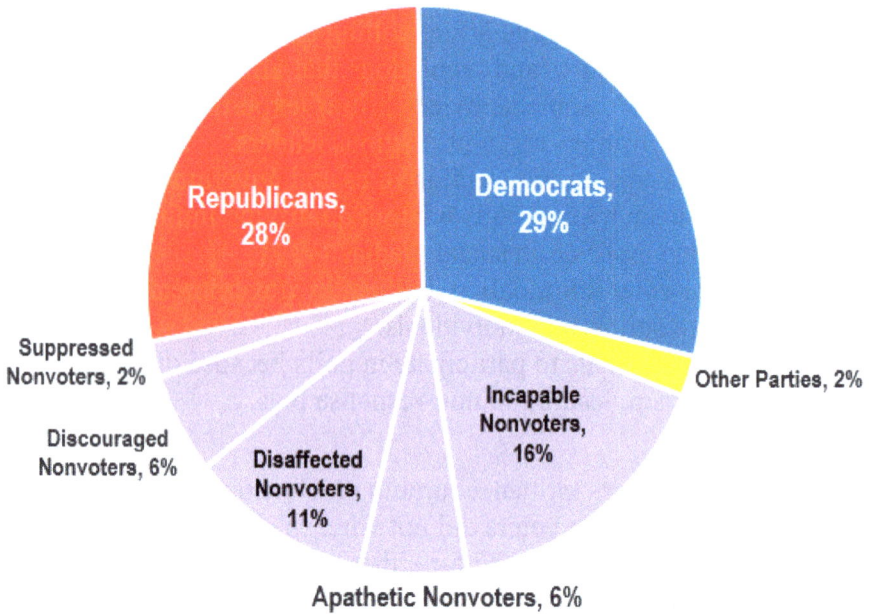

Republicans, 28%

Democrats, 29%

Suppressed Nonvoters, 2%

Other Parties, 2%

Incapable Nonvoters, 16%

Discouraged Nonvoters, 6%

Disaffected Nonvoters, 11%

Apathetic Nonvoters, 6%

Apathetic Nonvoters have no interest in politics; they are what most people believe typifies individuals who do not participate in elections.

Incapable Nonvoters are interested in politics but don't vote because of personal reasons, like illness, no transportation, or work requirements.

Disaffected Nonvoters are interested in the political process but are deterred by dissatisfaction with or ignorance about the candidates.

Discouraged Nonvoters would fulfill their responsibility to vote but are deterred by issues of convenience, like bad weather or long wait times.

Suppressed Nonvoters want to vote but are prevented by measures like gerrymmandering, ID requirements or limited polling times and places.

Figure 53. About 40% of eligible voters have not voted in U.S. Presidential elections since 2000.

Combining good polls with bad just to provide the appearance of improved credibility is misleading.

Polls are *instruments* for measuring human opinions. Like any measurement instrument, they have limitations. They have to be designed and used correctly. Even so, they may not have the capability to resolve small differences. This is a big issue in elections in which the differences are smaller than the MoE.

Political polls are a bit like blood pressure measurements. There are dozens of factors that can affect the results. The instrument (survey or sphygmomanometer) is

Surveys.

continually upgraded as knowledge of the phenomenon (opinions or blood pressure) and the technology for measuring it improves. The process of taking a measurement (polling or sphygmomanometry) is continually refined to control extraneous sources of variability. And, the subjects of the measurement (participants or patients) are understood better with more and better information (demographics or lifestyle choices) to put results into context. You don't abandon measuring the information (opinions or blood pressure) just because you don't completely believe the results. You fix the instrument, control the process better, and try again. There aren't any other options.

Pollsters will do everything they can to improve their surveys despite challenges. It's why they are in the profession.

Polling has been declared broken many times in the past, starting in 1948 and continuing most years since 2000. Each time, professional statisticians have investigated the causes of the discrepancies and improved how surveys are conducted. That's how science works (Chapter 9).

But, survey statisticians focus their attention on statistical methodologies. It's what they know. They can improve those. However, they can't affect the issues with elections and society that make polling appear to be broken.

Given the current state of U.S. politics, perhaps current polling methods have reached their *limit-of-detection.* Solutions have to come from outside the discipline of statistics. Polling isn't broken, it continues to mature as a science. Still, voters must come to appreciate the benefits and limitations of statistical surveys.

Don't be discouraged if your brain is exploding from all the information in this chapter about what is involved in creating a professional survey. Don't sweat the details. Just remember that creating sound surveys is complicated and difficult, even for professionals, and sometimes things go wrong. Every unexpected result isn't attributable to a nefarious conspiracy.

And that's why statistical surveying is the most difficult analysis in all of statistics and why it is seldom taught in Stats 101.

Sometimes, things that look simple are actually more complicated than you can imagine.

Surveys.

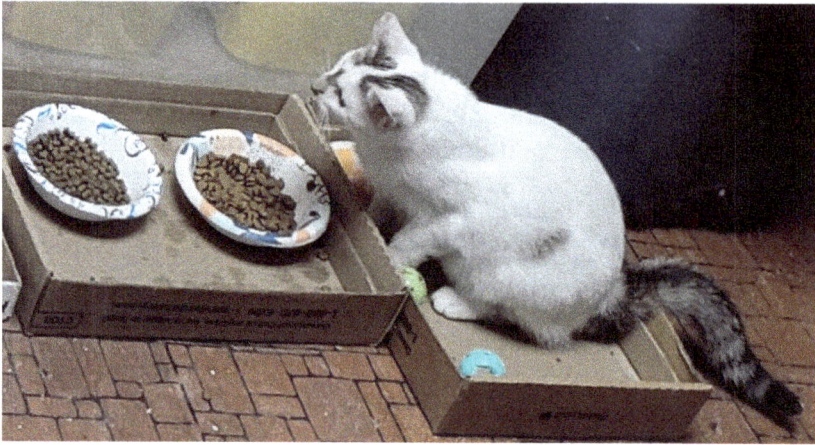

We should do a survey of which kibble all us cats like best and then do a statistical test of the results. We could call it "The Kibble Buffet Experiment."

Surveys.

CHAPTER 6. COMPARISONS.
Is There Really A Difference?

INTRODUCTION

It's *SIGNIFICANT!* That's what the media highlights to attract concerned readers. You've all seen it. Even dignified researchers high-five when their statistical tests pass the 0.05 threshold. And why not? Statistical testing has been the ultimate arbiter of all decisions involving numbers for over a century.

I'm significant!

But there's a darker side to statistical testing. For one, researchers can't get their works published unless they find *significance*. That puts them under a lot of pressure to get the right results. And, tests are complicated enough that they can be manipulated. Most problematic, though, is that there is a general lack of understanding about how statistical tests work. If statistical testing is going to be worth its billing, you have to understand how it works … and doesn't work.

That's what this chapter is about.

STATISTICAL COMPARISONS

In school, you probably had to line up by height now and then. That wasn't too difficult. There weren't too many individuals being lined up and they were all in the same place at the same time. An individual's place in line was decided by comparing his or her height to the heights of other individuals. The comparisons were visual; no measurements were made. Everyone made the same decisions about the height comparisons. You didn't need statistics to solve the problem. So why might you ever need statistics to compare heights?

The nice thing about statistical comparisons is that you don't have to measure the phenomenon in the entire population at the same place or the same time, and you can then make inferences about groups instead of just individuals or items. What may even be better is that if you follow statistical testing procedures, people who don't know any better will agree with your findings.

Testing.

Now for even more …

POPULATIONS

Statistics are primarily concerned about groups having some fundamental commonalities. These groups are called *populations*. Populations are more difficult to compare than just pairs of individuals because you have to identify what properties define the population and then measure the characteristics of the *phenomenon* that you want to compare. You've heard this before in Chapter 3, it's nothing new.

The advantage of a statistical comparison, though, is that you can extrapolate your results from a *sample* to the entire *population*. That's called *induction*.

For example, you might want to compare the heights of male high school freshmen in two different school districts. That would be called a *two-population test.* One population would be male high school freshmen in school district 1 and the other population would be male high school freshmen in school district 2. The phenomenon you want to compare is the heights of the two populations. It's not as easy as just visually comparing the heights of pairs of individuals because they are not located in the same place. You have to measure some, but not all, of the heights of the individuals in the two populations.

SAMPLES

You don't have to measure every individual in the two populations so long as you measure a *representative sample* of the individuals in each population. You can improve your chances of getting a representative sample by using the three Rs of variance control— *Reference*, *Replication*, and *Randomization* (Chapter 3).

How many individual heights would you need to measure? The *Law of Large Numbers* says that the more individuals you measure the closer your estimates will be to the true population values. But there's a complication. With too many measurements, statistical tests might detect *meaningless* differences. As with potato chips and middle managers, too many samples can be as bad as not enough. There's a right amount and it's a matter of *resolution*.

I'm still a kitten but I'm taller than any of the other cats.

Some people think the answer is 30 samples, which is often taught in Stats 101. However, that's a misunderstanding of two 1908 journal articles by William Gosset, a statistician and Head Brewer for Guinness, who published under the pseudonym **Student**. He wrote the first paper introducing the *t-*

Testing.

distribution and then wrote a second paper based on the first about an experiment that used 30 samples. People thought the second paper endorsed the use of 30 samples. It didn't. Oops.

As it turns out, the *precision* of estimates of a sample mean increases very rapidly up to about 10 samples then continues to increase, albeit at a decreasing rate, to over 100 samples. So, there's nothing extraordinary about using 30 samples but the belief has persisted as an old statisticians' tale.

STUDY PERSPECTIVES

Statistical analyses come from two perspectives—*observational* and *experimental*. (There are also *meta-analyses* that involve combining the findings from many independent studies in ways that make their findings comparable. Don't worry about these at this point.)

In an *observational study*, the researcher develops a *research question* about a *phenomenon* in a *population*, then samples the population using an appropriate statistical sampling scheme (Chapter 3). The samples selected from the population have *inherent characteristics* associated with them that are what's analyzed in the study. The characteristics might be biological (e.g., species, sex, age, habitat), chemical (e.g., composition, concentration), physical (e.g., size, construction, temperature), set in a certain time period or location, or any of an infinite number of other possibilities.

In an *experimental* study, the researcher develops a *research question* about a phenomenon in a population, then samples the population using an appropriate statistical sampling scheme. This is the same as in an observational study. However, the samples selected from the population are then *randomly assigned* to the experimental conditions being tested. The experimental conditions, called *treatments*, could be a pharmaceutical, a medical or veterinary procedure, an educational alternative, or any of an infinite number of other possibilities.

I'm like an experimental study— small, focused, and streamlined.

I'm like an observational study—collect everything and sort it all out later.

Both types of studies may have a *control group* in which the characteristics or treatments being studied are not present. The purpose of a control group is to test whether the characteristic or treatment is associated with a difference in the measurement of the phenomenon. It is an example of the *Reference* approach to evaluating variability (Chapter 3).

 Testing.

For example, say a school principal wanted to determine if there was a difference in grades between the males and the females in math classes. That would be an *observational* study because the subjects, the students, assign themselves to the sex groups. If the principal wanted to determine which of two approaches was better for teaching math, students would be assigned randomly to classes in which one of the two methods was taught. Then their grades on standardized tests would be compared. That would be an *experimental* study.

> *The division between those who try to learn about the world by manipulating it and those who can only observe it had led, in natural science, to a struggle for legitimacy. The experimentalists look down on the observers as merely telling uncheckable just-so stories, while the observers scorn the experimentalists for their cheap victories over excessively simple phenomena.*
>
> Richard C. Lewontin, 1929-2021, American evolutionary biologist, mathematician, geneticist, and social commentator, from **Sex, Lies, and Social Science** (1995).

DISTRIBUTIONS

If you were comparing the heights of two individuals, you would only be concerned with whether one individual's height is greater than, equal to, or less than the other individual's height. When you're comparing populations, there's not just one height but many, and you only know what some of the heights are (hopefully, a representative sample of them). That's where distributions come in.

In statistical testing, a *frequency distribution* refers to the number of times each value of the measured phenomena appears in the population. A bar chart of these values, with the values on the horizontal axis and their frequencies on the vertical axis, is called a *histogram* (see Chapters 3 and 4).

Sometimes histograms have many values located at the middle of the distribution and fewer values farther away from the center. They look like a bell and are called *bell curves*. Such bell-shaped curves are usually assumed to represent a *Normal probability distribution* (see Chapter 2). The center of the distribution of values is estimated by the *average*. The variability of the values, how far they stretch along the horizontal axis, is estimated by the *variance* or the *standard deviation*, the square root of the variance (see Chapter 3).

The thin ends of the bell curve are called the *tails*. The average and the variance of the values are called *parameters* of the distribution because they are in the mathematical formula that defines the form of the Normal distribution.

So why is the Normal distribution so important to statistics? It is because of the *Central Limit Theorem*. The Central Limit Theorem says that the distribution of

Testing.

averages of samples of a population will approximate a Normal distribution (if there are enough samples) no matter what the underlying frequency distribution of the population is. This means that the Normal distribution can be used as a basis for statistical comparisons for virtually any continuous-scale data. This is a very big deal.

Having a mathematical equation like the Normal distribution that can be used as a model of the frequency of phenomenon values in the population is advantageous because you can use the distribution model to represent the characteristics of the population.

THE JARGON OF TESTING

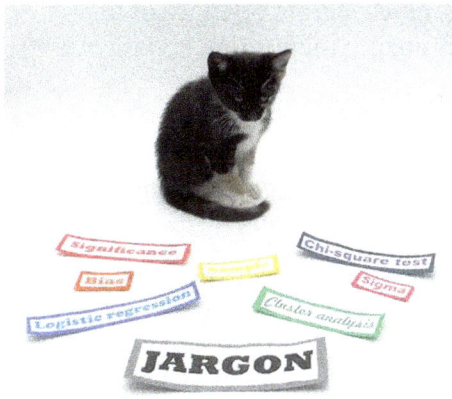

Still more jargon ?!?

To summarize:

Statistical comparisons, or *statistical tests* as they are usually called, involve *populations,* groups of individuals or items having some fundamental commonalities. The members of a population also have one or more characteristics, called *phenomena,* which are what is compared in the populations. You don't have to measure the phenomena in every member of the population. You can take a *representative sample*.

Once you have data on the *phenomenon* from the *representative sample* of the *population*, you calculate *descriptive statistics* (Chapter 3) and conduct *statistical tests*. Statistical tests can involve one population (comparing a population phenomenon to a constant), two populations (comparing a population phenomenon to the phenomenon in another population), or three or more populations. You can compare just one phenomenon (called *univariate tests*) or two or more phenomena (called *multivariate tests*).

Parametric statistical tests compare frequency distributions, the number of times individual values (or small intervals of values called *bins*) of the measured phenomena appear in the population. Most tests involve the *Normal distribution* in which the center of the distribution of values is estimated by the *average*, also called the *mean*. The variability of the distribution is estimated by the *variance* or the *standard deviation*, the square root of the variance. The mean and standard deviation are called *parameters* of the Normal distribution because they are in the mathematical formula that defines the form of the distribution.

Statistical comparisons consider both the *accuracy* (i.e., the difference between measurements of a phenomenon and the true values of the phenomenon in the

Testing.

population of individuals) and the *precision* (i.e., how consistent or variable the measurements of the phenomenon are). Formulas for statistical tests usually involve some measure of accuracy (involving some measure of *central tendency*) divided by some measure of precision (involving some measure of *dispersion*).

Most statistical tests focus on the extreme ends of the Normal distribution, called the *tails*. Tests of whether population means are equal are called *non-directional, two-sided,* or *two-tailed tests* because differences in both tails of the Normal distribution are examined. Tests of whether population means are less-than (<) or greater-than (>) each other are called *directional, one-sided,* or *one-tailed tests* because the difference in only one tail of the Normal distribution is examined.

Statistical tests that don't rely on the distributions of the phenomenon in the populations are called *nonparametric* tests. Nonparametric tests often involve converting the data to ranks and analyzing the ranks using the *median* and the *range*.

There's even more to statistical testing but those are the basic concepts and nomenclature. Table 25 provides a summary.

ASSUMPTIONS

> *The most misleading assumptions are the ones you don't even know you're making.*
>
> Douglas Adams, 1952-2001,
> English author

You make a few *assumptions* in conducting statistical tests. First, you assume that your population is real (i.e., not a *phantom population*) and that your samples of the population are representative of all the possible measurements, and that they are independent of each other. Then, if you plan to do a parametric test, you assume that the measurements of the phenomenon are *Normally distributed* and that the *variances are the same* in all the populations being compared. These and other assumptions are discussed below. But first, you need to know just a little about statistical models.

A statistical model is just an equation with five components:

🐾 *Dependent variable*. Called the y-variable, it is placed to the left of the equals sign in the model. It is the variable that is the focus of a statistical analysis. There is only one dependent variable in a statistical model (usually).

🐾 *Independent variable*. Called the x-variable, it is placed to the right of the equals sign. It is the variable that is used to predict or explain the dependent variable. There may be one or many independent variables in a statistical model.

I'm just the dependent variable. We're gonna need a bigger truck for the whole model.

Testing.

Table 25. Summary of basic concepts and nomenclature of statistical testing.

Test Specification	Why it is Important
Population	Groups of individuals or items having some fundamental commonalities relative to the phenomenon being tested. Populations must be definable and readily reproducible so that results can be applied to other situations.
Number of populations being compared	The number of populations determines whether a comparison can be a relatively simple 1- or 2-population test or a complex ANOVA test.
Phenomena	The characteristic of the population being tested. It is usually measured as a continuous-scale attribute of a representative sample of the population.
Number of phenomenon	The number of phenomenon determines whether a comparison will be a relatively simple univariate test or a complex multivariate test.
Representative sample	A relatively small portion of all the possible measurements of the phenomenon on the population selected in such a way as to be a true depiction of the phenomenon.
Sample size	The number of observations of the phenomenon used to characterize the population. The sample size contributes to the determinations of the type of test to be used, the size of the difference that can be detected, the power of the test, and the meaningfulness of the results.
Hypotheses	You start statistical comparisons with a research hypothesis of what you expect to find about the phenomenon in the population. The research hypothesis is about the differences between the categories of the variable representing the population. You then create a null hypothesis that translates the research hypothesis into a mathematical statement that is the opposite of the research hypothesis, usually written in term of no change or no difference. This is the subject of the test. If you do not reject the null hypothesis, you adopt the alternative hypothesis.
Distribution	Statistical tests examine chance occurrences of measurements on a phenomenon. These extreme measurements occur in the tails of the frequency distribution. Parametric statistical tests assume that the measurements are Normally distributed. If the distribution is different from the tails of a Normal distribution, the results of the test may be in error.
Directionality	Null hypotheses can be non-directional or two-sided (i.e., $\mu=0$), in which both tails of the distribution are assessed. They can also be nondirectional or one-sided (i.e., $\mu<0$ or $\mu>0$), in which only one tail of the distribution is assessed.
Assumptions	Statistical tests assume that the measurements of the phenomenon are independent (not correlated) and are representative of the population. They also assume that errors are normally distributed and the variances of populations are equal.
Type of test	Statistical tests can be based on a theoretical frequency distribution (parametric) or based on some imposed ordering (nonparametric). Parametric tests tend to be more powerful.
Test Parameters	Test parameters are the statistics used in the test. For t-tests using the Normal distribution, this involves the mean and the standard deviation. For F-tests in ANOVA, this involves the variance. For nonparametric tests, this usually involves the median and range.
Confidence	Confidence is 1 minus the false-positive error rate. The confidence is set by the person doing the test before testing as the maximum false-positive error rate they will accept. Usually, an error rate of 0.05 (5%) is selected but sometimes 0.1 (10%) or 0.01 (1%) are used, corresponding to confidences of 95%, 90%, and 99%..
Power	Power is the ability of a test to avoid false-negative errors $(1-\beta)$. Power is based on sample size, confidence, and population variance and is NOT set by the person doing the test, but instead, calculated after a significant test result..
Degrees of Freedom	The number of values in the final calculation of a statistic that are free to vary. For a t-test, the degrees of freedom is equal to the number of samples minus 1.
Effect Size	The smallest difference the test could have detected. Effect size is influenced by the variance, the sample size, and the confidence. Effect size can be too small, leading to false negatives, or too large, leading to false positives.
Significance	Significance refers to the result of a statistical test in which the null hypothesis is rejected. Significance is expressed as a p-value.
Meaningfulness	Meaningfulness is assessed by considering the difference detected by the test to what magnitude of difference would be important in reality.

- *Coefficients*. Values that the independent variables are multiplied by to weight the importance of the variable. There is one coefficient for each independent variable in the model.
- *Constant*. A single value used to adjust the level of the model at the y-axis.
- *Error term*. A value that represents the uncertainty in the model. Differences between model estimates and actual values are called *residuals*. Residuals are the focus of several of the assumptions.

Testing.

That's enough for now. Chapter 8 discusses models in detail.

REPRESENTATIVENESS

All statistical analyses assume that the sample of the population is representative of the population being investigated. Some statistics books don't discuss this as a basic assumption because it is viewed as more of a requirement than an assumption. But, obtaining representative samples of populations can be a challenge.

Unlike the other assumptions, failure to obtain a representative sample from a population under study would necessarily be a fatal flaw for any statistical analysis. You might not know it, though, because there's no good way to determine if a sample is representative of the underlying population. To do that, you would need to know all the pertinent characteristics of the population. But if you knew that about the population, you wouldn't need to bother with a sample.

We represent a diverse population.

As a consequence, representativeness has to be addressed indirectly by building randomization and variance control into the sampling program before it is undertaken. If randomization cannot be incorporated into the sampling procedure in some way, the only alternative is to try to evaluate how the sample might not be representative. This is seldom a satisfying exercise. Making statements like "the results are conservative because only the *worst cases* were sampled" are usually conjectural, qualitative, and biased. It's called the *worst-case fallacy* (Chapter 9).

ADDITIVITY AND LINEARITY

Statistical models are assumed to be additive and linear. *Additivity* refers to the right side of the model being the sum of coefficients times independent variables plus the constant. *Linearity* refers to the relationships between the dependent variable and the independent variables forming straight (i.e., linear) lines.

Additivity implies that the relationships between the independent variables and the dependent variable are not correlated with each other (i.e., are independent) so that the effects can be summed to estimate the overall effect. Additivity prohibits *interactions*, in which the effects of one independent variable depends on another.

Additivity would seem to preclude multiplicative models but they can be accommodated by using logarithms. This is because the product of variables can be transformed to the sum of the logarithms of the variables, thus making the logarithms an additive model. What additivity doesn't allow, for example, is one independent variable being an exponent of another.

Testing.

Linearity means that the relationships between the dependent variable and the independent variables are straight lines. Violations of this assumption can often be corrected by using *transformations* (Chapter 9) of the independent variable.

I guess we don't mean the same thing by "additivity."

There are two types of nonlinearity. *Intrinsically-linear* relationships, such as some simple *curvilinear* relationships, can be made into linear relationships through the use of *transformations*. *Intrinsically-nonlinear* relationships can't be adjusted using transformations but they can be analyzed using special types of analysis.

It's not always easy to distinguish i*ntrinsically-linear* from *intrinsically-nonlinear* relationships. In general, if the model coefficients are not part of a mathematical function, the relationship is probably *intrinsically-linear.* If the coefficients are part of a mathematical function with independent variables, the relationship is probably *intrinsically-nonlinear.*

Additivity is more often mentioned for statistical models involving hypothesis testing while linearity is more often mentioned for models involving regression. Statistical testing and regression are closely linked, but that is a story for Stats 102.

INDEPENDENCE

Another assumption common to statistical models is that the errors in the model, the *residuals*, are independent of each other.

Some introductory statistics textbooks describe this assumption in terms of the measurements of the dependent variable. There are two reasons for this. First, it's a lot easier for beginning students to understand, especially if they aren't familiar with the mathematical form of statistical models and the concept of model errors (residuals). Second, and more importantly, the two approaches to describing the independence assumption are equivalent. This is because a data value can be expressed as the sum of an inherent true value and some random error (Chapter 3). If you have controlled

We're all independent of each other.

   Testing.

all sources of extraneous variation, the data and the model errors should be identically distributed.

Say you were conducting a study that involved measuring the temperature of human subjects. Without your knowledge, a well-meaning assistant provides beverages in the waiting room—hot coffee and iced tea. When you plot a histogram of the human temperature data, you see three peaks (called *modes*) in the data frequency distribution, one centered at 98.6°F, another at a degree or so higher and a third a degree or so lower. Your data have violated the independence assumption.

Maybe I wouldn't be so good at running a coffee shop… no opposable thumbs.

The subjects who drank the coffee all had their temperatures influenced by the higher temperature of the coffee. The subjects who drank the iced tea all had their temperatures influenced by the lower temperature of the iced tea.

What are the chances you might notice this dependency? If you had a dozen or so subjects, the chances wouldn't be good. With 100 subjects, you might notice something. With 1,000 subjects, you would almost certainly notice the effect. (If you're providing beverages to 1,000 subjects, though, you might consider getting out of research and opening a coffee shop.)

Assessing independence also involves looking for *serial correlations* (i.e., correlations based on the order of the measurements), *autocorrelations* (i.e., correlations based on the times the measurements were taken), and spatial correlations (i.e., correlations based on the locations of the measurements).

A serial correlation is the correlation between data points with the previously-measured data points. For example, making measurements with an instrument that is drifting out of calibration will introduce a serial correlation.

Spatial or temporal dependence are often present in environmental data. For example, two soil samples located very close together are more likely to have similar attributes than two samples located very far apart. Likewise, two groundwater samples from a well collected a day apart are more likely to have similar

We may be littermates but we are totally independent.

 Testing.

attributes than two groundwater samples from the well collected a year apart.

When the independence assumption is violated, the calculated probability that a population and a fixed value (or two populations) are different will be underestimated if the correlation is negative, or overestimated if the correlation is positive. The magnitude of the effect is related to the degree of the correlation.

NORMALITY

Statistical comparisons require that errors in the model are Normally distributed. This amounts to assuming that the measurements themselves are Normally distributed.

This is a reasonable assumption because of the *Central Limit Theorem.* The CLT says that estimates of the mean of any collection of measurements will be Normally-distributed regardless of the frequency distribution of the measurement themselves, so long as enough samples are used to make the estimates.

Figure 54 shows a graph of a Normal distribution. The area under the curve represents the total probability of measured values occurring, which is equal to 1.0. Values near the center of the distribution, near the mean, have a large probability of occurring while values near the *tails* (the extremes) of the distribution have a small probability of occurring.

In statistical testing, the Normal distribution is used to estimate the probability that the measurements of the phenomenon will fall within a particular range of values.

To estimate the probability that a particular value will occur, you use the values of the mean and the standard deviation in the formula for the Normal distribution. Actually though, you never have to do that

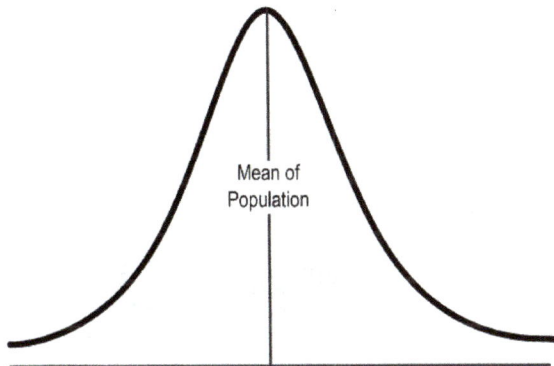

Figure 54. Graph of a Normal distribution.

because there are tables for the Normal distribution (and the t-distribution for small sample sizes). Even easier, the functions are available in many spreadsheet applications, like Microsoft Excel, so you don't have to perform any look-ups or calculations.

Statistical tests focus on the *tails* of the distribution where the probabilities are the smallest. It doesn't matter much if the measurements of the phenomenon near the mean exactly follow a Normal distribution so long as they do in the tails. The more the distribution of a measurement deviates from a Normal distribution in the tails, the more the calculations of test probabilities will be in error.

Testing.

The *t-distribution* can be used instead of the Normal distribution for small sample sizes. This is because the t-distribution is similar to the Normal distribution except that it has more area in its tails.

EQUAL VARIANCES

*Equal Variances (*also called *homoscedasticity* or *homoskedasticity)* means that the errors in a statistical model have the same variance for all values of the dependent variable.

For models involving grouping variables, the assumption means that all groups have about the same variance. Evaluating this assumption involves calculating the variances for each group and looking at the ratios of the sample sizes and the variances.

For continuous-scale measurements, equal variances means that the variance of the errors don't change across the entire scale of measurement. For example, in the case of a measurement instrument, the equal variances assumption requires that the error variance be about the same for measurements at the low, middle, and high portions of the instrument's range. Another example would be measurements made over many years. Improvements in measurement technologies could cause more recent measurements to be less variable (i.e., more precise) than historical measurements.

Violations of the equal variance assumption tend to affect statistical models more than do violations of the Normality assumption. Generally, the effects of violating the homogeneity-of-variances assumption will be small if the largest ratio of variances is near one and the sample sizes are about the same for all values of the independent variables. As differences in both the variances and the numbers of samples become large, the effects can be great.

Be sure to consider unequal variances.

Violations of equal variances can often be corrected using *transformations*. In fact, transformations that correct deviations from Normality will often also correct unequal variances. *Non-parametric statistics* are also used to address violations of this assumption.

PROCESS

The concept behind statistical testing is to determine how likely it is that a difference in two populations' parameters like the means (or a population parameter and a constant) could have occurred by chance. If the probability of the difference

Testing.

occurring is small enough to occur in the tails of the distribution, there is only a small probability that the difference could have occurred by chance.

Differences having a probability of occurrence less than a prespecified value are said to be *significant* differences. The prespecified value, which is the acceptable *false positive error rate*, **α**, may be any small percentage but is usually taken as five-in-a-hundred (0.05) or one-in-a-hundred (0.01), and sometimes ten-in-a-hundred (0.10), one-in-a-thousand (0.001), or one-in-a-million (0.000001).

Figure 55 is a flowchart of the process of planning, testing, and reporting statistical tests.

PLANNING

First, you plan the comparison by understanding the *populations* you will take a r*epresentative sample* of individuals from and measure the *phenomenon* on. Then you assess the *frequency distributions* of the measurements to see if they fulfill the Normality and equal variances assumptions.

TESTING

Second, you test the measurements by considering the *test parameters*, the *type of test*, the *hypotheses*, the test *directionality*, *degrees of freedom*, and any violations of *assumptions*.

Phases of Statistical Testing

Figure 55. Three phases of statistical testing.

The testing phase begins with a *research hypothesis,* a statement of what you expect to find about the phenomenon in the population. From there, you create a *null hypothesis* that translates the research hypothesis into a mathematical statement about the **opposite** of the research hypothesis.

The null hypothesis is usually written in terms of no change or no difference. For example, if you expect that the average heights of students in two school districts

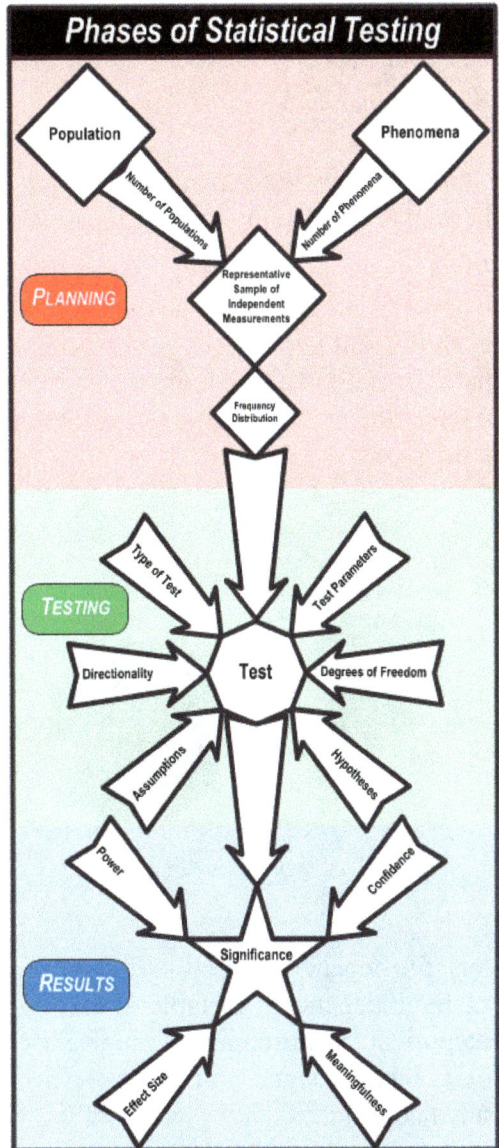

Testing.

will be different because of some demographic factors (your research hypothesis), then your null hypothesis would be that the means of the two populations are equal.

REPORTING

Third, you review the results by determining the *confidence*, *effect size,* and *power* of the test, and assessing the test's s*ignificance* and *meaningfulness*.

When you conduct a statistical test, the result does not mean you prove your hypothesis. Rather, you can only *reject* or *fail-to-reject* the null hypothesis. If you reject the null hypothesis, you adopt the *alternative hypothesis.* This would mean that it is more likely that the null hypothesis is not true in the populations. If you fail to reject the null hypothesis, it is more likely that the null hypothesis is true in the populations.

TESTS

It seems like there are hundreds of kinds of statistical tests. For the most part, though, they are just variations of the concept of accuracy-in-terms-of-precision. Here are a few of the most common statistical tests.

SIMPLE TESTS

What will my hypothesis test reveal?

The appropriateness of tests for a statistical comparison depends on the *measurement scales* of the data. Usually for parametric tests, the dependent variable (the measurements of the phenomenon) is continuous and the independent variable (the divisions of the populations being tested) is categorical. The grouping variables used as independent variables are called *effects.* In advanced designs, continuous-scale variables used as independent variables are called *covariates.*

Other *scales of measurement* for the dependent variable, like binary scales and restricted-range scales, require special tests or test modifications. Scales are a big part of why there seem to be so many different tests.

Table 26 highlights a few of the most common parametric statistical tests.

Statistical tests can be designed for a variety of situations. Most involve simple comparisons of means but there are also tests for comparing percentages, frequencies, ranks, and paired data. Tests can compare one population to a standard value or multiple populations to each other. Advanced *Analysis-of-Variance* (ANOVA) designs can do much more.

Testing.

Table 26. Common statistical designs and tests.

Populations	Grouping Variables	Comparison	Test
1	0	Population mean vs constant	z-test, t-test
	1	Population level 1 mean vs Population level 2 mean	z-test, t-test
	2 or more	Level 1 vs Level 2 etc.	ANOVA F-test
2	0	Population 1 mean vs Population 2 mean	z-test, t-test
	any	Level 1 vs Level 2 etc.	ANOVA F-test
3 or more	any	Level 1 vs Level 2 etc.	ANOVA F-test

Don't worry about remembering formulas for the statistical tests (unless a teacher tells you to). Most testing is done using software with the test formulas already programmed. If you need a test formula, you can always search the internet.

Z-TESTS AND T-TESTS

The *z-test* and the *t-test* have similar forms relating the difference between a population mean and a constant (*one-population test*) or two population means (*two-population test*) to some measure of the uncertainty in the population(s). The difference in the tests is that z-tests are for Normally-distributed populations where the variance is known while t-tests are for populations where the variance is unknown and must be estimated from the sample.

t-Tests depend on the number of observations made on the sample of the population. The greater this *sample size,* the closer the t-test is to the z-test. Adjustments of two-population t-tests are made when the sample sizes or variances are different in the two populations. These tests can also be used to compare paired (e.g., before vs after) data.

F-TESTS

Statistical tests are not all about means. In an Analysis of Variance, *ANOVA*, the variances in the groups (called *levels*) being compared are partitioned between variation associated with the *factors* (independent variables) in the design (called *model variation*) and random variation (called *error variation*).

ANOVA is conceptually similar to multiple two-population t-tests, but produces fewer type I (*false positive*) errors. While t-tests use t-values from the t-distribution, ANOVAs use F-tests based on the F-distribution.

I thought an F-test was about deciding if someone was feline friendly.

Testing.

An *F-test* is the ratio of the *between-groups variance* to the *within-groups variance*. The F-statistic can also be expressed as the ratio of the *model variance* to the *error variance*. Once again, the comparison is about accuracy (model) versus precision (error).

When there are only two means to compare, the t-test and the ANOVA F-test are equivalent according to the relationship $F = t^2$. When the F-statistic is close to 1, the null hypothesis cannot be rejected.

χ^2-TESTS

The *chi-squared test* is used to determine whether there is a significant difference between the *expected frequencies* and the *observed frequencies* in mutually exclusive categories of a *contingency table*. The test statistic is the square of the observed frequency minus the expected frequency divided by the expected frequency.

We're in an X shape and there are two of us, so we're being a χ^2.

NONPARAMETRIC TESTS

Nonparametric tests are also called distribution-free tests because they don't rely on any assumptions concerning the frequency distribution of the data. Instead, the tests use ranks or other imposed orderings of the data to identify differences. Table 27 highlights a few of the most common nonparametric statistical tests.

Table 27. Common nonparametric tests.

Dependent Variable Scale	Levels of Categorical Independent Variable	Test
Categorical	Percentage of the target population	binomial test
	2 matched groups	McNemar's test for 2x2 contingency tables
	2 or more independent groups	Chi-square test for contingency tables
	2 independent groups	Fisher's exact test
Continuous	2 matched groups	Wilcoxon rank-sum test / Wilcoxon sign-rank test
	2 independent groups	Mann-Whitney U test / Wilcoxon-Mann Whitney test
	2 or more matched groups	Friedman test
	2 or more independent groups	Kruskal-Wallace H test

CONFIDENCE INTERVALS

A *confidence interval* is the numerical interval around a statistic, most commonly the mean, that is estimated from a sample and has a certain confidence of including the true value for the population. Confidence intervals have an *Upper Confidence Limit* (UCL) and a *Lower Confidence Limit* (LCL).

Testing.

For a population mean:

Confidence Interval = Sample mean ±
t-value times (sample standard deviation divided by (square root of the number of samples))

Confidence intervals are comparable to some types of parametric tests that are calculated with means, standard deviations, and *degrees-of-freedom* (sample size minus 1, in most cases).

THE RING-TOSS GAME

Consider the analogy of a nearsighted man playing a ring-toss game at a carnival.

The peg that the man would toss a ring at represents the mean of the population. The location of the peg, representing the value of the population mean, is fixed. The man can't change it. The diameter of the peg represents the inherent variability of the population mean. The man can't change that either. The fuzziness with which the man sees the peg represents the additional variation associated with sampling, measurement, and environmental variability. The man can change that. If the man put on corrective lenses, it would represent his effort to control excess variation.

Confident kitten.

The ring that the man tosses represents the confidence interval. The diameter of the ring represents the size of the confidence interval (i.e., the UCL minus the LCL). The man is able to select from several different sizes of rings, representing different probabilities that the ring he tosses will cover the peg.

If the man wanted to be very confident that he could toss a ring over the peg, he would use a large ring resulting in fewer misses (i.e., when the ring doesn't cover the peg). Misses represent instances in which the confidence interval doesn't include the population mean. If the ring (the confidence interval) becomes too large, though, the game becomes meaningless. Thus, there must be some limits on how large the ring should be (how many misses are acceptable).

By convention, most statistical inferences, including confidence intervals, use a 95% confidence level, equivalent to a 5% miss rate in the ring-toss game. Sometimes either a 90% level (resulting in a smaller confidence interval and more misses) or a 99% level (resulting in a larger interval and fewer misses) is used. A 90% level would be more appropriate when the consequences of not including the true population value in the interval are relatively minor. Confirmatory inferences, on the other hand, often use a 99% confidence level. When in doubt, use 95%.

Testing.

The most important thing to take from this analogy is that just as the tossed ring tries to cover the stationery peg, the calculated interval tries to include the fixed population mean. It bothers statisticians when people say that the population mean falls within a confidence interval. It's a subtle but important point.

OTHER INTERVALS

Confidence intervals are the most commonly used type of statistical interval but not the only type. There are also *prediction intervals* and *tolerance intervals*.

- ❦ **Confidence intervals** provide a certain amount of confidence that a population parameter, like the *mean*, will be included within its limits.
- ❦ **Prediction intervals** provide a certain amount of confidence that an individual prediction from a statistical model will be included within its limits.
- ❦ **Tolerance intervals** provide a certain amount of confidence that certain proportions of a population will be included within its limits. (Note: Statistical tolerance intervals are different from engineering tolerances used to specify design acceptability of materials and processes.)

All three intervals are calculated as:

$$\text{Interval} = \text{Mean} \pm (\text{standard deviation} * \text{probability factor})$$

For *confidence intervals*, the probability factor is $t_{1-\alpha/2,n-1} * \sqrt{(1/n)}$
For *prediction intervals*, the probability factor is $t_{1-\alpha/2,n-1} * \sqrt{(1+1/n)}$
For *tolerance intervals*, the probability factor is
$z_{(1-p)/2} * \sqrt{(n+1)/n} * \sqrt{(n-1)/\chi^2_{(\alpha,n-1)}} * \sqrt{1+((n-3-\chi^2_{(\alpha,n-1)})/(2(n+1)^2))}$

For the same confidence and number of samples, *confidence intervals* will be more narrow than *prediction* or *tolerance intervals*. Don't worry about calculating any of the intervals, just be aware if a statistical presentation is using one of them because they mean very different things.

WHY USE INTERVALS

Statistical intervals are just another way to express the results of a statistical calculation. The interval provides a new perspective for looking at the estimated value by essentially substituting the interval range for the standard deviation.

For example, instead of reporting estimates of a population mean and standard deviation, an interval containing the range of possible values of the mean at a specified level of confidence could be reported. This removes the focus on a specific value for the mean, although this is often reported as well, and forces attention on the uncertainty of the estimate. Some people like this.

Testing.

For the same reason, some people dislike putting confidence limits around means they calculate. Limits show how imprecise data, and statistics calculated from them, actually are. They prefer one number to remember in making their decisions and uncertainty is an afterthought.

How the information is formatted is a matter of preference. Nevertheless, if you are going to make an informed decision, you have to know not just the evidence, but the reliability of the evidence as well.

> *Confidence interval. The time interval between when you graduate and when you get onto the next thing (e.g., job, college, professional school) and realize you don't really know anything.*
>
> C. Kufs, Confidence Interval, posted on 2-22-2012 at statswithcats.net.

ANOVAs

So what do you do if you have more than two populations or more than one phenomenon or some other weird combinations of data you want to compare? You use an *Analysis of Variance* (*ANOVA*).

ANOVA includes a very large variety of statistical designs used to analyze differences in groups. It is a generalization of the t-test of a factor (called *main effect* or *treatments* in ANOVA) to more than two groups (called *levels* in ANOVA).

In an ANOVA, the variances in the levels of factors being compared are partitioned between variation associated with the factors in the design (called *model variation*) and random variation (called *error variation*). ANOVA is conceptually similar to multiple two-population t-tests, but produces fewer type I (false positive) errors.

While t-tests use t-values from the t-distribution, ANOVAs use F-tests from the F-distribution. An F-test is the ratio of the model variation to the error variation. When there are only two means to compare, the t-test and the ANOVA F-test are equivalent according to the relationship $F = t^2$.

There are many types of ANOVA designs.

One-way ANOVA is used to test for differences among three or more independent levels of one effect. *Multi-way ANOVAs* (sometimes called *factorial ANOVAs*) are used to test for differences between two or more effects. A *two-way ANOVA* tests two effects, a *three-way ANOVA* tests three effects, and so on.

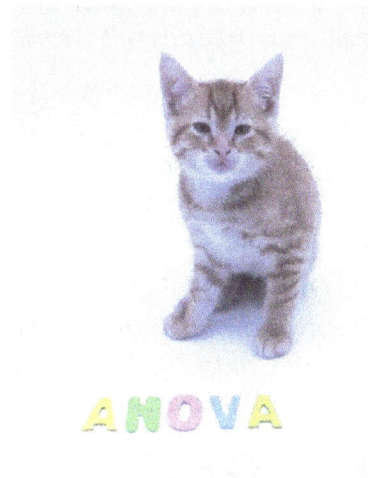

ANalysis Of VAriance.
It's what we kittens say
to impress cats.

Testing.

Multi-way ANOVAs have the advantage of being able to test the significance of *interaction effects*. Interaction effects occur when two or more effects combine to affect measurements of the phenomenon. Interactions are a violation of the *additivity assumption*, so they are not always welcome news. All that means, though, is that the estimates of the individual effects can't be summed to estimate the total effect.

There are also numerous other types of ANOVA designs, some of which are too complex to explain adequately in a single sentence, or even five. Here are a few of these designs.

ANOVAs for repeated measures (also called *repeated measures* or *within-subjects ANOVAs*) are used when the same subjects are used for each treatment effect, as in a *longitudinal study*.

ANOVAs for variance control use statistical design elements to control extraneous variance instead of just the three Rs of *reference*, *replication*, and *randomization* (Chapter 3). The significance of the design elements determined in the ANOVA is not important so long as they control variability in the *main effects*. If the design element is a nominal-scale variable, it is called a *block effect* and the model is called a *blocked ANOVA* or an *ANOVA with blocking*. If the design element is a continuous-scale variable, it is called a *covariate* and the model is called an *Analysis of Covariance (ANCOVA)*.

ANOVAs for random effects assume that the levels of a main effect are sampled from a population of possible levels so that the results can be extended to other possible levels. These models are relatively uncommon.

ANOVAs for multiple dependent variables, called *multivariate analysis of variance* (*MANOVA*) models are used when there is more than one *dependent variable* that characterizes the phenomenon.

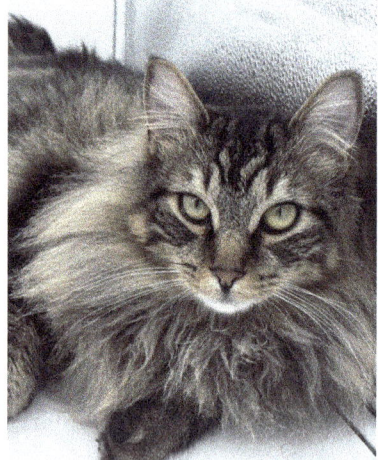

OK, that's enough of that.

PROBABILITY VALUES

Traditionally, *test statistics* were calculated and then compared to a standard listed in a table at the back of most statistics textbooks. If the test statistic was greater than the standard, it meant that the difference was larger than what might have been expected by chance, and was said to be *statistically significant*. If the test statistic was not greater than the standard, it meant that the null hypothesis could not be rejected so the difference was said to be *non-significant*.

Testing.

Today, statistical software reports exact probabilities, *p-values*, for statistical tests instead of relying on manual comparisons. It's another modern convenience that most people just don't seem to appreciate enough.

The *p value*, or calculated probability, is the probability of finding the observed, or more extreme, results when the *null hypothesis* (**H₀**) of a study question is true. The null hypothesis is usually an hypothesis of *no difference* between groups, that is defined before the start of an experiment. The *alternative hypothesis* (**H₁** or **H_A**) is the opposite of the null hypothesis, usually the status quo hypothesis you set out to investigate.

CONFIDENCE

The results of statistical tests are sometimes in error, but fortunately, you have some control over the rates at which errors occur. There are four possibilities for the results of a statistical test.

- **True Positive**. The statistical test fails to reject a null hypothesis that is true in the population.
- **True Negative**. The statistical test rejects a null hypothesis that is false in the population.
- **False Positive**. The statistical test rejects a null hypothesis that is true in the population. This is called a *Type I error* and is represented by α. The *Type I error rate* you will accept for a test is called the *confidence*, **1-α**. For example, a 95% confidence means that there is a 5% chance of making a Type I error (α **= 0.05**).
- **False Negative**. The statistical test fails to reject a null hypothesis that is false. This is called a *Type II error* and is represented by β. The ability of a particular comparison to avoid a *Type II error* is represented by **1-β** and is called the *power* of the test. Typically, *power* should be at least 0.8 for a 20% *Type II error rate*.

The term *significance level* (α, alpha) refers to a pre-chosen Type I error rate that you are willing to accept for the experiment. If the calculated *p value* for a test is less than the chosen significance level α, then the null hypothesis is rejected.

Rejecting a null hypothesis is what makes a test *significant*. However, significance does NOT imply that the results are *meaningful*. For example, a test might find a 10-unit difference between the means of two

> *A difference is a difference only if it makes a difference.*
>
> Darrell Huff, **How to Lie with Statistics**, p. 58.

populations to be *significant* even though the smallest size of a difference that can be measured is 100-units. *Significant* but not *meaningful*.

Testing.

Statistical tests have to be designed so that *significant* test results are also *meaningful*. This involves selecting an appropriate *sample size* (**n**) for a prespecified *significance level* (α) and the anticipated difference between the populations that will be tested. The challenge is that while the sample size and significance level are specified before the experiment is conducted, the difference between the populations isn't known until afterwards.

When the specifications of a nonsignificant statistical test are adjusted and repeated just to obtain a significant result, it is called *p-hacking*. p-hacking is a very bad thing to do because it compromises the randomness required by statistical procedures.

POWER

The *power* of a test is one minus the probability of *type II error* (β, *beta*). Power is only relevant when a null hypothesis is rejected. Ideally, *confidence* and *power* should be maximized, but they are related.

After you conduct the test, there are two pieces of information you need to determine—the sensitivity of the test to detect differences, called the *effect size*, and the *power* of the test. The *power* of the test depends on the *sample size*, the *confidence*, and the *effect size*.

The *effect size* provides insight into whether the test results are *meaningful*. *Meaningfulness* is important because a test may be able to detect a difference far smaller than what might be of interest, such as a difference in mean student heights less than a millimeter. Perhaps surprisingly, the most common reason for being able to detect differences that are too small to be meaningful is having too large a sample size.

I have the power ...
at least in a
significant statistical test.

More samples are not always better.

STATISTICAL TESTING IN ACTION

When you design a statistical test, you specify the hypotheses including the number of populations and directionality, the type of test, the confidence, and the number of observations in your representative sample of the population.

From the sample, you calculate the mean and standard deviation. You calculate the test statistic and compare it to standard values in a table based on the distribution. If the test statistic is greater than the standard value, you reject the null hypothesis.

Testing.

When you reject the null hypothesis, the comparison is said to be *significant*. If the test statistic is less than the standard value, you *fail to reject* the null hypothesis and the comparison is said to be *nonsignificant* or *not significant*.

Statistical software now provides exact probabilities that the null hypothesis is false so tables are no longer necessary.

Here are a few examples of what the process of statistical testing looks like for comparing a population mean to a constant.

Imagine you are analyzing the average height of male high school freshmen in Minneapolis school district #1. You want to know for an article you are writing for the school newspaper how their average height compares to the average height of 9th to 11th century Vikings (their mascot). Turn-of-the-century Vikings were typically about 5'9" or 69 inches (172 cm) tall, based on internet research.

This comparison doesn't need to be too rigorous. The only possible negative consequence to the test is it being reported as breaking news by podcasters. You'll accept a false positive rate (i.e., 1-confidence, α) of 0.10.

ONE POPULATION TESTS

All z-tests and t-tests involve either one or two populations and only one phenomenon. The population is represented by the nominal-scale, independent variable. The measurement of the phenomenon is the dependent variable, which can be measured using an ordinal, interval, or ratio scale. The formula for a t-test is:

Follow this example.

$$\text{t-value} = \frac{\text{mean of measurements from a population minus constant or mean of second population}}{\dfrac{\text{standard deviation of measurements from the population(s)}}{\sqrt{\dfrac{\text{number of measurements from the population} - 1}{}}}}$$

For this one-population test, you would be comparing the average of the measurements in the population of high schoolers to a constant. You do this using the formula for a one-population *t-test value* (or a *z-test value*) to calculate the t-value for the test.

The Normal distribution and the t-distribution are symmetrical so it doesn't matter if the numerator of the equation is positive or negative.

Testing.

Then compare that calculated value to a table of values for the t-distribution (for the appropriate number of *tails*, the *confidence* (**1-α**), and the *degrees of freedom* (the number of samples of the population minus 1). If the calculated t-value is larger than the table t-value, the test is *significant*, meaning that the mean and the constant are statistically different. If the table t-value is larger than the calculated t-value, the test is *not significant*, meaning that the mean and the constant are statistically the same.

NONDIRECTIONAL TESTS

Say you don't know many freshmen boys but you don't think they are as tall as Vikings. You certainly don't think of them as rampaging Vikings, although hormonally, they may be. They're younger so maybe they're shorter. Then again, they've grown up having better diets and medical care so maybe they're taller.

I think the Freshman and the Vikings were about the same height.

You make your research hypothesis that Freshmen are not likely to be the same height as Vikings.

The null hypothesis you want to test is:

Height of Freshmen = Height of Vikings

which is a nondirectional test. If you reject the null hypothesis, the alternative hypothesis:

Height of Freshmen ≠ Height of Vikings

is probably true of the Freshmen.

Say you then measure the heights of 10 freshmen and you get:

63.2, 63.8, 72.8, 56.9, 75.2, 70.8, 68.0, 64.0, 61.4, 65.2

(I used only ten values because I didn't want to type so much.) The measurements average 66 inches with a standard deviation of 5.3 inches. The t-value would be equal to:

(Freshmen height - Viking height) / (standard deviation / (√number of samples))

t-value = (66 inches - 69 inches) / (5.3 inches / (√10 samples))

t-value = -1.790

Testing.

Ignore the negative sign; it doesn't matter in this case because the test is non-directional.

In this comparison, the calculated t-value (1.79) is less than the table t-value (**t(2-tailed, 90% confidence, 9 degrees of freedom) = 1.833**) so the comparison is *not significant*. The comparison might look something like the diagram shown in Figure 56.

There is no statistical difference in the average heights of Freshmen and Vikings. Both are around 5'6" to 5'9" tall. That isn't to say that there weren't 6'0" Vikings, or Freshmen, but as a group, the Freshmen are about the same height as a band of berserkers. I'm sure that there are high school principals who will agree with this.

Obtaining t-Values
Remember, when obtaining t-values from a table or spreadsheet formula:
Confidence = $1 - \alpha$
Degrees of freedom = $n - 1$
$t_{(\alpha \text{ for 1-tails})} = t_{(\alpha/2 \text{ for 2-tails})}$

When you get a nonsignificant test, it's a good practice to conduct a *power analysis* to determine what protection you had against false negatives. For a t-test, this involves rearranging the t-test formula to solve for **t_beta**:

$$t_{beta} = (\sqrt{(\text{sample size})}/\text{standard deviation}) * \text{difference} - t_{alpha}$$

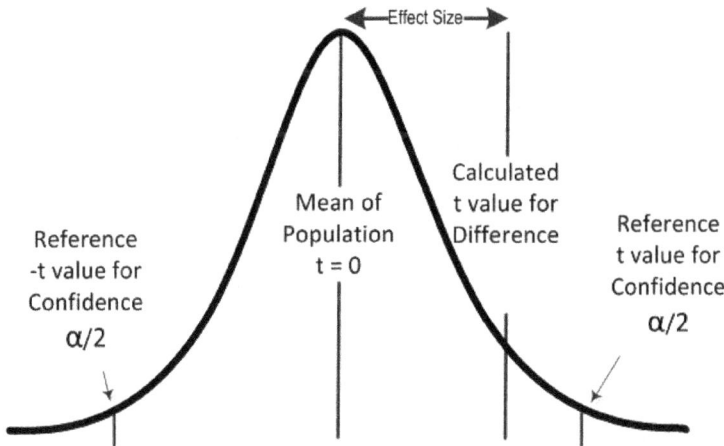

NON-SIGNIFICANT ONE POPULATION, NONDIRECTIONAL, T-TEST

Figure 56. Diagram of a non-significant one-population nondirectional t-test.

Testing.

The t_{alpha} is for the confidence you selected, in this case 90%. Then you look up the t-value you calculated to find the probability for beta. It's a cumbersome but not difficult procedure. In this example, the calculated t_{beta} would have been 1.24 so the power would have been 88%. That's not bad. Anything over 80% is usually considered acceptable.

Most statistical software will do this calculation for you. You can increase power by increasing the sample size or the acceptable Type 1 error rate (decrease the confidence) BEFORE conducting the test.

So if everything were the same (i.e., mean of students = 66 inches, standard deviation = 5.3 inches) except that you had collected 30 samples instead of 10 samples, the calculation would be:

t-value = (69 inches—66 inches) / (5.3 inches / (√30 samples))

t-value = 3.10

$t_{(2\text{-tailed, 90\% confidence, 29 degrees of freedom})}$ = 1.699

If you had collected 100 samples:

t-value = (69 inches—66 inches) / (5.3 inches / (√100 samples))

t-value = 5.66

$t_{(2\text{-tailed, 90\% confidence, 99 degrees of freedom})}$ = 1.660

These comparisons are both significant, and might look something like the diagram shown in Figure 57.

More samples give you better *resolution*.

DIRECTIONAL TESTS

Now say, in a different reality, you know that many of those freshmen boys grew up on farms and they're pretty buff. You even think that they might just be taller than the Vikings of a millennium ago. Therefore, your research hypothesis is that Freshmen are likely to be taller than the war-faring Vikings. The *null hypothesis* you want to test is:

Height of Freshmen ≤ Height of Vikings

which is a *directional test*. If you reject the null hypothesis, the alternative hypothesis:

Height of Freshmen >Height of Vikings

is probably true of the Freshmen.

So you measure the heights of 10 freshmen and get:

72.4, 71.1, 75.4, 69.0, 75.7, 73.3, 76.0, 58.8, 70.4, 78.6

Testing.

SIGNIFICANT ONE POPULATION,
NONDIRECTIONAL, T-TEST

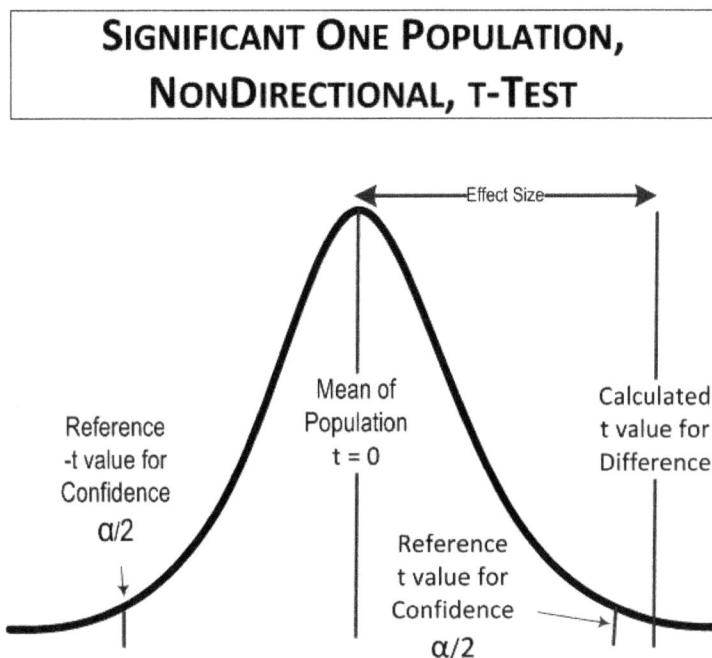

**Figure 57. Diagram of a significant
one-population nondirectional t-test.**

The measurements average 71.2 inches with a standard deviation of 5.3 inches. The t-value would be equal to:

(Freshmen height - Viking height) / (standard deviation / (√number of samples))

t-value = (72 inches - 69 inches) / (5.3 inches / (√10 samples))

t-value = 1.790

In this comparison, the table t-value you would use is for a one-tailed (directional) test at 90% confidence for 10 samples is

$t_{(1\text{-tailed}, \alpha = 0.1, 9 \text{ degrees of freedom})}$ = 1.383

For comparison, the values of $t_{(2\text{-tailed}, 0.90 \text{ confidence}, 9 \text{ degrees of freedom})}$, which was used in the first example, and $t_{(1\text{-tailed}, 0.95 \text{ confidence}, 9 \text{ degrees of freedom})}$ are both equal to 1.833. The reason those two t-values are equal is that the one-tailed value only considers half of the t-distribution area compared to a two-tailed test, which considers both tails. That means that if you use a directional test you can have a greater confidence (90% versus 95%).

The table t-value you would use, $t_{(1\text{-tailed}, 0.9 \text{ confidence}, 9 \text{ degrees of freedom})}$, is equal to 1.383. which is smaller than the calculated t-value, 1.790, so the comparison is *significant*. The comparison looks something like the diagram shown in Figure 58.

Testing.

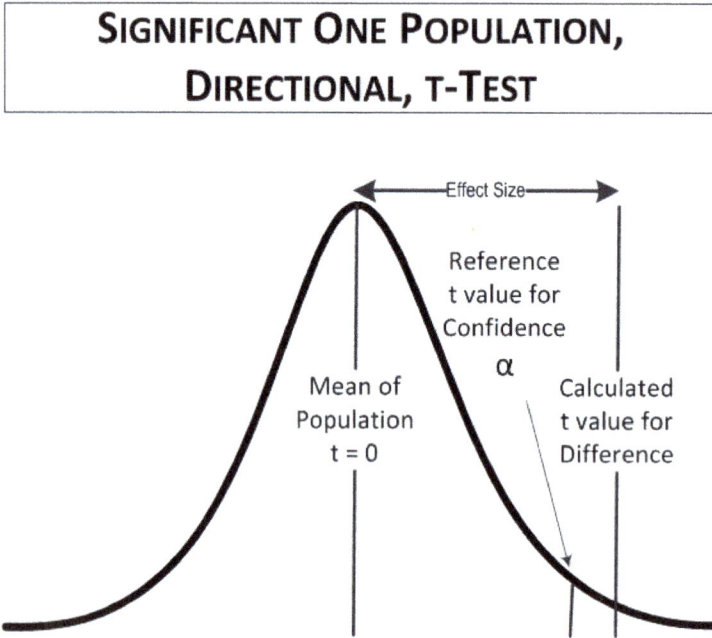

SIGNIFICANT ONE POPULATION, DIRECTIONAL, T-TEST

Figure 58. Diagram of a significant one-population directional t-test.

In this comparison, the Freshmen are on average at least 3 inches taller than their frenzied Viking ancestors. Genetics, better diet, and healthy living win out.

But what if the farm boys averaged only 71 inches:

(Freshmen height - Viking height) / (standard deviation / (√number of samples))

t-value = (71 inches - 69 inches) / (5.3 inches / (√10 samples))

t-value = 1.193

The table t-value you would use, $t_{(1\text{-tailed, 0.9 confidence, 9 degrees of freedom})}$, is equal to 1.383. which is larger than the calculated t-value, 1.193, so the comparison is *not significant*. The comparison looks something like the diagram shown in Figure 59.

And that's what one-population t-tests look like. Now for some two-population tests.

TWO-POPULATION TESTS

In a two-population test, you compare the average of the measurements in the first population to the average of the measurements in the second population, using the formula:

$$t = \frac{\overline{x}_1 - \overline{x}_2}{\sqrt{\left(\frac{\left((n_1-1)*s_1^2\right) + \left((n_2-1)*s_2^2\right)}{(n_1-1 + n_2-1)}\right)\left(\frac{1}{n_1} + \frac{1}{n_2}\right)}}$$

Testing.

NON-SIGNIFICANT ONE POPULATION, DIRECTIONAL, t-TEST

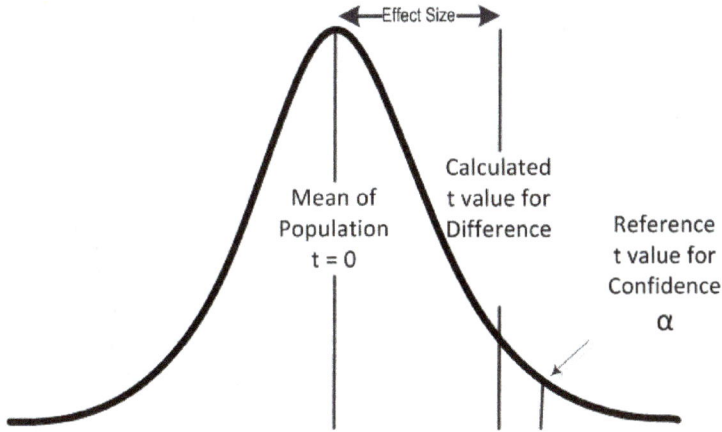

Figure 59. Diagram of a non-significant one-population directional t-test.

This is more complicated than the formula for a one-population test because you can have different standard deviations and different numbers of measurements in each population.

Here's what's happening. The numerator (top part of the formula) is the same in both t-test formulas. The leftmost term in the denominator calculates a weighted average of the variances, called a *pooled variance*.

$$s^2_{Pooled} = \frac{((n_1-1)*s^2_1)+((n_2-1)*s^2_2)}{(n_1-1+n_2-1)}$$

I'm balanced down here.

If the number of measurements taken of the two populations is the same, the test design is said to be *balanced*. If the variances of the measurements in the two populations are the same, the leftmost term in the denominator reduces to s^2. So, the formula for a balanced two-population t-test with equal variances is:

$$t = \frac{\overline{x}_1 - \overline{x}_2}{s\sqrt{\frac{2}{n}}}$$

 Testing.

I'm balanced up here.

Much more simple but not as useful as the more complicated formula. You might be able to control the number of samples from the populations but you can't control the variances.

Once you calculate a t-value, the rest of the test is similar to a one-population test. You compare the calculated t to a t-value from a table or other reference for the appropriate number of tails, the confidence (**1-α**), and the degrees of freedom (the number of samples minus 1).

If the calculated t-value is larger than the table t-value, the test is *significant*, meaning that the means are statistically different. This is shown in Figure 60. If the table t-value is larger than the calculated t-value, the test is *not significant*, meaning that the means are statistically the same. This is shown in Figure 61.

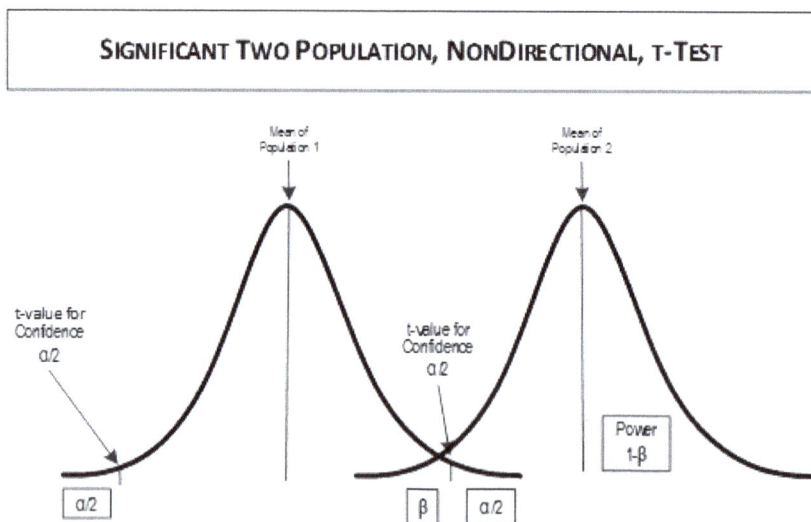

Figure 60. Diagram of a significant two-population nondirectional t-test.

WHICH CLASS TO TAKE?

Now imagine this. You're a sophomore statistics major at Faber College and you need to sign up for the dreaded Stats 102 class. The class is taught in the Fall and the Spring by two different instructors (Dr. Statisticus and Prof. Modearity) as either three, one-hour sessions on Mondays, Wednesdays, and Fridays, or as two, hour-and-a-half sessions on Tuesdays and Thursdays. You wonder if it makes a difference which class you take.

Testing.

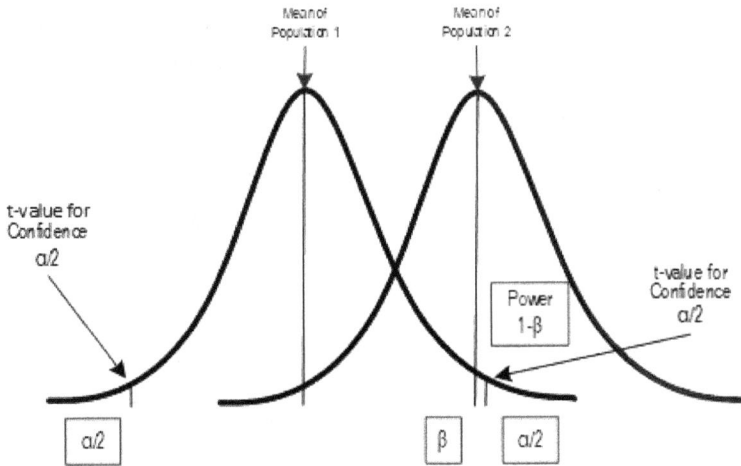

Figure 61. Diagram of a non-significant
two-population nondirectional t-test.

Having completed Stats 101, you know enough to do your own research. You start by getting the grades from the classes that were taught last year. Table 28 shows the data.

What class should you take to have the best chance to get a high grade?

Table 28. Grades for the example Stats 102 classes.

STATS 102	Fall Semester				Spring Semester			
	Dr. Statisticus		Prof. Modearity		Dr. Statisticus		Prof. Modearity	
	MWF	TuTh	MWF	TuTh	MWF	TuTh	MWF	TuTh
Students	96	95	92	93	93	93	90	94
	93	93	90	88	87	91	88	89
	89	90	84	87	86	89	86	86
	87	88	81	85	85	89	84	86
	85	87	80	81	84	88	83	84
	83	86	78	80	83	86	81	80
	82	82	73	78	82	84	77	79
	78	78	69	68	79	77	73	72
	75				76		68	
					69			
Mean	85	87	81	82	82	87	81	84
SD	6.6	5.2	7.2	7.1	6.2	4.6	6.8	6.3
n	9	8	8	8	10	8	9	8

Testing.

DATA

Dr. Statisticus gave out higher grades than Prof. Modearity in both semesters. He gave higher grades for the two-day classes than the three-day classes. Prof. Modearity gave out higher grades in the Spring and for the two-day classes. But, are these results having such small differences both significant and meaningful?

On the other side of the coin, only one person flunked (grade below 75) Dr. Statisticus' classes but six people flunked Prof. Modearity's classes. Three students flunked in the Fall while four students flunked in the Spring. Two people flunked Tuesday-Thursday classes and five people flunked Monday-Wednesday-Friday classes. This is complicated.

Table 29. Differences between the example Stats 102 classes.

	Semester		Instructor		Days	
	Fall	84.0	Dr. Statisticus	85.6	MWF	82.4
	Spring	83.5	Prof. Modearity	82.0	TuTh	85.2
Difference	0.5		3.5		2.7	

Looking at the averages (Table 29), you think that taking Dr. Statisticus' Tuesday-Thursday class in the fall would be your best bet. However, is a two or three point difference worth the class conflicts and scheduling hassles you might have? Does it really matter? Maybe it's time for some statistical testing.

These would be two-population tests because you have to compare two semesters, two instructors, and two class lengths in separate tests. You have no solid expectations for what the best semester, instructor, or class days would be despite your first glance at the averages for the classes.

I've looked on the other side of all these coins but I haven't found anything.

You decide to accept a false positive rate (i.e., 1-confidence, α) of 0.05. Your null hypotheses are:

$$\mu_{\text{FALL SEMESTER}} = \mu_{\text{SPRING SEMESTER}}$$
$$\mu_{\text{DR. STATISTICUS}} = \mu_{\text{PROF. MODEARITY}}$$
$$\mu_{\text{MWF}} = \mu_{\text{TuTh}}$$

SEMESTERS

Now for some calculations. First, the semesters.

$$\bar{X}_{\text{FALL SEMESTER}} = 84.0$$

Testing.

$\overline{X}_{\text{SPRING SEMESTER}}$ = 83.5
$n_{\text{FALL SEMESTER}}$ = 33
$n_{\text{SPRING SEMESTER}}$ = 35
$S^2_{\text{FALL SEMESTER}}$ = 49.7 (S = 7.05)
$S^2_{\text{SPRING SEMESTER}}$ = 41.7 (S = 6.46)

$$t = \frac{\overline{X}_{\text{Fall}} - \overline{X}_{\text{Spring}}}{\sqrt{\left(\dfrac{\left((n_{\text{Fall}}-1)s^2_{\text{Fall}}\right) + \left((n_{\text{Spring}}-1)s^2_{\text{Spring}}\right)}{(n_{\text{Fall}}-1 + n_{\text{Spring}}-1)}\right)\left(\dfrac{1}{n_{\text{Fall}}} + \dfrac{1}{n_{\text{Spring}}}\right)}}$$

$$t = \frac{84.0 - 83.5}{\sqrt{\left(\dfrac{\left((33-1)49.7\right) + \left((35-1)41.7\right)}{(33-1 + 35-1)}\right)\left(\dfrac{1}{33} + \dfrac{1}{35}\right)}}$$

$$t = 0.33$$

And the tabled value is:

$$t_{\text{(2-tailed, 0.05 confidence, 65 degrees of freedom)}} = 1.997$$

So there is no statistically significant difference between the Fall semester classes and the Spring semester classes, as shown in Figure 62.

You could do these calculations in Excel with the formula:

$$=\text{T.TEST(array1,array2,tails,type)}$$

Where type=3 is a t-test for two-samples with unequal variances.

There are also online sites for the calculations, such as evanmiller.org, from which the graphics for this example were produced.

INSTRUCTORS

Now for the instructors:

$\overline{X}_{\text{DR. STATISTICUS}}$ = 85.4
$\overline{X}_{\text{PROF. MODEARITY}}$ = 82.0
$n_{\text{DR. STATISTICUS}}$ = 35
$n_{\text{PROF. MODEARITY}}$ = 33
$S^2_{\text{DR. STATISTICUS}}$ = 37.5 (S = 6.12)
$S^2_{\text{PROF. MODEARITY}}$ = 48.5 (S = 6.96)

Mom is the best teacher.

   Testing.

Question: Does the average value differ across two groups?

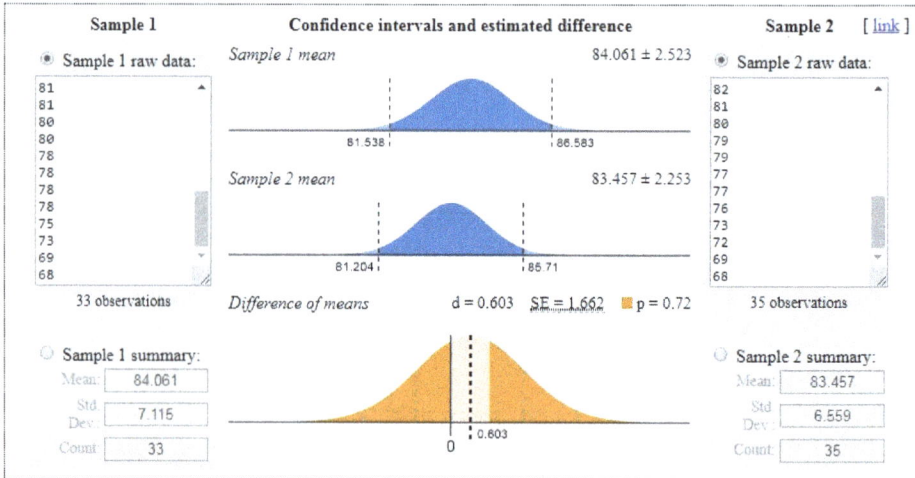

Verdict: No significant difference

Hypothesis: ● $d = 0$ ○ $d \leq 0$ ○ $d \geq 0$

Confidence: [＿＿＿＿＿＿＿＿] 95%

If the experiment is repeated many times, the confidence level is the percent of the time each sample's mean will fall within the confidence interval.

Figure 62. t-test for the difference between classes held during the Fall semester and the Spring semester.

$$t = \frac{\bar{X}_{Statisticus} - \bar{X}_{Modearity}}{\left(\frac{\left((n_{Statisticus}-1)s^2_{Statisticus}\right) + \left((n_{Modearity}-1)s^2_{Modearity}\right)}{(n_{Statisticus}-1 + n_{Modearity}-1)}\right)\left(\frac{1}{n_{Statisticus}} + \frac{1}{n_{Modearity}}\right)}$$

$$t = \frac{85.4 - 82.0}{\left(\frac{((35-1)37.5) + ((33-1)48.5)}{(35-1 + 33-1)}\right)\left(\frac{1}{35} + \frac{1}{33}\right)}$$

$$t = 2.08$$

And the tabled value is:

$$t_{(2\text{-tailed, }0.05\text{ confidence, }66\text{ degrees of freedom})} = 1.996$$

Testing.

Question: Does the average value differ across two groups?

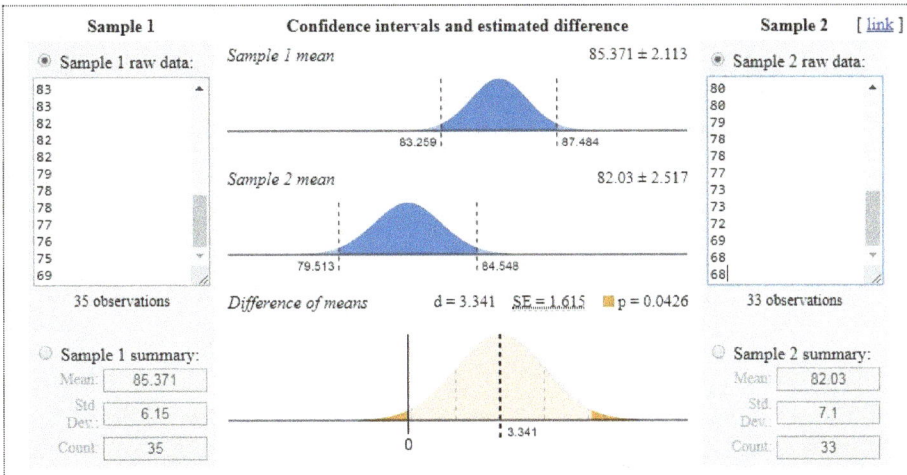

Figure 63. t-test for the difference in instructors.

So there is a statistically significant difference between instructors as shown in Figure 63. Dr. Statisticus gives higher grades than Prof. Modearity.

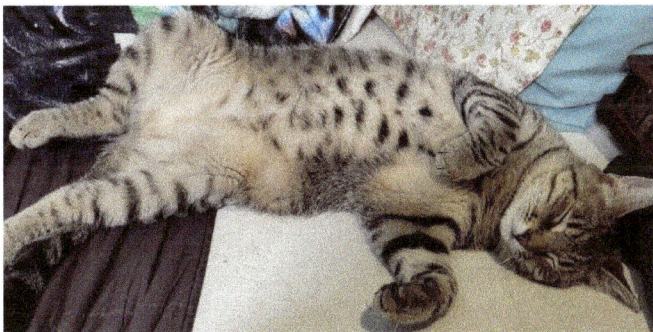

Yes, I'm still following ya.

DAYS OF THE WEEK

Now for the days of the week:

$$\bar{X}_{MWF} = 82.4$$
$$\bar{X}_{TUTH} = 85.2$$
$$n_{MWF} = 36$$
$$n_{TUTH} = 32$$

Testing.

S^2_{MWF} = 47.8 (S = 6.91)
S^2_{TuTH} = 39.4 (S = 6.28)

$$t = \frac{\bar{X}_{MWF} - \bar{X}_{TuTh}}{\left(\frac{\left((n_{MWF}-1)s^2_{MWF}\right) + \left((n_{TuTh}-1)s^2_{TuTh}\right)}{(n_{MWF}-1 + n_{TuTh}-1)}\right)\left(\frac{1}{n_{MWF}} + \frac{1}{n_{TuTh}}\right)}$$

$$t = \frac{82.4 - 85.2}{\left(\frac{\left((36-1)47.8\right) + \left((32-1)39.4\right)}{(36-1 + 32-1)}\right)\left(\frac{1}{36} + \frac{1}{32}\right)}$$

$$t = -1.68$$

So there is no statistically significant difference between the one-hour classes on Mondays, Wednesdays, and Fridays and the hour-and-a-half classes on Tuesdays and Thursdays. This is shown in Figure 64.

Question: Does the average value differ across two groups?

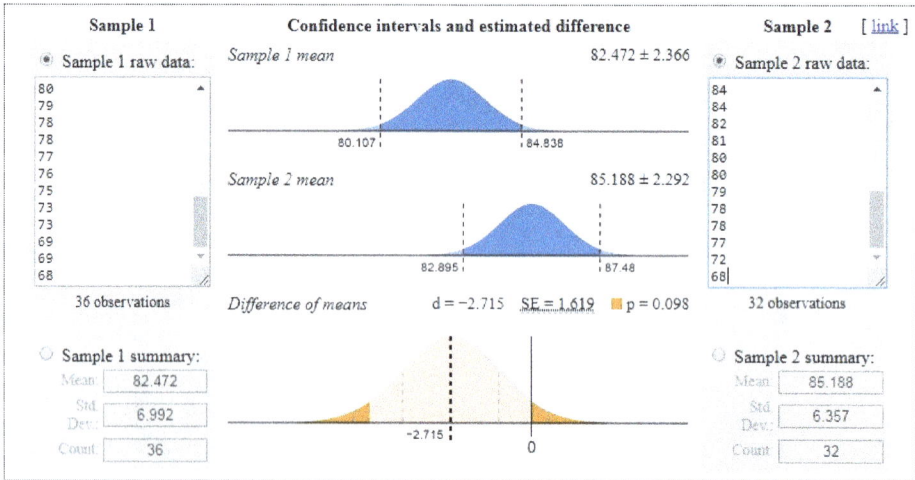

Figure 64. t-test for the difference between days.

Testing.

		Mean	SD	n
Semester	**Fall Semester**	84.0	7.1	33
	Spring Semester	83.5	6.5	35
	Calculated t-value		0.33	
			One Tail	**Two Tail**
	Table t-value		1.67	2.00
	Exact Probability		0.37	0.74
		Mean	**SD**	**n**
Instructor	**Dr. Statisticus**	85.4	6.1	35
	Prof. Modearity	82.0	7.0	33
	Calculated t-value		2.08	
			One Tail	**Two Tail**
	Table t-value		1.67	2.00
	Exact Probability		0.02	0.04
		Mean	**SD**	**n**
Days	**MWF**	82.4	6.9	36
	TuTh	85.2	6.3	32
	Calculated t-value		1.68	
			One Tail	**Two Tail**
	Table t-value		1.67	2.00
	Exact Probability		0.05	0.10

Table 30. Summary of statistical comparisons of instructor, semester, and schedule.

DECISION

Table 30 provides a summary of the three tests.

So take Dr. Statisticus' class whenever it fits in your schedule.

We're all going to take Stats 102.

 Testing.

ANOVA APPROACH

There's another, more efficient way to analyze this data set—ANOVA. The design would be called a *three-way design* because there are three factors: instructor; semester; and days of the week. Figure 65 is a block diagram of the ANOVA design for the example.

The ANOVA approach has several advantages over the three t-tests. It is more efficient even though there are more statistical tests being conducted because the tests are conducted in a hierarchical manner. If an overall test of an effect is nonsignificant, the follow-up tests of the effect levels are not conducted.

The ANOVA would also allow *interaction* effects to be tested. It controls for error rates inflated by the number of t-tests. And, it usually has greater statistical power to detect differences.

Looks like a multi-way ANOVA design.

The downsides are that you have to have knowledge and experience with ANOVA designs and appropriate software.

Using an ANOVA approach would allow the design to be expanded in a variety of ways. For example, more than two instructors or more than two semesters could be compared. More than two populations could be compared, such as other departments in Faber College that taught Stats 102 or even other colleges.

If the scores for the students were recorded every month of the semester, the data could be analyzed with a *Repeated-Measures ANOVA*. If a survey were conducted of

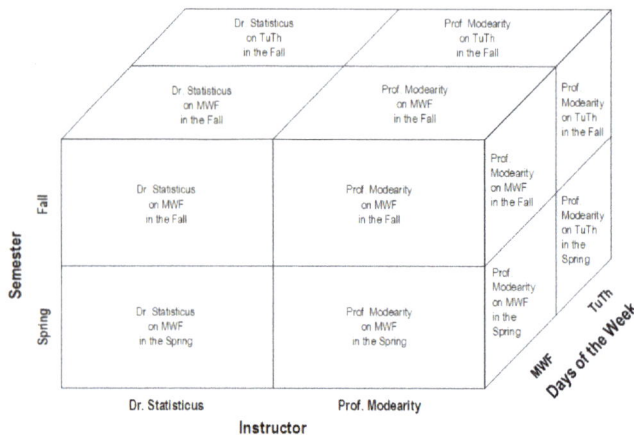

Figure 65. ANOVA design for Stats 102 example.

Testing.

the students' overall satisfaction with the courses, the additional metric could be analyzed as a *Multivariate Analysis of Variance* (*MANOVA*).

The students' year in college (freshman, sophomore, junior, or senior, an ordinal-scale measure) could be added as an effect to control variance, called a *blocking factor*, making it a *blocked ANOVA*. Students' GPA (grade point average, a continuous scale measure) could also be added to control variance as a *covariate*, making it an *Analysis of Covariance* (*ANCOVA*) design.

If only Dr. Statisticus and Prof. Modearity were levels of the instructor main-effect, it would be called a *fixed-effect design*. However the instructors main-effect could be specified as a *random effect* allowing other instructors to be considered part of the population that included Dr. Statisticus and Prof. Modearity. Then inferences could be made to any instructor who taught Stats 102. If a design includes both fixed and random effects, it is called a *mixed-effects* design.

WHAT TO LOOK FOR IN STATISTICAL TESTS

Believe it or not, there's quite a bit more to learn about all of these topics if you go further in statistics. There are special t-tests for proportions, regression coefficients, and samples that are not independent (called *paired-sample t-tests*). There are tests based on other distributions besides the Normal and t-distributions, such as the binomial and chi^2 distributions. There are also quite a few nonparametric tests, usually based on ranks. And, of course, there are many topics on the mathematics end and on more metaphysical concepts like *meaningfulness*.

I can't wait until I'm old enough to learn to drive.

Statistical testing is more complicated than portrayed by some people but it's still not as formidable as, say, driving a car.

You might learn to drive as a teenager but not discover statistics and statistical testing until college. Both statistical testing and driving are full of intricacies that you have to keep in mind.

In testing you consider an issue once, while in driving you must do it continually. When you make a mistake in testing, you can go back and correct it. If you make a mistake in driving, you might get a ticket or cause an accident.

After you learn to drive a car, you can go on to learn to drive motorcycles, trucks, buses, and racing vehicles. After you learn simple hypothesis testing, you can go on to learn ANOVA, regression, and many more advanced techniques. So if you think you can learn to drive a car, you can also learn to conduct a statistical test.

Testing.

If you need to evaluate whether a statistical test has been done correctly, you'll usually have to look at the original report on the analysis. The analysts probably got the type of test, the sample size, the test setup (i.e., number of populations, directions, null and alternative hypotheses) and probabilities all correct. Look for issues with violations of the test assumptions. Consider the size of the difference in the means being tested. Is it meaningful? Finally, if the analysis involves an ANOVA, count the number of tests being conducted on all the groups. Too many tests is a recipe for disaster because some are likely to be misleading just by chance.

If you need to conduct your own statistical test, consider taking that Stats 101 course or getting a friendly statistician to help you out.

I'm a cat in a Cat.

Don't be discouraged, statistical testing is complicated. Both the concepts and the jargon are intense. But like driving, it's all a matter of practicing … hopefully without causing any irreparable damage. Don't worry about formulas and calculations (except in Stats 101), there are plenty of websites available to help you out.

Testing.

They idolize me because I understand ANOVA.

Testing.

Statistical testing is really complicated.
I hope data relationships are easier.

Testing.

CHAPTER 7. RELATIONSHIPS.
As Complicated As You Might Think.

INTRODUCTION

You've probably seen scatter plots (Chapter 4) showing data points plotted for two variables with a smooth line drawn through them indicating a *trend*. The plot may also have shown the *equation* for the line and a *correlation coefficient* (or *coefficient of determination*). If you're lucky, the presentation didn't mention anything about causation. This is the kindergarten of data relationships.

When you study data relationships further, you'll learn that there is much, much more to be aware of. There are more kinds of relationships and there are more ways to characterize them. There are relationships that aren't linear and there are ways to determine if they are noteworthy. And, there are even ways to judge whether causation is actually involved.

I think I'm going to need help.

That's what this Chapter is about.

CATEGORIES OF DATA RELATIONSHIPS

There are four ways that data can be related to each other if there is any relationship at all. Everybody has heard about *causation* but data relationships may also reflect *influence* or *association* or even mere *coincidence*. Table 31 provides a summary of these categories. Within these categories, there are at least 27 possible types of data relationships, not even counting the coincidental, *spurious relationships*.

CAUSES

A *cause* is a condition or event that directly triggers, initiates, makes happen, or brings into being another effect, condition or event. A cause is a sine qua non; without a cause, an effect will not occur. Causes are directional. A cause must precede its effect.

INFLUENCES

An *influence* is a condition or event that changes the manifestation of an existing condition or event but does not cause it to happen. Influences can be direct or

Relationships.

Table 31. Categories of data relationships.

CATEGORIES OF RELATIONSHIPS			
TYPE OF RELATIONSHIP	**PATTERNS**	**DIRECTION**	**FREQUENCY**
CAUSE — A condition or event that directly triggers, initiates, makes happen, or brings into being another condition or event.	Without a cause a consequent cannot occur.	Causes are directional. A cause must precede its consequent.	True causal relationships are rare.
INFLUENCE — A condition or event that changes the manifestation of another condition or event.	Can be direct or mediated by a separate condition or event.	May exist at any time before or during the influenced condition or event. Influences may be unidirectional or bidirectional.	Influential relationships are relatively common.
ASSOCIATION — Two conditions or events that appear to change in a related way. Associations can be spurious or real.	May exist at any time before, during, or after the associated condition or event.	Have no effect on each other and may not exist in different populations or in the same population at different times or places.	Very common.

mediated by a separate condition or event. Influences may exist at any time before or after the influenced condition or event. Influences may be unidirectional or bidirectional.

ASSOCIATIONS

Associations are conditions or events that appear to change in a related way. Any variables measuring such conditions or events may also appear to change in a related way and are said to be associated.

Associations can be spurious or real. Associations may exist at any time before or after the associated condition or event. Unlike causes and influences, associated variables have no effect on each other and may not exist in different populations or in the same population at different times or places.

Associations are commonplace. Many observed correlations may just be associations. Influences and causes are less common but, unlike associations, they can be supported by the science or other principles on which the data are based.

Relationships.

The strength of a correlation coefficient is not related to the type of relationship. Causes, influences, and associations can all have strong as well as weak correlations depending on the efficiency of the variables being correlated and the pattern of the relationship.

TYPES OF DATA RELATIONSHIPS

I'm an influencer.

A topology of data relationships is important because it helps people to understand that not all relationships reflect a *cause.* This is where number crunching ends and statistical-thinking shifts into high-gear. Be prepared.

Here are nine relationships that involve events or conditions termed *A*, *B*, and *C*.

A ➜ B

DIRECT RELATIONSHIP

Direct relationships are easy to understand. If there are no statistical obfuscations, direct relationships should exhibit a high degree of correlation. In practice, though, not every relationship is direct or simple. Some are downright complex.

Most discussions of correlation and causation focus on the simple, direct relationship that one event or condition, *A*, is related to a second event or condition, *B*. The relationship proceeds in only one direction.

For example, gravitational forces from the Moon and Sun cause ocean tides on the Earth. *A* causes *B* but *B* does not cause *A*. Another direct relationship is that age influences height and weight. Age doesn't cause height and weight but we tend to grow larger as we age so *A* influences *B*. *B* does not influence *A*.

A ⟷ B

FEEDBACK RELATIONSHIP

In a feedback relationship, *A* and *B* are linked in a loop. *A* causes or influences *B*, which then causes or influences *A*, and so on. Feedback relationships are bidirectional. They will be correlated. For example, poor performance in school or at work (*A*) creates stress (*B*) which degrades performance further (*A*) leading to more stress (*B*) and so on.

Relationships.

COMMON-CAUSE RELATIONSHIP

In a common-cause relationship, a third event or condition, *C*, causes or influences both *A* and *B*. For example, hot weather (**C**) causes people to wear shorts (*A*) and drink cool beverages (*B*). Wearing shorts (*A*) doesn't cause or influence beverage consumption (*B*), although the two are associated by their common cause. A plot of this data will show that *A* and *B* are correlated, but the correlation represents an underlying association rather than an influence or a cause. Another example is the influence obesity has on susceptibility to a variety of health maladies.

A → C → B

MEDIATED RELATIONSHIP

In a mediated relationship, *A* causes or influences *C* and *C* causes or influences *B* so that it appears that *A* causes or influences *B*. *A* and *B* will be correlated. For example, rainy weather (*A*) often induces people to go to their local shopping mall for something to do (*C*). While there, they shop, eat lunch, and go to the movies thus providing the mall with increased revenues (*B*). In contrast, snowstorms (*A*) often induce people to stay at home (*C*) thus decreasing mall revenues (*B*). Bad weather doesn't cause or influence mall revenues directly but does influence whether people visit the mall.

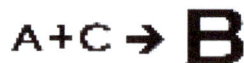

A+C → B

STIMULATED RELATIONSHIP

I need a cool drink on this hot day.

In a stimulated relationship, A causes or influences B but only in the presence of C. Stimulated relationships may not appear to be correlated using a *Pearson correlation coefficient* but may appear to be correlated using a *partial correlation coefficient*. There are many examples of this pattern, such as metabolic and chemical reactions involving enzymes or catalysts.

Relationships.

A+C ➜ B

SUPPRESSED RELATIONSHIP

In a suppressed relationship, A causes or influences B but not in the presence of C. As with stimulated relationships, suppressed relationships may only appear to be correlated using a partial correlation coefficient. Medicine has many examples of suppressed and stimulated relationships. For example, pathogens (A) cause infections (B) but not in the presence of antibiotics (C). Some drugs (A) cause side effects (B) only in certain at-risk populations (C).

A ➜ B
A ➜ B

INVERSE RELATIONSHIP

In inverse relationships, the absence of *A* causes or influences *B*, OR the presence of *A* minimizes *B*. Correlation coefficients for inverse relationships are negative. For example, vitamin deficiencies (*A*) cause or influence a wide variety of symptoms (*B*).

We're gonna need a bigger cart soon.

A ➜
A ➜
A ➜ B

THRESHOLD RELATIONSHIP

In threshold relationships, *A* causes or influences *B* only when *A* is above a certain level. For example, rain (**A**) causes flooding (**B**) only when the volume or intensity is very high. These relationships aren't usually revealed by correlation coefficients.

Relationships.

COMPLEX RELATIONSHIP

In complex relationships, many *A* factors or events contribute to the cause or influence of *B*. Numerous environmental processes fit this pattern. For example, a variety of atmospheric and astronomical factors (*A*) contribute to influencing climate change (*B*). Even many correlation coefficients may not explain this type of relationship; it takes more involved statistical analyses.

Table 32 provides a summary of types of data relationships.

It's just a coincidence that I'm laying on the air conditioning vent during this heat wave.

SPURIOUS DATA RELATIONSHIPS

The fourth category of relationship is the *spurious relationship* in which *A* appears to cause or influence *B,* but really does not. Often the reason is that the relationship is based on *anecdotal evidence* that is not valid more generally. Sometimes spurious relationships may be some other kind of relationship that isn't understood. Table 33 summarizes six other reasons why spurious relationships are so common.

MISUNDERSTOOD RELATIONSHIPS

The science behind a relationship may not be understood correctly. For example, doctors used to think that spicy foods and stress caused ulcers. Now, there is greater recognition of the role of bacterial infection. Likewise, hormones have been found to be the leading cause of acne rather than diet (i.e., consumption of chocolate and fried foods).

MISINTERPRETED STATISTICS

There are many examples of statistical relationships being interpreted incorrectly. For example, the sizes of homeless populations appear to influence crime. Then again, so do the numbers of museums and the availability of public transportation. All of these factors are associated with urban areas, but not necessarily crime.

Relationships.

Table 32. Types of data relationships.

TYPES OF RELATIONSHIPS	
BASED ON EVENTS OR CONDITIONS DESIGNATED AS **A, B,** OR **C**	
DEFINITION OF RELATIONSHIP	EXAMPLES
DIRECT — *A* influences *B*.	Gravitational forces from the Moon and Sun (**A**) influence ocean tides on the Earth (**B**). Age (**A**) influences height and weight (**B**).
FEEDBACK — *A* and *B* are linked in a loop; *A* influences *B* which influences *A* and so on.	Poor performance in school or at work (**A**) creates stress (**B**) which degrades performance further (**A**) leading to more stress (**B**) and so on.
COMMON — A third event or condition, *C*, influences both *A* and *B*.	Hot weather (**C**) influences people to wear shorts (**A**) and drink cool beverages (**B**). Poor health (**C**) influences people to be more susceptible to other health maladies (**A**) and (**B**).
MEDIATED — *A* influences *C* and *C* influences *B* so that it appears that *A* influences *B*.	Rainy weather (**A**) often influences people to go to their local shopping mall for something to do (**C**). While there, they shop, eat lunch, and go to entertainment venues thus providing the mall with increased revenues (**B**). Snowstorms (**A**) often influence people to stay at home (**C**) thus decreasing mall revenues (**B**).
STIMULATED — *A* influences *B* but only in the presence of *C*.	Some metabolic and chemical reactions (**A**) and (**B**) require enzymes or catalysts (**C**) to occur.
SUPPRESSED — *A* influences *B* but not in the presence of *C*.	Pathogens (**A**) cause infections (**B**) but not in the presence of antibiotics (**C**). Some drugs (**A**) cause side effects (**B**) only in certain at-risk populations (**C**).
INVERSE — The absence of *A* influences *B*.	Vitamin deficiencies (**A**) influence a variety of symptoms (**B**).
THRESHOLD — *A* influences *B* only when *A* is above a certain level.	Rain (**A**) causes flooding (**B**) only when the volume or intensity is very high.
COMPLEX — Many *A* factors or events contribute to the influence of *B*.	A variety of atmospheric and astronomical factors (**A**) contribute to influencing climate change (**B**). Numerous other environmental processes also fit this pattern.

MISINTERPRETED OBSERVATIONS

Incorrect reasons are sometimes attached to real observations. Many old-wives' tales are based on credible observations. For example, the notion that hair and nails continue to grow after death is an incorrect explanation for the legitimate observation that they appear to increase in size as the skin around them dehydrates. Apophenia refers to seeing patterns or meaning in uncorrelated data. One form of apophenia is

Relationships.

Table 33. Examples of spurious relationships.

SPURIOUS RELATIONSHIPS **A** APPEARS TO CAUSE OR INFLUENCE **B**, BUT DOES NOT.		
REASON FOR FALSE IMPRESSION		*EXAMPLE*
MISUNDERSTOOD RELATIONSHIPS	The science behind a relationship may not be understood correctly.	Medicine is notorious for advancing spurious relationships based on anecdotal evidence. Doctors used to think that spicy foods and stress caused ulcers. Now, there is greater recognition of the role of bacterial infection. Likewise, hormones have been found to be the leading cause of acne rather than diet (i.e., consumption of chocolate and fried foods).
MISINTERPRETED STATISTICS	There are many examples of statistical relationships being interpreted incorrectly.	The sizes of homeless populations appear to influence crime. Then again, so do the numbers of museums and the availability of public transportation. All of these factors are associated with urban areas but not necessarily crime.
MISINTERPRETED OBSERVATIONS	Incorrect reasons are attached to real observations.	Many old wives' tales are based on credible observations. For example, the notion that hair and nails continue to grow after death is an incorrect explanation for the legitimate observation.
URBAN LEGENDS	Some urban legends have a basis in truth and some are pure fabrications, but they all involve spurious relationships.	In South Korea, it believed that sleeping with a fan in a closed room will result in death.
BIASED ASSERTIONS	Some spurious relationships are not based on any evidence, but instead, are claimed in an attempt to persuade others of their validity.	The claim that masturbation makes you have hairy palms is not only ludicrous but also easily refutable. Likewise, almost any advertisement in support of a candidate in an election contains some sort of bias, such as cherry picking.
COINCIDENCES	Some events happen by pure chance, or appear to happen by pure chance.	Serendipity (being in the right place at the right time), happenstance (it was destiny), and many varieties of mental sensations like déjà vu.

pareidolia in which a random visual pattern is interpreted to be meaningful, such as seeing faces in clouds, toast, and rock formations.

URBAN LEGENDS

Some urban legends have a basis in truth and some are pure fabrications, but they all involve spurious relationships.

For example, In South Korea, it is believed that sleeping with a fan in a closed room will result in death. Some people may die in their sleep in a closed room but it wouldn't be attributable to a fan.

In the U.S., parents are warned every year to screen their children's Halloween candy for foreign objects or substances placed by malicious strangers to harm random children. Verified cases of adulterated candy are extremely rare but have caused parents to reject all homemade treats and even restrict or eliminate trick-or-treating.

It's in National Geographic so it must be true.

Relationships.

BIASED ASSERTIONS

Some spurious relationships are not based on any evidence, but instead, are claimed in an attempt to persuade others of their validity.

For example, the claim that masturbation makes you have hairy palms is not only ludicrous but also easily refutable because everybody in the world does not have hairy palms. Likewise, almost any advertisement in support of a candidate in an election contains some sort of bias, such *as cherry picking*.

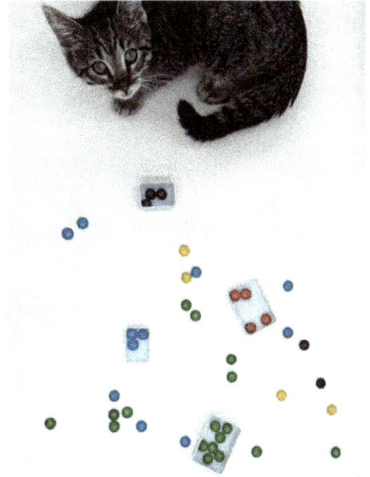

COINCIDENCES

Mother Nature has a wicked sense of humor. Sometimes she makes things happen in ways that make some of us think they are meaningfully related.

Are you going to check my M&M's for razor blades or are you going to steal them?

There is serendipity (being in the right place at the right time) and happenstance (it was destiny). Then there are the many varieties of mental sensations: déjà vu (fleeting feeling of having an experience before); déjà vécu (feeling of having a longer-term experience before); déjà senti (feeling of having a specific thought before); déjà visité (feeling of having been somewhere before); déjà entendu (feeling of having heard something before); déjà fait (feeling of having done something before); déjà arrive (feeling that something already happened); déjà dit (feeling of having said something before); déjà goûté (feeling of having tasted something before); déjà lu (feeling of having read something before); déjà rencontré (feeling of having met someone before); déjà rêvé (feeling of having dreamt something before); déjà presenti (feeling that something is about to happen); jamais vu (feeling that something familiar is unfamiliar); and presque vu (mental lightbulb that doesn't light).

Weird things do happen. Don't necessarily believe every *correlation coefficient* you calculate. Replicate and verify.

CORRELATIONS

I wonder if this will help my presque vu.

The idea that two variables might be related numerically dates to the 1880s, when Sir Francis Galton brought together concepts already over half a century old into the notions of correlation and regression. A decade later, Karl Pearson created an index to quantify the strength of the

Relationships.

relationship between two continuous-scale data series. Today, that index is called *Pearson's r*, or just the *correlation coefficient*.

Yes, weird things do happen.

The formula for the Pearson correlation coefficient is fairly nasty. The numerator estimates a degree of association of two variables, *x* and *y*, by summing the products of the centered variables (i.e., the mean of each variable subtracted from each original measurement). Then, the denominator adjusts the scales of the variables to have equal units by taking the product of the sum-of-the-squares of the deviations for each variable. Thus, **r** is the *standardized* sum of cross-product of two centered variables. Whew!

This is amazing when you consider that Pearson created his correlation coefficient over a century ago, before the invention of candy corn, fly swatters, and volleyball.

Since Pearson's work, other types of coefficients have been developed for variables that use other scales of measurement. Don't worry, you'll never have to use any of the formulas (unless your Stats 101 instructor tells you to). There are websites and free apps that will do the calculations from your raw data for you.

TYPES OF CORRELATION

So, if you have a dataset with more than one variable, you'll want to look at a correlation coefficient, usually Pearson's **r**. The Pearson correlation coefficient is used for linear relationships when both variables are measured on a continuous (i.e., interval or ratio) scale. Values of **r** can range from -1.0 to +1.0:

- 🐾 -1.0 represents a perfect correlation of data points on a line having a negative slope.
- 🐾 0.0 represents no linear relationship between the variables.
- 🐾 +1.0 represents a perfect correlation of data points on a line having a positive slope.

There are several variations of the Pearson correlation coefficient. The *multiple correlation coefficient*, denoted by **R**, indicates the strength of the relationship between a *dependent variable* and two or more *independent variables*. The *partial correlation coefficient* indicates the strength of the relationship between a dependent

$$r = \frac{\sum (x_i - \bar{x})(y_i - \bar{y})}{\sqrt{\sum (x_i - \bar{x})^2 \; \sum (y_i - \bar{y})^2}}$$

r = correlation coefficient
x_i = values of the x-variable in a sample
$\bar{x}$ = mean of the values of the x-variable
y_i = values of the y-variable in a sample
$\bar{y}$ = mean of the values of the y-variable

Relationships.

variable and an independent variable with the effects of other independent variables held constant. The *adjusted* or *shrunken correlation coefficient* indicates the strength of a relationship between variables after correcting for the number of variables and the number of data points.

There are also correlation coefficients for variables measured on noncontinuous scales. The Spearman **R**, for instance, is computed from ordinal-scale ranks.

Table 34 summarizes eleven types of correlation coefficients.

WHAT'S A *GOOD* CORRELATION?

When you look at a correlation coefficient, you might wonder, is that good? Is a correlation of -0.73 good but not a correlation of +0.58? Just what is a good correlation and what makes a correlation good?

As it turns out, what is considered a good correlation depends on who you ask and what they need it for. A chemist calibrating a laboratory instrument to a standard would want a correlation coefficient of at least 0.98. An engineer conducting a regression analysis of a treatment process might look for a correlation coefficient between 0.6 and 0.8. A biologist conducting an ANOVA of the size of field mice living in two different habitats might hope for a correlation coefficient of at least 0.2.

I think a good correlation coefficient is 9 ... or maybe that's the number of lives I have.

Is 0.2 really a good correlation or does a good correlation have to be at least 0.6 or even 0.98? As it turns out, the chemist, the engineer, and the biologist each have different experimental situations, different objectives, and different types of data. Those correlations were all good for those uses. So, the *meaningfulness* of a correlation coefficient depends, in part, on the expectations of the person using it.

How do you know what value of a correlation coefficient you should expect for it to be good for your situation? You might find some guidance from a statistician or a prior experiment, or you could calculate the square of the correlation coefficient, called the *coefficient of determination*, *R-square*, or R^2.

R-square is an estimate of the proportion of variance in the dependent variable that is accounted for by the independent variable(s). It is used commonly to put the strength of the relationship between variables into an understandable context so that alternative statistical models can be compared.

Relationships.

Table 34. Types of correlation coefficients.

Type of Correlation		First Variable	Second Variable	Comments
Pearson		One variable; continuous scale	One variable; continuous scale	This is the correlation coefficient you learned about in your introductory statistics course. It is used to quantify the relationship between two variables.
	Multiple	One variable; continuous scale	More than one variable; continuous scales	Used to quantify the relationship between a single variable and a set of several variables.
	Partial	One variable; continuous scale	More than one variable; continuous scales	Used to quantify the relationship between two variables while holding the effects of a set of other variables constant.
	Canonical	More than one variable; continuous scales	More than one variable; continuous scales	Used to quantify the relationship between two sets of several variables.
Spearman Rho, Kendall Tau		Interval or ordinal scale	Ordinal scale	
Polyserial		Interval scale	Ordinal scale	
Polychoric		Ordinal scale	Ordinal scale	Used to quantify the relationship between two variables that are not measured on continuous scales.
Point-biserial		Continuous scale	Binary scale	
Biserial		Interval scale	Binary scale	
Rank biserial		Ordinal scale	Binary scale	
Phi, tetrachoric		Binary scale	Binary scale	

Decide what proportion of variance you are looking for a relationship to account for. That would be your target R^2. Then take the square root of that proportion. That is the **r** you would be looking for.

You might be able to decide how good your correlation is from a gut feel for how much of the variability you wanted a relationship to account for. For example, correlation coefficient values between approximately -0.3 and +0.3 account for less than 9 percent of the variance in the relationship between two variables, which might indicate a weak or non-existent relationship. Values between -0.3 and -0.6 or +0.3 and +0.6 account for 9 percent to 36 percent of the variance, which might indicate a weak to moderately-strong relationship. Values between -0.6 and -0.8 or +0.6 and +0.8 account for 36 percent to 64 percent of the variance, which might indicate moderately-strong to strong relationship. Values between -0.8 and -1.0 or +0.8 and

Relationships.

+1.0 account for more than 64 percent of the variance, which might indicate a very strong relationship.

You can find other guidelines and ways to interpret correlation coefficients on many websites.

WHY YOU DON'T GET THE CORRELATION YOU EXPECT

Analysts often expect that a large value for a correlation coefficient, either positive or negative, means that there is a noteworthy relationship between two variables. This is not always the case. Furthermore, a small correlation may not always mean there is no relationship between the variables.

The size of a correlation will depend on several factors that need to be considered, including the type of relationship and the data used to characterize it. This is important because analysts may devote all their time investigating large correlations while disregarding relationships that have smaller correlations but greater importance.

Did I do that?

STATISTICAL REASONS

There are several statistical reasons for unexpected correlations:

Relationships. Correlation coefficients assume that the relationship between two variables is linear. Nonlinear relationships result in smaller than expected correlation coefficients. A scatter plot of the variables can usually confirm this problem, which can often be corrected with a *data transformation*.

Outliers. The strength of a correlation coefficient can be deflated or inflated by outliers. A scatter plot can usually confirm the presence of outliers although deciding how to treat them may be more problematic.

Excessive uncontrolled variance. Sometimes, data points that appear to be outliers may just be instances of excess variance. Excess variance is probably the most common cause of smaller than expected correlations. Usually, excess variance is the result of a lack of adequate control in data generation.

Inappropriate sample. Data points that look like outliers or excess variance may be *sham samples*. Sham samples are not representative of the population being analyzed, and so, confound any calculated statistics.

Relationships.

Hidden trends. Data may include trends hidden in subpopulations (*Simpson's paradox*). These hidden trends can detract from the calculated correlation of the overall data set.

Inefficient metrics. Variables used in the analysis may not be appropriate for investigating the phenomenon in question. That is, the issue is the metric not the phenomenon. As a consequence, the strength of a relationship will be smaller than expected.

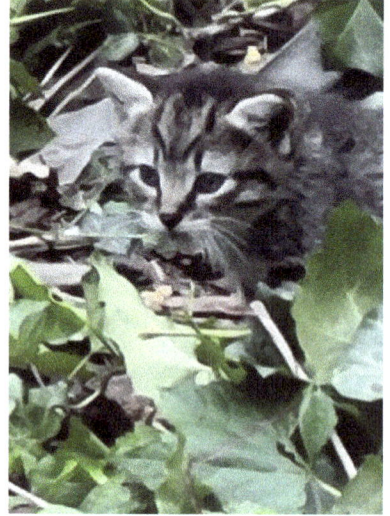

Consequently, any evaluation of a correlation should include looking at the coefficient's sign and size, a scatter plot of the relationship, and a statistical test of significance. There's a lot more information in data relationships than can be expressed by a single statistic.

If there are no statistical issues with a dataset, it's important to also consider what types of relationships between the variables are possible.

I'm a hidden kitten.

RELATIONSHIP REASONS

When analysts see a large correlation coefficient, they begin speculating about possible reasons. They'll naturally gravitate toward their initial hypothesis (or preconceived notion, their *confirmation bias*) which set them to investigate the data relationship in the first place. Because hypotheses are commonly about causation, they often begin with this least-likely type of relationship. Besides causation, relationships can also reflect *influence* or *association*.

OTHER REASONS

Correlation coefficients have a few other pitfalls to be aware of. Here are three.

First, the value of a *multiple* or *partial correlation coefficient* may not necessarily be reliable from subsample to subsample even if the coefficients are significantly different from zero. That's because the calculated values will tend to be inflated if there are many variables but only a few data pairs. Calculating a *shrunken correlation coefficient* will compensate for this obstacle.

Second, a large correlation isn't necessarily a good thing. If you are developing a statistical model and find that your predictor variables are highly correlated with your dependent variable, that's great. But if you find that your predictor variables are highly correlated with each other, that's not good. It's called *multicollinearity*.

Relationships.

Multicollinearity increases the variability in calculated statistics for a statistical model.

Third, if you're calculating many correlation coefficients from a large dataset, you might find that the number of data pairs is different for each calculation because of missing data. Some statisticians believe it is acceptable to compare correlations calculated with different numbers of data pairs. Other

I never expected that.

statisticians believe it is unwarranted, nonsensical, dishonest, fraudulent, heinous, and sickeningly evil. And you thought politics was contentious.

How to Verify That Good Is Really *Good*

Other than just considering the value of a correlation coefficient in context, there are two things you can do to assure yourself that a correlation really is good. You can plot the data and conduct a statistical test.

Scatter Plots

You should always plot the data used to calculate a correlation to ensure that the coefficient adequately represents the relationship.

The magnitude of **r** is very sensitive to the presence of nonlinear trends and outliers. Nonlinear trends in the data cause the magnitude of the relationship to be underestimated. You can often use data transformations to straighten any nonlinear patterns you see.

Outliers (i.e., data values not representative of the population) that are located perpendicular to the data trend cause the relationship to be underestimated. Outliers parallel to the data trend cause the relationship to be overestimated.

Statistical Tests

Every calculated correlation coefficient is an estimate. The real value may be somewhat more or somewhat less. You can conduct a statistical test (Chapter 6) to determine if the correlation you calculated is different from zero. If it's not, there is no evidence of a relationship between your variables.

This test looks at the absolute value of the correlation coefficient and the number of data pairs used to calculate it. The larger the value

I'm out here all alone. I must be an outlier.

Relationships.

of the correlation and the greater the number of data pairs, the more likely the correlation will be significantly different from zero.

Figure 66 displays the number of data pairs that are needed for a correlation coefficient to be significantly different from zero. For example, a correlation of 0.5 would be significantly greater than zero based on about 11 data pairs but a correlation of 0.1 wouldn't be significantly different from zero with 380 data pairs.

Number of Data Pairs Needed for a Correlation Coefficient
to be Significantly Different from Zero

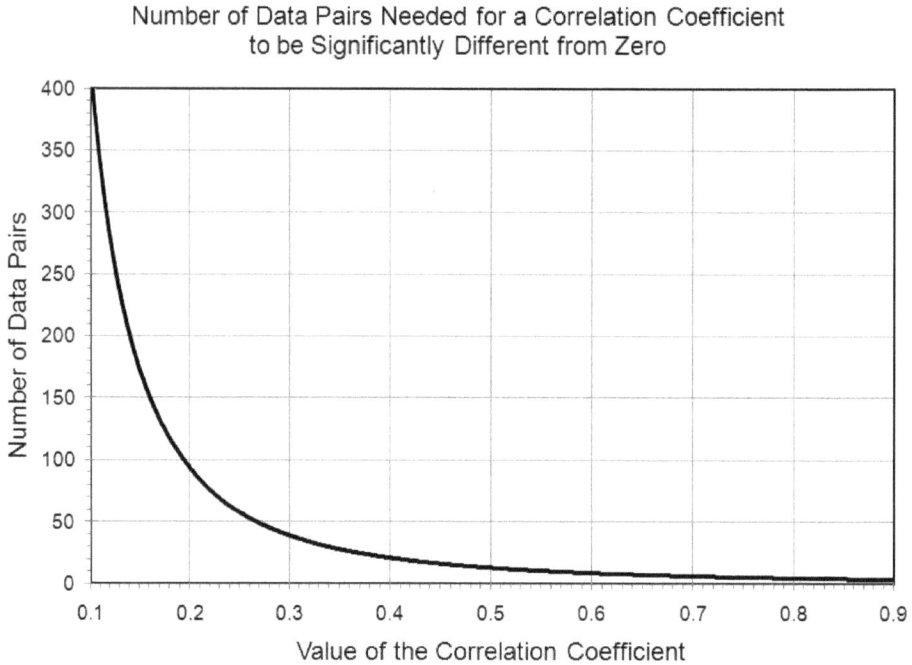

Figure 66. Significance of correlation coefficients.

That's why statistical software outputs the number of data pairs and the test probability with each correlation. With some software, you can also calculate a *confidence interval* around your estimate of **r** to see if the interval includes the value you set as a goal. But one way or the other, you have to consider the variability of your calculated estimate to decide if the correlation is good.

SUMMARY

So what makes a good correlation depends on what your expectations are, the value of the estimate, whether the estimate is significantly different from zero, and whether the data pairs form a linear pattern without any unrepresentative outliers. You have to consider correlations on a case-by-case basis. Remember too, though, that "no relationship" may also be an important finding.

Relationships.

Table 35 provides a summary of what to look for and where to look.

Table 35. What to look for in correlation coefficients.

What to Look For		Data to Use	Where to Look			
			Sign of r	Value of r	Plot	Test
Nature of Relationship	Is the trend linear?	All data from pairs of variables			X	
Direction of Relationship	Is the trend slope positive (rising) or negative (decreasing)?		X		X	
Strength of Relationship	How closely related are data points for two variables?			X	X	X
Outliers	Are there data points that are not representative of the population trend?				X	
Hidden Trends	Are there correlations in some data groupings but not others?	Data from pairs of variables from a single relevant group	X	X	X	X
Multi-collinearity	Are independent variables correlated with each other?	All data from pairs of independent variables	X	X	X	X
Auto-correlation	Are values of a dependent variable correlated with the order in which the variable was measured	A single dependent variable and the order the data were generated	X	X	X	X

HOW TO TELL IF CORRELATION IMPLIES CAUSATION

Everyone agrees that correlation is not the same as causation. However, those two words—correlation and causation—have generated quite a bit of discussion.

You've undoubtedly have heard the admonition:

Correlation does not imply causation.

This dictum dates back to the 1850s but was resurrected by Huff a hundred years later. Perhaps what Huff should have said was:

Correlation does not ALWAYS imply causation.

After all, sometimes it does. Consider that *where there's smoke there's fire* was first coined in the 1550s. So perhaps for those 300 years, people weren't so cautious about claiming causality. That was during the Spanish Inquisition, so maybe …. To be fair, Huff didn't have the internet to rely on in the 1950s to sort all this out. He lived in the halcyon days before trolls and flame wars on social media. He didn't anticipate that his cautious reminder would become weaponized.

Relationships.

In reality, calculated correlation coefficients are innocent bystanders in debates over *causation*, *influence*, and *association*. With all the statistical and relationship nuances that can affect their interpretation, it's a wonder that they are so often offered alone as determinants of causality and influence. As with all statistics, correlation coefficients need to be interpreted in the context provided by other types of information.

Now brace yourself for this: just as a large value for a correlation coefficient does not always imply causation, a real causal relationship does not always guarantee a large value for a correlation coefficient. In other words:

Causation does not always imply correlation.

The reason for this is twofold. First, and most fundamentally, the relationship has to be *direct*. Other types of data relationships—*feedback*, *common-cause*, *mediated*, *stimulated*, *suppressed*, *inverse*, *threshold*, and *complex*—don't always exhibit strong correlations. Second, correlations may be suppressed for statistical reasons, like *outliers*, *excessive uncontrolled variance*, *hidden data trends*, *inappropriate samples*, and *inefficient metrics*.

That's why you can't rely on correlation alone to demonstrate causation. Still, causality is an important concept to evaluate in many aspects of life.

WHY CAUSALITY MATTERS

No one gets perturbed if you say two conditions or events are correlated but even suggest that one causes the other and you'll hear the clichéd admonition if not even harsher criticism. It's not easy to prove causality, though, so there must be a reason for putting in the effort.

Causality is a big deal. It needs to be spelled out.

Here are the *Six Ps* for why you might want to establish causality.

- 🐾 *Promote* the relationship to reap benefits, such as between agricultural methods and crop production or pharmaceuticals and recovery from illnesses.
- 🐾 *Prevent* the cause to avoid harmful consequences, such as airline crashes and manufacturing defects.
- 🐾 *Prepare* for unavoidable harmful consequences, such as natural disasters, like floods.
- 🐾 *Prosecute* the perpetrator of the cause, as in law, or lay blame, as in politics.
- 🐾 *Pontificate* about what might happen in the future if the same relationship occurs, such as in economics.

🐈 Relationships.

- 🐾 ***Probe*** for knowledge based on nothing more than curiosity, such as why cats purr.

So how can you tell if correlation does in fact imply causation?

CRITERIA FOR CAUSALITY

Sometimes it's next to impossible to convince skeptics of a causal relationship. Sometimes it's even tough to convince your supporters. Developing criteria for causality has been a topic of concern for centuries. In medicine, several sets of criteria have been proposed over the years.

One widely cited set is the criteria described in 1965 by Austin Bradford Hill, a British medical statistician. *Hill's criteria for causation* specify the minimal conditions necessary to accept the likelihood of a causal relationship between two measures as:

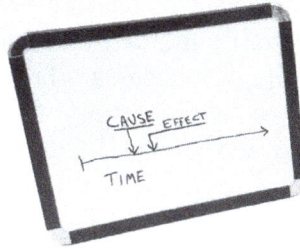

Causation requires one thing to come before another, but that's not all.

- 🐾 ***Strength***. A relationship is more likely to be causal if the correlation coefficient is large and statistically significant.
- 🐾 ***Consistency***. A relationship is more likely to be causal if it can be replicated.
- 🐾 ***Specificity***. A relationship is more likely to be causal if there is no other likely explanation.
- 🐾 ***Temporality***. A relationship is more likely to be causal if the effect always occurs after the cause.
- 🐾 ***Gradient***. A relationship is more likely to be causal if a greater exposure to the suspected cause leads to a greater effect.
- 🐾 ***Plausibility***. A relationship is more likely to be causal if there is a plausible mechanism linking the cause and the effect.
- 🐾 ***Coherence***. A relationship is more likely to be causal if it is compatible with related facts and theories.
- 🐾 ***Experiment***. A relationship is more likely to be causal if it can be verified experimentally.
- 🐾 ***Analogy***. A relationship is more likely to be causal if there are proven relationships between similar causes and effects.

These criteria are sound principles for establishing whether some condition or event causes another condition or event. No individual criterion is foolproof, however. That's why it's important to meet as many of the criteria as possible. Still, sometimes causality is unprovable.

🐈 Relationships.

THREE STEPS TO DECIDE IF CORRELATION IMPLIES CAUSATION

Hill's criteria can be thought of as aspects of the process of *critical thinking* (Chapter 9) for deciding if a relationship involves causation. The criteria don't all have to be met to suggest causality and some may not even be possible to meet in every case. The important point is to consider the criteria in a careful and unbiased process.

STEP 1. CHECK THE METRICS.

The admonition that *correlation-does-not-imply-causation* is used to remind everyone that a correlation coefficient may actually be characterizing a non-causal influence or association rather than a causal relationship. A large correlation coefficient does not necessarily indicate that a relationship is causal. On the other hand, saying that correlation is a necessary but not sufficient condition for causality, or in other words, causation cannot occur without correlation, is also not necessarily true. There are quite a few reasons for a lack of correlation.

So, before you get too excited about some causal relationship, make sure the correlation is statistically legitimate. You can't assess the relationship's *gradient* (i.e., the sign of the correlation coefficient) and *strength* (i.e., the value of the correlation coefficient) if the correlation is erroneous. Make sure to:

- 🐾 Use metrics (variables) that are appropriate for quantifying the relationship. For example, don't use an *index* that is a ratio of the other metric in the relationship.
- 🐾 Use an appropriate correlation coefficient based on the scales of the relationship metrics.
- 🐾 Confirm that the samples are representative of the population being analyzed and that the relationship is linear (or you are using nonlinear methods for the analysis).
- 🐾 Make sure that there are no outliers or excessive uncontrolled variance.

The *gradient* of most causal relationships is positive. Inverse relationships will have a negative gradient. The strength of causal relationships could be almost anything; it depends on what you expect. If you don't know what to expect, look at the *coefficient of determination.* R-square is an estimate of the proportion of variance shared by two variables. It is used commonly to interpret the strength of the relationship between variables. Be aware, though, that even causal relationships may show smaller than expected correlations.

STEP 2. EXPLAIN THE RELATIONSHIP.

If you are comfortable with the *gradient* and *strength* of the correlation coefficient, the next step is to define the *pattern* of the relationship. The correlation may not be of any help in exploring the pattern of the relationship because data plots for

Relationships.

different patterns can look similar. Nonetheless, there's no sense expending more effort if the correlation is in any way suspect.

First, check for *temporality* in the data. If the cause doesn't always precede the effect then either the relationship is a feedback relationship or is not causal. If cause and effect are not measured simultaneously, *temporality* may be obscured.

Next, try to determine what pattern of relationship is likely. This is not easy but it's also not a permanent determination. If you are uncertain, start with either a *direct* or an *inverse* relationship, which can be determined from data plots. Then as you study the relationship further, you can assess whether the relationship may be based on *feedback, common-source, mediation, stimulation, suppression, threshold,* or *multiple complexities*.

A different kind of feedback relationship.

Consider your relationship in terms of Hill's criteria of *plausibility, coherence, analogy*, and *specificity*. *Plausibility* and *coherence* are perhaps the easiest of the criteria to meet because it is all too easy to rationalize explanations for observed phenomena. They may also rely on related theories that can change over time. *Analogy* is a bit more difficult to meet but not impossible for a fertile mind. However, analogous relationships may appear to be similar but in fact be attributable to very different underlying mechanisms. Narrow-minded people rely on *specificity* in their arguments. Then again, relationships may have no other likely explanation because a phenomenon is not well understood.

STEP 3. VALIDATE THE EXPLANATION.

Perhaps the most important of Hill's criteria are *experiment* and *consistency*, both of which involve conducting additional research studies to confirm and explore the causality. These studies may be either experimental or observational depending on the natures of the population and the phenomenon.

In an *experimental* study, researchers decide what conditions the subjects (the entities being experimented on) will be exposed to and then measure variables of interest. In an *observational* study, researchers observe subjects that possess the conditions being assessed and then measure variables of interest.

Both types of designs have their challenges. Researchers may not be able to manipulate the conditions under study in an experiment because of cost, logistics, or ethical concerns. Observational studies may be subject to *confounding conditions* that interfere with the interpretation of results. Consequently, verifying that a relationship is causal is much easier said than proven.

Relationships.

If you're serious about proving there is a causal relationship between two conditions or events, you have to verify the relationship using an appropriate and effective research design. Such an experiment usually requires a mathematical model of the relationship, a testable hypothesis based on the model, incorporation of variance-control measures, collection of suitable metrics for the relationship, and an appropriate analysis.

An appropriate analysis may be statistical or deterministic. If the experiment verifies the relationship, especially if it can be consistently replicated by independent parties, there will be solid proof of causality and any spurious relationships will be disproved. The two problems are that this validation can involve considerable effort and that not every relationship can be verified experimentally.

IMPLYING CAUSALITY

> *The plural of anecdote is not data.*
>
> Roger Brinner, American economist.

Hill's criteria were developed for medicine. Medical research may start with anecdotal observations and progress to statistical collection of data. Add demographics and patterns of occurrence may become apparent. Then the patterns are assessed to look for coherent, plausible explanations and analogues.

Some medical hypotheses can be tested and analyzed statistically. Pharmaceutical effectiveness is an example. Psychological and agricultural relationships can often be tested. Other relationships that can't be manipulated must be analyzed based on observations. Epidemiological studies are examples. Without being able to rely on the *experiment* and *consistency* criteria, causality can only be argued using the weaker *plausibility*, *coherence*, *analogy*, and *specificity* criteria. This is also true with natural phenomena, like landslides and earthquakes.

Some conditions are unique or the underlying knowledgebase is insufficient to explain the phenomenon convincingly, so even the *plausibility*, *coherence*, *analogy*, and *specificity* criteria aren't useful. Economic and political relationships often fall into this category.

I think I'm pretty unique.

So, if you hear someone claim that a relationship is causal, consider how Hill's criteria might apply before you believe the assertion.

WHAT TO LOOK FOR IN DATA RELATIONSHIPS

Data relationships are most commonly shown in *bivariate plots,* plots of one variable versus another, each with its own scale on a separate axis. You might know them as *two-way plots*, *x-y plots*, *2D plots*, *scatter plots*, *line plots*, or another of their many monikers. *Dependent variables* are customarily plotted on the vertical y-axes against the *independent variables* on horizontal x-axes.

Relationship plots can show a variety of interesting and meaningful elements. Graphed data can display patterns—*shocks*, *steps*, *shifts*, *cycles*, and *clusters*—or trends—*linear*, *curvilinear*, *nonlinear*, or *complex* trends involving different dimensions (*temporal*, *spatial*, *categorical*, *hidden*, or *multivariate*). Sometimes the most interesting revelations you can garner from a dataset are the ways that it doesn't fit expectations, like *censoring*, *heteroscedasticity*, or *outliers*.

Linear *trends* are easiest to discern followed by simple curves. The more complex the curve, the harder it is to identify in data. Trends may also be discontinuous. Patterns are often harder to recognize than trends because they may be manifested in only some of the data whereas trends usually are formed by all the data.

Here's what to look for in relationship plots.

There's a lot more to relationships than I thought.

PATTERNS

There are five patterns to look for in graphs of data relationships:

- Shocks
- Steps
- Shifts
- Cycles
- Clusters.

Examples of these patterns are shown in Figure 67.

Shocks are seemingly random excursions far from the main body of data. They are *outliers* but they often recur, sometimes in a similar way, suggesting a common though sporadic cause. Some shocks may be attributed to an intermittent malfunction in the measurement instrument. Sometimes they occur in pairs, one in the positive direction and another of similar size in the negative direction. This is often because of missed reporting dates for business data.

Relationships.

Steps are periodic increases or decreases in the body of the data. Steps progress in the same direction because they reflect a progressive change in conditions. If the steps are small enough, they can appear to be a linear trend and be analyzed as such.

Shifts are increases and/or decreases in the body of the data. They are like steps but shifts tend to be longer than steps and don't necessarily progress in the same direction. Shifts reflect occasional changes in conditions. The changes may remain or revert to the previous conditions, making them more difficult to analyze with linear models.

Cycles are increases and decreases in the body of the data that usually appear as a waveform having fairly consistent amplitudes and frequencies. Cycles reflect periodic changes in conditions, often associated with time, such as daily or seasonal cycles. Cycles cannot be analyzed effectively with linear models. Sometimes different cycles add together making them more difficult to recognize and analyze.

Clusters are groups of data points that are mostly separate. They may be circular or elongated. They may partially overlap. Clusters may indicate some underlying reason for data points to be associated. This is important when the goal of an analysis is *classification*.

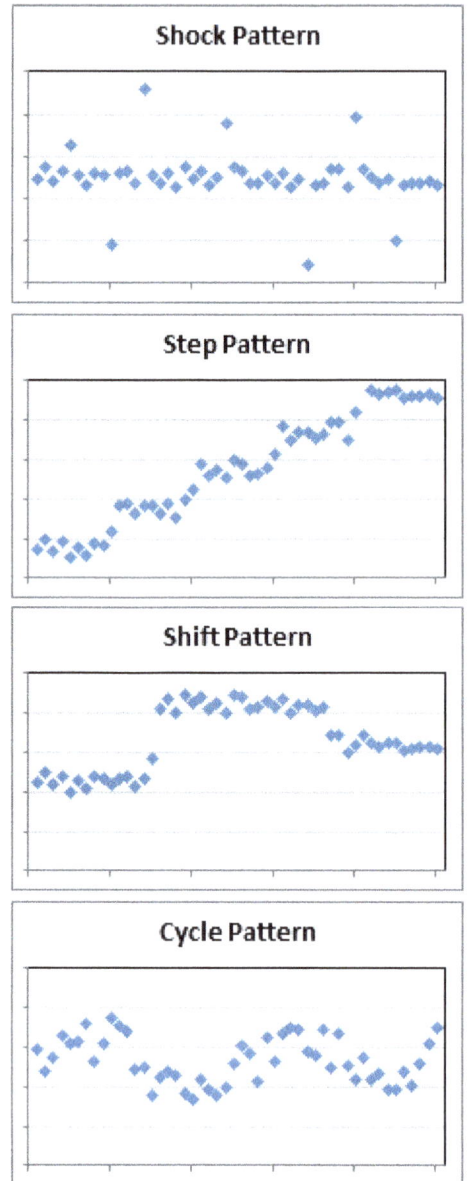

Figure 67. Types of patterns in statistical graphics.

TRENDS

Simple *trends* can be easier to identify because they tend to be formed by all the data and because they are more familiar to most data analysts. Again, graphs are the best place to look for trends.

Linear trends are easy to see; the data form a straight line. *Curvilinear* trends can be more difficult to recognize because they don't follow a simple path. With some experience and intuition, however, they can be identified. *Nonlinear* trends look similar to curvilinear trends but they require more complicated nonlinear models to

Relationships.

We're nonoverlapping, circular clusters.

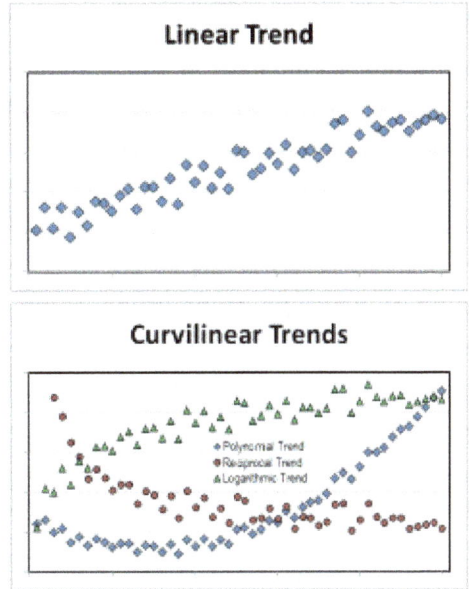

Linear Trend

Curvilinear Trends

Figure 68. Types of simple trends in statistical graphics.

We're overlapping, elongated clusters.

analyze. Some curvilinear trends can be analyzed with linear models with the use of transformations. Examples of simple trends are shown in Figure 68.

There are also more complex trends involving different dimensions, including:

- 🐾 Temporal
- 🐾 Spatial
- 🐾 Categorical
- 🐾 Hidden
- 🐾 Multivariate.

Temporal trends can be more difficult to identify because time-series data can be combinations of shocks, steps, shifts, cycles, and linear and curvilinear trends. The effects may be seasonal, superimposed on each other within a given time period, or spread over many different time periods. Confounded effects are often impossible to separate, especially if the data record is short or the sampled intervals are irregular or too large.

I'm ready to go on a vacation.

Relationships.

Spatial trends present a different twist. Time is one-dimensional; at least as we now understand it. Distance can be one-, two-, or three-dimensional. Distance can be in a straight line (i.e., as the crow flies) or along a path (e.g., travel routes).

Defining the location of a unique point on a two-dimensional surface (i.e., a plane) requires at least two variables. The variables can represent coordinates (northing/easting, latitude/longitude) or distance and direction from a fixed starting point. At least three variables are needed to define a unique point location in a three-dimensional volume, so a variable for depth (or height) must be added to the location coordinates.

Looking for spatial patterns involves interpolation of geographic data using one of several available algorithms, like *moving averages, inverse distances, splines, trend surfaces,* or *geostatistics.*

Categorical trends are no more difficult to identify than any trend except you have to break out categories to do it. One thing you might see when analyzing categories is *Simpson's paradox*. The paradox occurs when trends appear in categories of the data that are different from the trends that appear in the overall group.

Hidden trends are trends that appear only in segments of the data and not the entire dataset. For example, they may appear in only a finite range of an independent variable. You may be able to detect these trends in graphs if you have enough data but they are often too subtle to recognize. If you look too hard you may see patterns that aren't really there (*apophenia*).

Multivariate trends add a layer of complexity to most trends which are bivariate because they involve an added dimension. Still, you look for the same things, patterns and trends, only you have to examine at least one additional dimension. The extra dimension may be shown on a third axis or by some other way of representing data, like labeling, icons, sizes, or colors.

Examples of complex trends are shown in Figure 69.

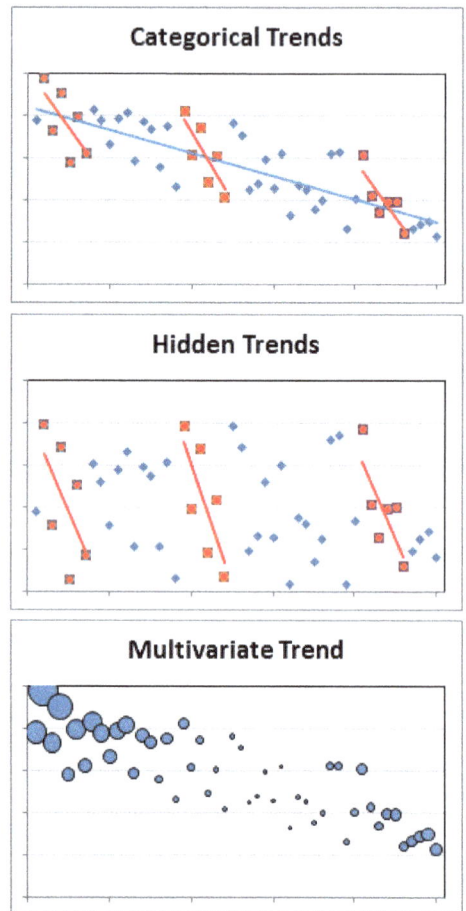

Figure 69. Types of complex trends in statistical graphics.

Relationships.

ANOMALIES

Sometimes the most interesting revelations you can garner from a graph are the ways that the data don't fit expectations. Three things to look for are:

- ❧ Censoring
- ❧ Heteroscedasticity
- ❧ Outliers.

Censoring is when a measurement is recorded as <value or >value, indicating that the measurement instrument was unable to quantify the real value.

For example, the real value may be outside the range of a meter, counts can't be approximated because there are too many or too few, or a time can only be estimated as before or after.

Censoring is easy to detect in a dataset because the data values are qualified with either < or >.

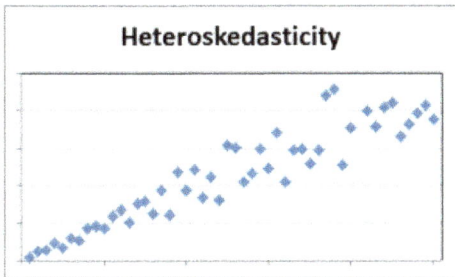

Figure 70. Example of heteroskedasticity (heteroscedasticity) in a scatterplot.

Heteroscedasticity is when the variability in a variable is not uniform across its range. This is important because *homoscedasticity* (the opposite of heteroscedasticity, also called equal variances) is assumed by probability statements in parametric statistics. Look for differing thicknesses of trends on plots of data variances. An example of a heteroscedastic trend is shown in Figure 70.

Influential observations and *outliers* are the data points that don't fit the overall trends and patterns or fall far away from a cluster of data. Finding anomalies isn't that difficult. Look back at the examples of scatter plots and line plots in Chapter 4. What is often difficult is determining why they are anomalous and deciding what to do with them in an analysis.

Figure 71 illustrates some examples of three types of outliers to look for.

In-line outliers may seem like a good thing because they will greatly inflate the estimate of R^2 for a model. You have to judge fairly whether the observation is legitimate or not before you incorporate it into a model.

Cross-trend outliers are the opposite. They will greatly reduce the estimate of R^2. Again, you have to judge fairly whether the observation is legitimate before you decide to delete it.

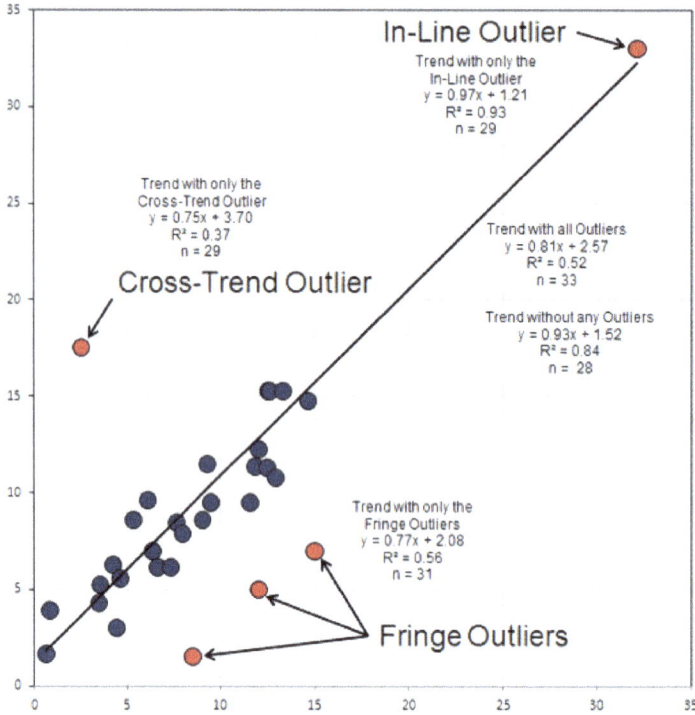

**Figure 71. Effects of outliers on
the characterization of data trends.**

Fringe outliers are the most challenging. They are harder both to identify and to decide what to do with since they are likely to share most of the characteristics of the rest of the data points.

*We have a great relationship
even when we're asleep*

Relationships.

Don't be discouraged if something you thought you understood—correlation—is a lot more complicated than you imagined. Now that you've heard about some of these intricacies, you are better prepared to understand and assess the legitimacy of statistical relationships even if you don't remember all the details.

I'm flabbergasted. Nobody ever explained all this stuff about types of relationships and correlations.

Relationships.

I'm in a relationship with a calico model.
Is that what we're talking about here?

Relationships.

CHAPTER 8. MODELS.
Putting A Face On A Relationship.

INTRODUCTION

Once you acknowledge a data relationship, your next step might be to try to model it. There are many ways to do that. As you might expect, the process of creating a statistical model of data is complex. It requires more knowledge and experience than kittens can provide. Still, if you understand the basics of statistical models, their components, and how they might go wrong, you'll be able to spot common faults that others may overlook.

That's what this Chapter is about.

MODELS

What's the first thing you think of when you hear the word *model*? The plastic dolls and model airplanes you used to play with? A fashion model? The model of the car you drive? The person who is your role model? But what do any of those things have to do with data analysis? Read on; you're about to find that statistical analyses begin and end with models.

BY ANY OTHER NAME

So, what do a globe, a mannequin, and the Normal distribution have in common? They are all called *models*. They are all representations of something, usually an ideal or a standard.

I need this model to pet my head.

Customarily, models are used to:

- 🐾 *Display* what they represent (e.g., model airplanes) or are associated with (e.g., fashions).
- 🐾 *Substitute* for incomplete real-world data, such as using the Normal distribution as a surrogate for a sample distribution.
- 🐾 *Manipulate* their components to learn more about the things they represent (e.g., scientific models for planetary motion).

Models can be true physical representations, approximate (or at least as good as practicable), or simplified, even cartoonish compared to what they represent. They can be about the same size, bigger, or most typically, smaller, whatever makes them

Models.

Table 36. Examples of models of different relative size and accuracy.

		Relative Size of Model		
		Larger than Actual	Same as Actual	Smaller than Actual
Accuracy of Representation	True	Oversized exhibition models for industrial trade shows	Cadavers used in medical schools	Ant Farms
	Approximate	Anatomical models used in colleges and medical schools	NASA flight simulators	Army Corps of Engineers models of waterways
	Simplified	Molecular models used in education	ResusiAnnie (dummies) used for CPR instruction	Architectural models

easiest to handle. Table 36 provides some examples of models having different sizes and accuracies.

Models often represent physical objects but they can also be written, drawn, or consist of mathematical equations or computer programming. In fact, using equations and computer code can be much more flexible and less expensive than building a physical model.

Models can represent a wide variety of phenomena, including conditions such as weather patterns, behaviors such as customer satisfaction, and processes such as widget manufacturing.

MODELS ARE EVERYWHERE

Whether you know it or not, you deal with models every day. Your weather forecast comes from meteorological models. Mannequins are used to display how fashions may look on you. Blueprints are drawn models of objects or structures to be built. Examples are plentiful.

Humans, in particular, are modeled all the time because of our complexity. Children play with dolls as models of playmates. Mannequins are simplified models of fashion models, who in turn, are models of people who might wear a fashion designer's wares. Posing models provide reference points for artists. Crash test dummies reveal how the human body might react in an automobile accident. Medical researchers use laboratory

Good thing these trucks are only models.

animals in place of humans for basic research. Medical schools use donated cadavers

Models.

Some models don't seem to be very good, like this model of a mouse. But sometimes, models are created with desirable features that are accentuated for further manipulation.

as models, very good ones as it turns out, of human anatomy. So, there should be nothing unfamiliar or intimidating about models.

Whether it is a physical scale-model of a hydroelectric dam or a mathematical model of weather patterns, a model is nothing more than a tool used to stimulate the imagination by simulating an object or phenomenon.

The model airplane takes its young pilot looping through the blue skies of a summer day. Globes teach geography and orreries teach planetary motion. The mannequin shows the bride-to-be how beautiful she'll look in the gown at her wedding. The concept car unveiled today gives consumers an idea of what they may be driving in a few years.

The National Hurricane Center uses over a dozen mathematical models to forecast the intensities and paths of tropical storms and to help understand the complex dynamics of hurricanes.

It should come as no surprise, then, that scientists, engineers, and mathematicians use models, especially virtual models, all the time. It may be surprising, though, that virtual models are also used extensively in business, economics, politics, and many other fields. Nevertheless, there is a mystique associated with modeling, especially the mathematical variety.

Some people believe that models are infallible and unchanging. Some believe that models are impossibly complex and necessarily unfathomable. Some believe that models are sophisticated delusions for obfuscating real data. In reality, none of these opinions is correct, at least entirely.

Some models are really simple compared to the original.

CLASSIFICATION OF MODELS

There are many ways that models are classified, so the catalog of models shown in Figure 72 isn't unique. The models may be described with different terms or broken out to greater levels of detail. It is also common for models to be hybrids of several different types. Examples include mash-ups of analytical and stochastic components used to analyze phenomena such as climate change and subatomic particle physics.

Models.

Nevertheless, the catalog should give you some examples of the variety of ways models help us explore our lives.

PHYSICAL MODELS

Your first exposure to a model was probably a physical model like a baby pacifier or a plush animal, and later, a doll or a toy car. From then, you've seen many more, from ant farms to anatomical models in school. You probably even created your own models with Legos and kits for plastic models of cars, planes, and ships. You may have built a Pinewood Derby racer or fabricated your own Halloween costumes. They are all representations of something else.

Physical models aren't used often for research and advanced applications because they are difficult and expensive to build and calibrate to a realistic experience. Flight simulators, hydrographic models of river systems, and reef aquariums are well-known examples, though.

CONCEPTUAL MODELS

Models can also be expressed in words and pictures. These are used in virtually all fields to convey mental images of some mechanism, process, or other phenomenon that was or will be created. Blueprints, flow diagrams, geologic fence diagrams, anatomical diagrams are all *conceptual models*. So are the textual descriptions that go with them. In fact, you should always start with a simple text model before you embark on building a complex physical or mathematical model.

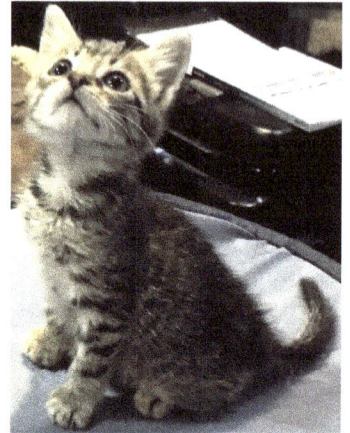

Will I be able to create a model if I read this Chapter?

MATHEMATICAL AND COMPUTER MODELS

Mathematical models can be either *theoretical* (i.e., derived mathematically from scientific principles) or *empirical* (i.e., based on experimental observations).

THEORETICAL MODELS

Theoretical models are based on scientific laws and mathematical derivations. For example, celestial movements and radioactive decay are phenomena that can be evaluated using theory-based models. To calibrate a theoretical model, the form of the model (i.e., the equation) is fixed and the inputs are adjusted so that the calculated results adequately represent actual observations. Both theoretical models and deterministic-empirical models provide solutions that presume that there is no uncertainty. These solutions are termed *exact* (which does not necessarily imply *correct*). Exact models have a single solution for a given set of inputs.

Models.

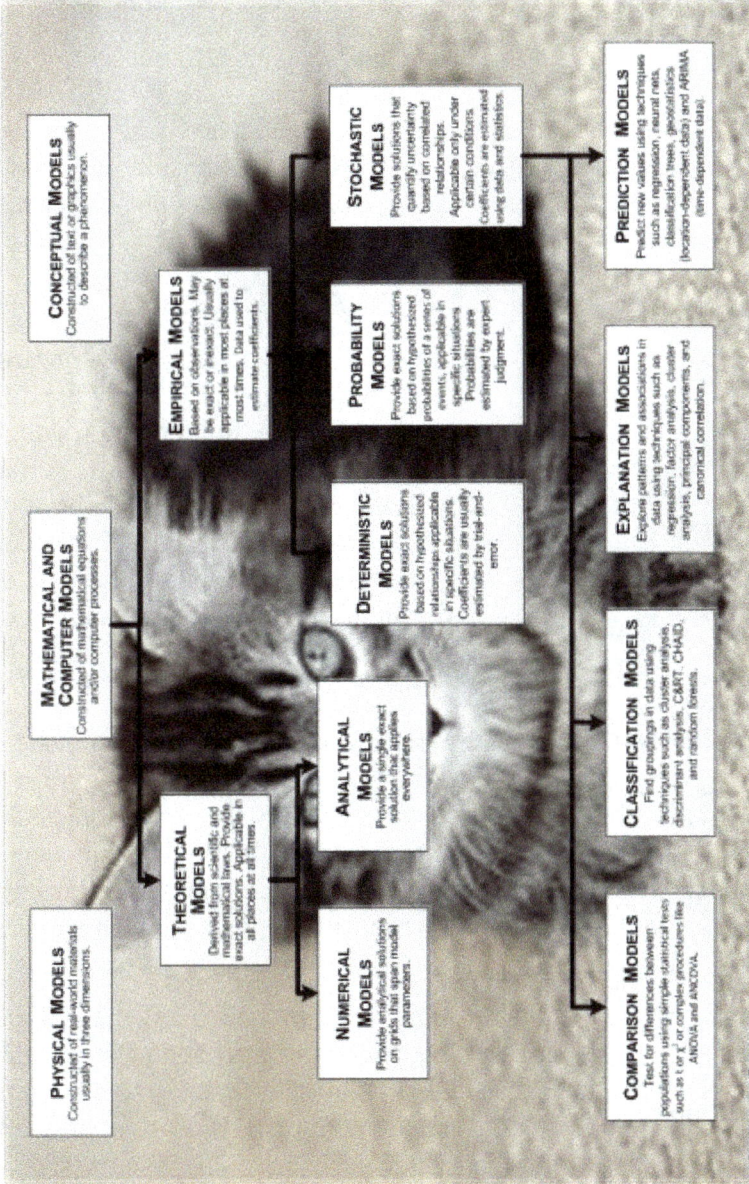

Categories of Models

CONCEPTUAL MODELS
Constructed of text or graphics usually to describe a phenomenon.

MATHEMATICAL AND COMPUTER MODELS
Constructed of mathematical equations and/or computer processes.

PHYSICAL MODELS
Constructed of real-world materials usually in three dimensions.

THEORETICAL MODELS
Derived from scientific and mathematical laws. Provide exact solutions. Applicable in all places at all times.

EMPIRICAL MODELS
Based on observations. May be exact or inexact. Usually applicable in most places at most times. Data used to estimate coefficients.

STOCHASTIC MODELS
Provide solutions that quantify uncertainty based on correlated relationships. Applicable only under certain conditions. Coefficients are estimated using data and statistics.

PROBABILITY MODELS
Provide exact solutions based on hypothesized probabilities of a series of events, applicable in specific situations. Probabilities are estimated by expert judgment.

DETERMINISTIC MODELS
Provide exact solutions based on hypothesized relationship, applicable in specific situations. Coefficients are usually estimated by trial-and-error.

ANALYTICAL MODELS
Provide a single exact solution that applies everywhere.

NUMERICAL MODELS
Provide analytical solutions on grids that span model parameters.

PREDICTION MODELS
Predict new values using techniques such as regression, neural nets, classification trees, geostatistics (location-dependent data) and ARIMA (time-dependent data).

EXPLANATION MODELS
Explore patterns and associations in data using techniques such as regression, factor analysis, cluster analysis, principal components, and canonical correlation.

CLASSIFICATION MODELS
Find groupings in data using techniques such as cluster analysis, discriminant analysis, CART, CHAID, and random forests.

COMPARISON MODELS
Test for differences between populations using simple statistical tests such as t or χ² or complex procedures like ANOVA and ANCOVA.

Figure 72. Categories of models for data analysis.

Models.

ANALYTICAL MODELS

Analytical models are mathematical equations derived from scientific laws that produce exact solutions that apply everywhere. For example, F (force) = M (mass) times A (acceleration) and E (energy) = m (mass) times c^2 (speed of light squared) are analytical models. Many concepts in classical physics can be modeled analytically.

NUMERICAL MODELS

Numerical models are mathematical equations that have a time parameter. Numerical models are solved repeatedly, usually on a grid, to obtain solutions over time. This is sometimes called a *dynamic model* (as opposed to a *static model*) because it describes time-varying relationships.

Does light move faster if you are closer to it?

EMPIRICAL MODELS

Empirical models differ from theoretical models in that the model is not necessarily fixed for all instances of its use. Rather, empirical models are developed for specific situations from measured data. Model formulation and calibration are simultaneous. However, the selection of the form of the equation and the inputs used in an empirical model are usually based on related theories. Models developed using statistical techniques are examples of empirical models.

Empirical models can be deterministic, probabilistic, stochastic, or sometimes, a hybrid of the three. They are developed for specific situations from measured data. Empirical models differ from theoretical models in that the model is not necessarily fixed for all instances of its use. There may be multiple reasonable empirical models that can apply to a given situation.

Deterministic-empirical models presume that a specific mathematical relationship exists between two or more measurable phenomena (as do theoretical models) that will allow the phenomena to be modeled without uncertainty under a given set of conditions (i.e., the model's inputs and assumptions). Biological growth models are examples of deterministic empirical models.

Both theoretical models and deterministic empirical models provide *exact* solutions, having no uncertainty. In contrast, *stochastic-empirical models* presume that changes in a phenomenon have a random component. The random component allows

Models.

stochastic empirical models to provide solutions that incorporate uncertainty into the analysis.

DETERMINISTIC MODELS

Deterministic empirical models presume that a mathematical relationship exists between two or more measurable phenomena (as do theoretical models) that will allow the phenomena to be modeled without uncertainty (or at least, not much uncertainty, so that it can be ignored) under a given set of conditions. The difference is that the relationship isn't unique or proven. There are usually assumptions. Biological growth and groundwater flow models are examples of deterministic-empirical models

The vet is going to check our biological growth to see if we fit his model.

PROBABILITY MODELS

Probability models are based on a set of events or conditions all occurring at once. In probability, it is called an *intersection of events*. Probability models are *multiplicative* because that is how intersection probabilities are combined (Chapter 2).

STOCHASTIC MODELS

Stochastic-empirical models, also called *statistical models*, presume that changes in a phenomenon have a random component. The random component allows stochastic empirical models to provide solutions that incorporate uncertainty into the analysis. In fact, the model equation is generated by quantifying and minimizing errors (i.e., uncertainty). Statistical models rely on uncertainty while theoretical models largely ignore it. Stochastic models include lottery picks, weather, and many problems in the behavioral, economic, and business fields.

Statistical models can be used to compare, classify, explain, and predict.

Comparison models. In *statistical comparison models*, the dependent variable is a grouping-scale variable (one measured on a nominal scale). The independent variables can be either grouping, continuous, or both. Simple hypothesis tests (Chapter 6) include:

- χ^2 *tests* that analyze cell frequencies on one or more grouping variables.
- *t-tests* **and** *z-tests* that analyze independent variable means in two or fewer groups of a grouping variable.

Analysis-of-Variance (ANOVA) models (Chapter 6) compare the means of independent variables for two or more groups of a dependent grouping variable. *Analysis-of-Covariance* (ANCOVA) models compare independent-variable means

Models.

for two or more groups of a dependent grouping variable while controlling for one or more continuous variables. Multivariate ANOVA and ANCOVA compare two or more *dependent* variables using multiple independent variables. There are many more types of ANOVA model designs.

Classification models. Classification and identification models also analyze groups.

Clustering models identify groups of similar cases based on continuous-scale variables. Clustering models do not have a dependent variable; all the variables used to identify data groupings are treated the same. There need be no prior knowledge or expectation about the nature of the groups.

There are several types of cluster analysis, including *hierarchical clustering*, *K-Means clustering*, *two-step clustering*, and *block clustering*. Often,

We're in our clusters.

the clusters or segments are used as inputs to subsequent analyses. Clustering models are also known as *segmentation models*.

Discriminant analysis models have a nominal-scale dependent variable and one or more continuous-scale independent variables. They are usually used to explain why the groups are different, based on the independent variables, so they often follow a cluster analysis.

Logistic regression is analogous to *linear regression* but is based on a binary or ordinal dependent variable. Often, models for calculating probabilities use a binary (0 or 1) dependent variable with logistic regression.

There are many analyses that produce *decision trees*, which look a bit like organization charts. *Classification and Regression Trees* split categorical dependent variables into groups based on continuous or categorical-scale independent variables. All splits are binary. *Chi-square Automatic Interaction Detectors* (CHAID) generate decision trees that can have more than two branches at a split. A *Random Forest* consists of a collection of simple tree predictors.

We're definitely in two different classifications.

Explanation models. *Explanation models* aim to explain associations within or between sets of variables. With explanation models, you select enough variables to address all the theoretical aspects of the phenomenon, even to the point of having

Models.

some redundancy. As you build the model, you discover which variables are extraneous and can be eliminated.

Factor Analysis (FA) and *Principal Components Analysis* (PCA) are used to explore associations in a set of variables where there is no distinction between dependent and independent variables. These two types of statistical analysis:

- Create new metrics, called *factors* or *components*, which explain almost the same amount of variation as the original variables.
- Create fewer factors/components than the original variables so further analysis is simplified.
- Require that the new factors/components be interpreted in terms of the original variables. They often make more conceptual sense so subsequent analyses are more intuitive.
- Produce factors/components that are statistically independent (uncorrelated) so they can be used in regression models to determine how important each is in explaining a dependent variable.

Canonical Correlation Analysis (CCA) is like PCA only there are two sets of variables. There is no dependency in the two sets of variables, one is termed simply the *left variables* and the other is the *right variables*. Pairs of components, one from each group, are created that explain independent aspects of the dataset.

Regression analysis is also used to build explanation models. In particular, regression using principal components as independent variables is popular because the components are uncorrelated and not subject to *multicollinearity*.

Prediction models. Some models are created to *predict* new values of a dependent variable, *forecast* future values of a time-dependent (autocorrelated) variable, or *extrapolate* more distant values of a spatially-dependent (regionalized) variable. In prediction models, accuracy tends to come easy while precision is elusive.

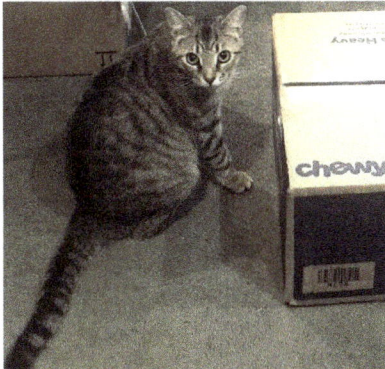

I can predict what's in this box.

Regression is the most commonly used technique for creating prediction models. Transformations are used frequently. Prediction models usually keep only the variables that work best in making a prediction, and they may not necessarily make a lot of conceptual sense.

Neural Networks is a predictive modeling technique inspired by the way biological nervous systems process information. The technique involves interconnected nodes or layers that apply predictor variables in different ways, linear and nonlinear, to all or some of the dependent variable values. Unlike most modeling techniques, neural networks can't be articulated so they are not useful for explanation purposes.

Models.

TIME-DEPENDENT MODELS

There are a variety of types of *time-dependent models,* including *smoothing models, time-series regression models, ARIMA models, decomposition models,* and *spectral models.*

Smoothing models use prior time-series measurements to produce estimates of later time-series measurements. A *moving-average model* is a simple time-series model in which a certain number of prior measurements are averaged together to produce a later measurement. The number of measurements in a moving-average model can be limited to accentuate variability or increased to eliminate perturbations. Weighted averages can be used to emphasize the contribution of more recent data. There are many types of smoothing models. Some models include linear, exponential, or seasonal trends. Some models are additive and some are multiplicative.

I'm caught in a neural net.

Linear regression can also be used to develop time-series models. The advantage of this method is that the software is readily available. Five types of functions are used in *time-series regression models*—linear *trends,* sinusoidal *cycles, lags* and *leads, differences,* and non-temporal predictors. Like smoothing models, time-series regression models can be arduous to fit properly. Their advantage however is that more people are familiar with the concepts of regression analysis than with other types of time-series modeling.

Autoregressive integrated moving average (ARIMA) models are specially designed for time-series data and require specialized software. *Autoregressive* terms in an ARIMA model refer to a prior measurement in a time series. *Moving-average* terms refer to an average of some number of prior measurements. *Integrated* terms refer to the difference between two prior measurements in a time series.

A big difference between regression and ARIMA models is that regression models are based on the number of time periods after an arbitrary start date whereas ARIMA models are based on the time difference between data points. Regression models do not require that all data be collected at regular intervals whereas ARIMA models do. As a consequence, ARIMA models require specialized software. While they are more difficult to develop, ARIMA models do tend to produce better forecasts than other types of time-series models.

Decomposition models attempt to separate out time-series trends, cycles, lags, and differences, analyze each component separately, and then combine the results into a

Models.

Time measurements are easy to understand but they are hard to model.

composite model. The Census Bureau and the stock market use this approach frequently.

Spectral models attempt to fit cyclic functions, particularly sine and cosine functions, to time-series data. This type of model usually requires hundreds of observations, and is used often in physics and astronomy.

LOCATION-DEPENDENT MODELS

Location measurements can represent one, two, or even three dimensions, namely length, width, and height or depth. In one dimension, location measurements can be analyzed like time measurements with one big difference. With time measurements, a point in time can only be influenced by earlier points in time. With location measurements, a point in space can be influenced by points anywhere near it. In two and three dimensions, location measurements may be represented by *maps* or *cross-sections*. Both maps and cross-sections can be contoured, which involves placing lines on a representation of the area to show where equal values of the measurement might be (see Chapter 4).

The first step in the process of contouring is to create a grid of estimated values derived from the original data. This process is called *gridding* or *interpolation*. Commonly used interpolation algorithms include *moving averages, inverse-distance averages, splines, trend-surfaces*, and *geostatistics*.

There is a time and a place for everything.

Moving Average Interpolation. A moving average value for a grid node is produced by taking a simple average of a predetermined number of data values located near the grid node. Moving average surfaces tend to be very smooth and will emphasize trends. They will not produce a surface that extends outside the range of the original data.

Inverse-Distance Interpolation. Inverse-distance interpolation produces grid values by weighting data values by the reciprocal of the distance from data points to grid nodes. The farther away a data point is from a grid node the less it contributes to the node's calculated value. Inverse distance surfaces tend to be less smooth than moving average surfaces, will emphasize both trends and anomalies, but will not produce a surface that extends beyond the original data.

 Models.

Spline Interpolation. Splines are mathematical functions that minimize the change in a surface's slope. Spline surfaces tend to be smoother than inverse distance surfaces and will emphasize trends rather than anomalies. Splines can also produce surfaces that extend outside the range of the original data.

Trend-Surface Interpolation. Unlike other gridding methods that rely on the distance and direction between sample locations, trend-surface models use location coordinates as independent variables and are fit like any regression model.

We're too close together for inverse-distance interpolation to be very effective.

Trend surfaces have the advantage of being able to use rotated and translated coordinates, other types of mathematical functions of the geographic coordinates, and non-location variables to achieve a better fit, all at the cost of a lot more effort. In general, trend surfaces tend to be very smooth and will emphasize trends even well outside the range of the original data. They are not used as commonly as other interpolation techniques.

Geostatistical Interpolation. Geostatistical interpolation is like inverse-distance interpolation except that the gridding process takes into account the spatial autocorrelation of the data (see *variograms* in Chapter 4). To produce a geostatistical surface, first the autocorrelation structure of the data is characterized in a process called *variogramming* (or *semivariogramming* to geostatistical purists). Grid values are interpolated based on the correlation structure in a process called *kriging* (an *eponym*). Geostatistical surfaces will emphasize both trends and anomalies.

There are other interpolation algorithms, data search controls, and other ways to customize a contoured surface depending on the available contouring software. How data are contoured will have a profound effect on how readers interpret reality.

THE *CORRECT* MODEL

There are many ways to model a phenomenon. Experience helps you to judge which model might be most appropriate for the situation. If you need some guidance, follow these steps.

🐾 **Step 1.** Start at top of the Catalog-of-Models figure. Decide whether you want to create a physical, mathematical, or conceptual model. Whichever you choose, start by creating a brief conceptual model in text so you have a mental picture of what your ultimate goal is and can plan for how to get there. If your goal is a physical or full blown conceptual model, do the research you'll need to identify appropriate materials and formats.

Models.

🐾 **Step 2.** If you want to select a type of mathematical model, start on the second line of the Catalog-of-Models figure and decide whether your phenomenon fits best with a theoretical or an empirical approach. If there are scientific or mathematical laws that apply to your phenomenon, you'll probably want to start with some type of theoretical model. If there is a component of time, particularly changes over time periods, you'll probably want to try developing a numerical model.

Some models are good and some models aren't.

Otherwise, if a single solution is appropriate, try an analytical model.

🐾 **Step 3.** If your phenomenon is more likely to require data collection and analysis to model, you'll need an empirical model. An empirical model can be probabilistic, deterministic, or stochastic. Probability models are great tools for thought experiments. There are no wrong answers, only incomplete ones. Deterministic models are more of a challenge. There needs to be some foundation of science (natural, physical, environmental, behavioral, or other discipline), engineering, business rules, or other guidelines for what should go into the model. More often than not, deterministic models are overly complicated because there is no way to distinguish between components that are major factors versus those that are relatively inconsequential to the overall results. Both probability and deterministic models are often developed through panels of experts using some form of the Delphi process.

Delphi process, huh?

🐾 **Step 4.** If you need to develop a stochastic (statistical) model, answer the five questions in Figure 73 to pick the right tool for the job.

🐾 **Step 5.** Consider adding hybrid elements. Don't feel constrained to only one type of component in building your model. For instance, maybe your statistical model would benefit from having deterministic, probability, or other types of terms in it. Calibrate your deterministic model using regression or another statistical method. Be creative.

Models.

Figure 73 doesn't include the hundreds of types of statistical models available today. It can't, new approaches are developed all the time. It is a good overview, though, of some traditional techniques to illustrate the process of identifying approaches to building a statistical model. It incorporates the numbers and scales of the variables and the aims of the analysis in the decision process. And, it fits on a single page.

STATISTICAL REGRESSION MODELS

Even if you never took an introductory statistics course you've probably heard the term *regression*. Statistical regression is a way to create an equation for a line that characterizes the most likely trend in a set of data paired on two variables.

In a regression model, the variable that characterizes the phenomenon to be modeled is called the *criterion variable*, *response variable*, *outcome variable*, *y-variable*, or most commonly, the *dependent variable*. Variables that are used to test, predict, or explain the dependent variable in the model are called *predictor variables*, *explanatory variables*, *grouping variables, x-variables* or most commonly, *independent variables*.

The most common method for fitting a regression line is called *least-squares*. The method finds a line in which the total deviations between the data pairs and the line are a minimum. The deviations are squared to account for deviations above and below the line, hence the term *least squares*.

Regression is a special word in statistics; it doesn't mean going backwards.

There are a variety of linear regression techniques. *Simple regression* involves one dependent variable and one independent variable. *Multiple regression* involves one dependent variable and more than one independent variable. Other techniques include *polynomial regression* and *logistic regression*.

Regression assumes that the variables are measured on continuous scales. Differences between the data and the regression line (i.e., *residuals*), are assumed to be independent, Normally distributed, and homoscedastic.

Additional details of statistical regression are described in scores of textbooks and websites. You'll hear a lot about it in Stats 101.

Models.

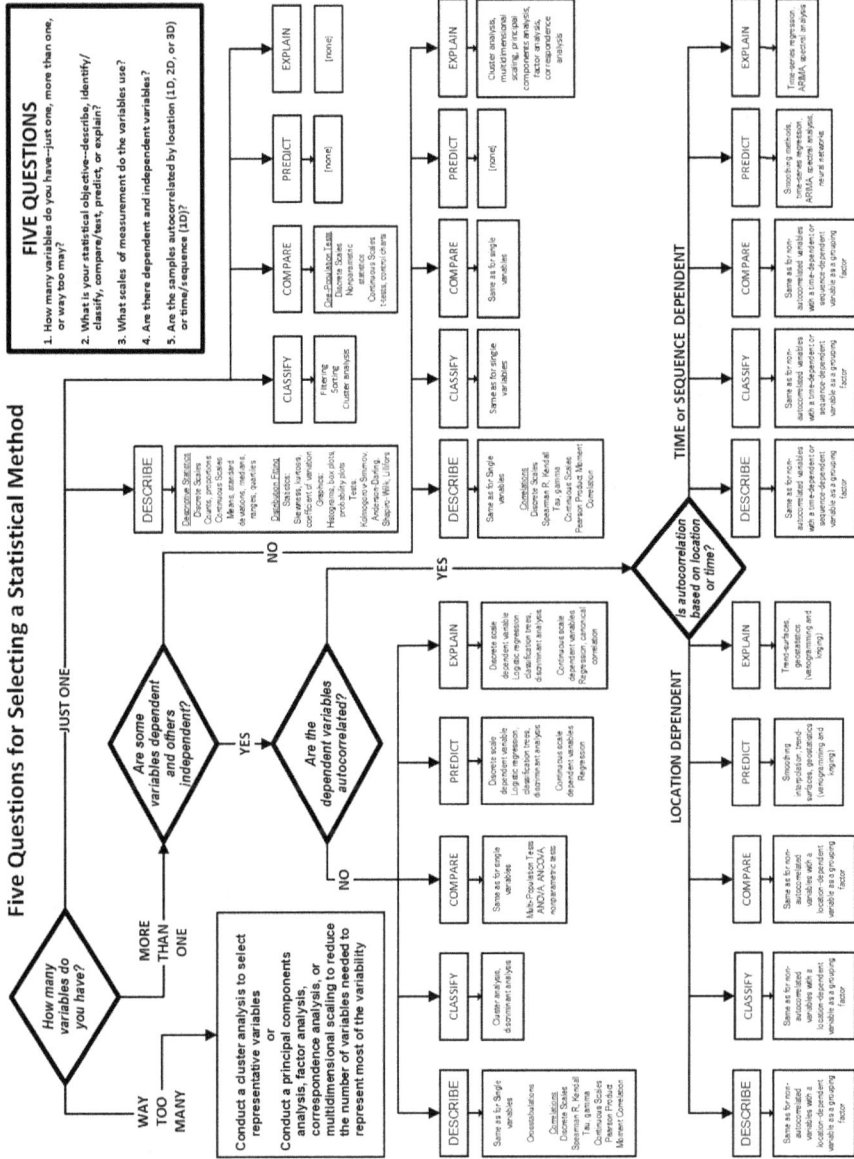

Figure 73. Decision tree for selecting statistical methods.

Models.

PARTS OF REGRESSION MODELS

A complicated and time-consuming aspect of model building involves selecting the components of the model—the variables, the samples, and the data. A prototype model is represented as:

$$\begin{matrix} \text{Dependent variable} \\ \text{that characterizes} \\ \text{the phenomenon} \end{matrix} = \begin{matrix} \text{Independent variable(s) that test, predict,} \\ \text{or explain the dependent variable} \end{matrix}$$

By convention, the dependent variable is always placed to the left of the equals sign, and the independent variables are placed to the right.

This representation says that the information in the dependent variable can be obtained from the information in the independent variable(s). Usually, though, the independent variables in a model won't all be equally important for describing a dependent variable. Each independent variable has to be weighted by multiplying it by an adjustment factor to account for the differences. The adjustment factors also correct for the independent variables being measured in different units or even scales-of-measurement.

So, a more detailed representation of a model would be:

$$\begin{matrix} \text{Dependent} \\ \text{variable} \end{matrix} = \begin{matrix} \text{Variable 1 Adjustment Factor * Independent variable 1 +} \\ \text{Variable 2 Adjustment Factor * Independent variable 2 +} \\ \text{... and so on ... +} \\ \text{Model Adjustment Factor} \end{matrix}$$

This says that the information in a dependent variable can be expressed as the sum of independent variables, which have been adjusted to account for their scales-of-measurement and for their contributions to the model, plus an adjustment factor for the entire model not related to a specific independent variable.

If all of the adjustment factors are constants, though not necessarily the same value, you have a *linear* model. The values for the adjustment factors are determined by the technique being used to calibrate the model, usually *least-squares*.

If the value of a dependent variable is equal to the sum of the adjustment factors times the values of the independent variables, plus the model constant, the model is called *exact* or *deterministic*

> *All information looks like noise until you break the code.*
>
> Hiro in Neal Stephenson's
> **Snow Crash** (1992)

Even with all those adjustment factors, though, sometimes the independent variables can't quite reproduce the values of the dependent variable, so there are errors. Add an error term to the model and you have a *statistical model*:

Models.

$$\text{Dependent variable} = \begin{array}{l} \text{Variable 1 Adjustment Factor * Independent variable 1 +} \\ \text{Variable 2 Adjustment Factor * Independent variable 2 +} \\ \text{... and so on +} \\ \text{Model Adjustment Factor +} \\ \text{Error} \end{array}$$

To be more concise, the terms in the model can be represented by letters and rewritten as:

$$y = a_0 + a_1x_1 + a_2x_2 + \dots a_nx_n + e$$

where:

> **y** is the dependent variable that characterizes the phenomenon.
>
> **x₁** through **xₙ** are the independent variables that test, predict, or explain the dependent variable.
>
> **a₀** is the Model Adjustment Factor.
>
> **a₁** through **aₙ** are the Variable Adjustment Factors. *a₁* through *aₙ* are constants called *coefficients* or *parameters* of the model. If **a₁** through **aₙ** aren't constants, it is a *nonlinear model*.
>
> **e** is the error term, which characterizes the uncertainty in the model.

The **y** and the **x**s are the variables you create and measure on your samples. The **a**s and the **e** are the constants that the statistical procedure estimates. That's how you get to a statistical model.

To add a little more perspective, if you have only one dependent variable, only one independent variable, and no error, the model reduces to:

$$y = a + bx$$

which you may remember from high school algebra as the equation of a straight line where **a** is the y-intercept and **b** is the slope of the line.

So, mathematical models really aren't so mysterious and shouldn't induce too much terror.

VARIABLES

The first step in assembling a set of variables for an analysis is to identify the concepts or aspects of the phenomenon you want to investigate. Concepts include hypotheses and theories as well as ideas, suppositions, beliefs, assertions, and premises, which may be less definitive or accepted. These concepts will come from the relationships known and supposed about the phenomenon. The reasons for doing this are that concepts can be multifaceted

Mathematical models shouldn't induce terror. They're just equations.

Models.

and linked to other concepts creating a framework of relationships underlying the phenomenon.

In traditional research, this is what a literature search is supposed to be for. Literature searches, though, are considered by some to be an academic activity not applicable to analyses done on the job. Not true. The process of thinking through what you want to measure is essential.

Once there are specific ideas to explore, ways they could be measured are identified. This search starts with conventional metrics, the ones everyone would recognize and know how they were determined. Then, any other ways to measure the concept directly are considered. From there, indirect measures or *surrogates* that could be used in lieu of a direct measurement are examined.

There are a lot of things to research about the world.

If there are no other options, the feasibility of developing a new measure based on theory is explored. Developing new measures or new scales-of-measurement is more difficult for the experimenter and less understandable for reviewers than using established measures, so they aren't a preferred strategy.

Finally, appropriate scales-of-measurement for the possible measures are selected. The difficulty in generating and using the data is a consideration. For example:

- *Qualities* are usually more difficult to measure accurately and consistently than quantities because there are more complex judgments involved.
- *Counts* are straightforward when they involve simple judgments as to what to count. Some judgments, such as species counts, can be relatively complex because you have to be able to identify the species before you can count it. Counts have no decimals and no negative numbers.
- *Amounts* are usually more difficult to measure than counts because the judgment process is more complex. Amounts have decimals but no negative numbers unless losses are admissible.
- *Ratio measures*, such as concentrations, rates, and percentages, are usually more difficult to measure than amounts because they involve two or more amounts. Ratio measures have both decimals and negative numbers.

Such changes can create problems when historical and current data are combined because variance differences attributable to evolving measurement systems introduce *heteroscedasticity*, which can produce misleading statistics. Appropriate controls to limit extraneous variability should be established.

The objective and ultimate use of the model must be taken into account. For example, if the aim of the model is to *predict* some dependent variable, quantitative

Models.

independent variables would usually be preferable to qualitative variables because they would provide more scale resolution. While a quantitative variable could be transformed into a less finely-divided scale or even a qualitative scale, it's impossible to go in the other direction. If you want your prediction model to be simple and inexpensive to use, don't select predictors that are expensive and time-consuming to measure.

Redundancy should be built into the variables if there is more than one way to measure a concept. Sometimes, one variable will display a higher correlation with a model's dependent variable or help explain outliers or analogous measurements in a related metric.

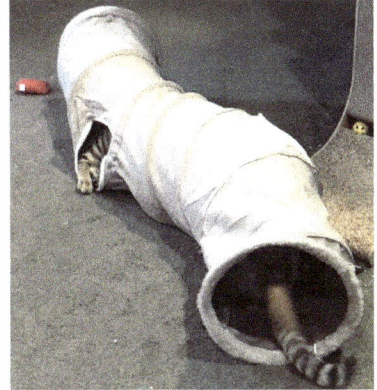

For example, redundant measures are often included in opinion surveys (Chapter 5) by using differently worded questions to solicit the same information. One question might ask "Did you like [something]?" and then in a later question ask "Would you recommend [something] to your friends?" or "Would you buy [something] again in the future?" to assess consistency in a respondent's opinion about a product.

You may not be able to decide which metric will be most effective in a model, so include them all and analyze them later to see which fits best.

Redundant variables can be a good check on data quality, but they all wouldn't necessarily go into the model. That would likely result in *multicollinearity*, correlations between independent variables, which is not a good thing.

DEPENDENT VARIABLES

To build a model, as many dependent variables as are needed to characterize the phenomenon are selected. Usually, statistical models have only one dependent variable. These are called *univariate* statistical models. If only one dependent variable is needed, that's great. It will make for a fairly straightforward analysis.

If more than one dependent variable is needed to describe a phenomenon, the model is called a *multivariate* statistical model, which is more complex to construct.

I'm a dependent kitten right now but I'll grow up to be an independent cat.

(Note: Some statisticians, particularly in the social sciences, refer to statistical procedures that involve either more than one *independent variable* or more than one *dependent variable* as multivariate. Other statisticians reserve this term for models

 Models.

involving multiple dependent variables because the complexity of the analysis is far greater if there are multiple dependent variables then if there are just multiple independent variables.)

If more than one dependent variable is needed, the number should be limited. Here are a few things that are often done to reduce the number of candidate dependent variables.

- **Focus on Separate Aspects of the Phenomenon.** Some phenomena are very complex or at least multifaceted. The number of dependent variables to be considered can be minimized by focusing on just one aspect of the phenomenon.
- **Narrow the Objective.** If the study is too broad, the aims of the study can be narrowed or the project can be divided into parts and conducted sequentially as subprojects.
- **Focus on Hard Information.** *Hard information* involves measurements of tangible, observable demonstrations as opposed to measurements of intangible beliefs or opinions. The number of dependent variables can sometimes be minimized by using only hard information.
- **Focus on Direct Information.** *Direct information* involves measurements specifically of the phenomenon being investigated, as opposed to measurements of factors associated with the phenomenon. The number of dependent variables can sometimes be minimized by using only variables that directly measure the phenomenon rather than its effects.

I'm not dropping out of this model.

- **Eliminate Correlated Variables.** If several candidate dependent variables are highly intercorrelated, the number of dependent variables can be minimized by using only the best and eliminating the rest.
- **Create Multiple Models.** If there is more than one dependent variable, separate models can be created for each one. This is like subdividing the objectives—not optimal but sometimes a necessary evil.
- **Conduct a Factor Analysis.** The number of dependent variables can sometimes be reduced using factor analysis, principal-components analysis, or another *variable-reduction technique* to combine the multiple variables into one.

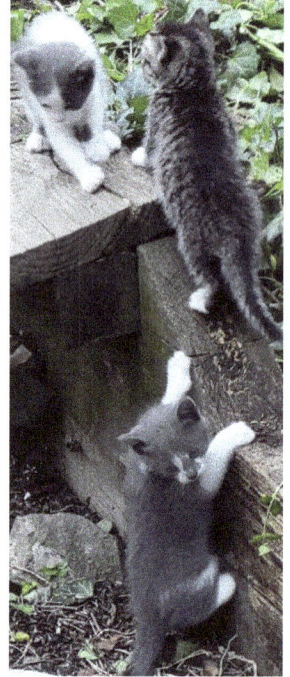

If none of these things can be done, the analysis will probably have to be multivariate.

Models.

INDEPENDENT VARIABLES

The selection of independent variables will hinge on what the model will be used for. Here are a few considerations that are used for identifying candidate measures and scales:

Variables for characterizing, classifying, identifying, and explaining. Variables to address all the theoretical aspects of the phenomenon, even to the point of having some redundancy, should be selected. Sometimes two differently measured or differently scaled variables that address the same theoretical concept will make dissimilar contributions to the model. When the model is calibrated, extra variables drop out.

Compared to my kibble, Charlie's food is delicious.

Variables for comparing. Only what is important to know should be tested, not everything under the sun. The number of variables should be kept to an absolute minimum or the analysis will become intractable. Conventionally recognized variables and scales should be used rather than creating new ones. This facilitates replication studies.

Variables for predicting. Prediction models should use variables and scales that are relatively inexpensive and easy to create or obtain otherwise the predictions will cost more to generate than they are worth. For example, if the model will be used repeatedly, say to make monthly forecasts, the model inputs need to be simple enough that all the data could be generated in a couple of days at most. If the inputs were so complex that they take months to generate, the model couldn't be used as intended.

Variables should stress precision. Accuracy tends to come easy while precision is elusive. Prediction models usually keep only the variables that work best in making a prediction, so the number of variables initially selected isn't that important. However, the more variables included in the initial conceptual model, the more work it is to winnow out the ones not needed.

Sugar will create the model and Dean will test it.

It's not a good idea to include a great number of predictors just to see what might turn up. This strategy could capitalize on the chance of one variable working great for the sample but not for new samples. Professional statisticians often split large datasets into portions used to create the model (*training datasets*) and test it independently (*testing dataset*).

Models.

Some variables have several possible scales. If these extra scales are related to each other by a linear algebraic relationship, only one is kept. This is because the variables will be perfectly correlated, and thus, will add no new information to the model.

Pick the scale that will give you the best resolution.

For example, in measuring temperature in degrees Fahrenheit, temperature in degrees Celsius doesn't also have to be measured because °C = 5/9(°F − 32). The scale that provides the best resolution is usually selected. In the example of temperature, Fahrenheit-scaled thermometers can be read with greater precision than Celsius-scale thermometers because they have smaller divisions. Digital thermometers that display decimal places would be better.

If two measures have unrelated scales or can be measured differently, all of them should be kept. The best measures are kept when the model is calibrated. For example, pH could be measured using pH paper, a field meter, or a lab titration. If a concept to be evaluated with a model were a person's size, both a height scale and a weight scale could be used. However, weight in both pounds and kilograms wouldn't be needed because the two scales are linearly related (1 kg = 2.2 lbs). Weight measured by different mechanisms—a balance beam, a strain gauge, a spring scale, or even a circus weight-guesser—could be kept because they use different techniques to measure weight (although they would probably be highly correlated).

SAMPLES AND DATA

Samples must represent the population to be analyzed. A lot of thought must go into defining the population and finding samples that will fairly represent that population. All those mental maneuvers go into a comprehensive sampling plan that details things like resolution, the number of samples, and the sampling scheme. So the sampler, the person who will generate the data, should never stray from the carefully thought-out plan. Field technicians, for instance, shouldn't move sampling locations so that they don't have to walk so far from their truck. Doctors shouldn't assign their friends to experimental groups that will get preferential treatment. Survey interviewers shouldn't concentrate on attractive members of the opposite sex. You get the idea.

Then there's also the process of generating the data to consider. As much as a plan aims to minimize variance with *reference*, *replication*, and *randomization*, there will always be opportunities at the point of data collection to improve the process. A

Models.

dropped meter may require recalibration that's not called for in the sampling plan. A surveyor can correct a map with an incorrectly located sampling point. An accountant can adjust misclassified entries in financial records.

Data analysts are mostly powerless to make such corrections and clarifications. They have to puzzle over the cause of outliers and missing data. The knowledge and experience of the people collecting the data have to be relied on. So when it comes to data, samplers should be given very clear instructions concerning what they can and cannot do to

I'm tired. I'm collecting the sample right here.

ensure the quality of the data, minimize variance, and achieve the intent, if not the letter, of the sampling plan. Document everything and preserve it in *metadata*. You never know what might become important.

MODEL DIAGNOSTICS

If you think building a statistical model is as easy as checking a box on an Excel or Tableau scatter plot, you're in for a big surprise. There are over a dozen diagnostic statistics that professional statisticians rely on whenever they build a model.

For a candidate model that has only one set of variables to assess, the diagnostic statistics provide information on how well the model fits the data. However, if more than one combination of independent variables is of interest, the diagnostic statistics help determine which of the models is preferable.

You won't need to understand how the following eighteen diagnostic statistics work, only that they exist and should be consulted whenever a serious statistical model is being developed.

OVERALL MODEL

The three most commonly used criteria for evaluating the overall quality of a model are the *coefficient of determination*, the *standard error of estimate*, and the *F-test*.

This is getting really complicated.

🐾 **Coefficient of Determination.** *R-square* or R^2 is a measure of how well the pattern of the model fits the pattern of the data, and hence, is a measure of *accuracy*. Some

Models.

statisticians believe that R-square is overused and flawed because it always increases as terms are added to a model. Whine. Whine. Whine.

- **Standard Error of Estimate.** The *SEE* takes into account the number of samples (more is better) and the number of variables (fewer is better) in the model, and is in the same units as the dependent variable. It is a measure of how much scatter there is between the model and the data, and hence, is a measure of *precision*.

- **F-test.** The *F-test* is a statistical test of whether the R-square value is different from zero. The F-value will vary with the numbers of samples and terms in the model. Many statisticians start by looking at the results of the F-test as a determiner of whether to keep the model, and then look at the R-square and SEE.

Evaluating models doesn't end with R-square, SEE, and F-test. There are five other diagnostic tools for evaluating the overall quality of statistical models, including diagnostic graphics and statistics.

Diagnostic Graphs for evaluating regression models include:

- **Plot of Observed vs. Predicted.** A graph with observed values on the y-axis and predicted values on the x-axis, in which data points should plot close to a straight 45-degree line passing through the origin of the axes. Some analysts switch the axes.

- **Plot of Observed vs. Residuals.** A graph with observed values on the y-axis and residuals (predicted values minus observed values) on the x-axis, in which data points should plot randomly around the origin of the axes. Some analysts switch the axes.

- **Histogram of Residuals.** A frequency distribution of the model's residuals that should approximate a Normal distribution.

Other diagnostic statistics used in evaluating an overall regression model include:

- **AIC and BIC.** The *Akaike's Information Criterion* and the *Bayesian Information Criterion* are statistics for comparing alternative models.

- **Mallows' Cp Criterion.** *Mallows' Cp Criterion* is a relative measure of inaccuracy in the model given the number of terms.

This will be a good tool to test my hunting skills.

Models.

Normally-distributed residuals are good.

MODEL COMPONENTS

Once a small number of alternative models is selected, statistical diagnostics are used to evaluate the components of the statistical model.

REGRESSION COEFFICIENTS

Statistical software calculates two types of regression coefficients. The *unstandardized regression coefficients* are the a_0 through a_n terms in the model. These are the values used to calculate a prediction of the y variable from the values of the x variables. The *standardized regression coefficients* are equal to the *unstandardized regression coefficients* divided by the *standard errors* of the coefficients. Standardized regression coefficients, also called *Beta coefficients*, are used to compare the relative importance of the independent variables.

You can tell the difference between the unstandardized and the standardized coefficients if they are not labeled clearly by looking for a coefficient for the constant in the model. There is no standardized coefficient for the constant term in the model but there is an unstandardized coefficient.

VARIABLES

Three diagnostic statistics are used to assess the importance of the variables in the model.

- **t-tests and probabilities.** Used to verify that the regression coefficients are different from zero. The t-values may change significantly depending on what other terms are in the model. The probabilities for the tests are used to include or discard independent variables.
- **Variance Inflation Factors.** *VIFs* are measures of how much the model's coefficients change because of correlations between the independent variables (i.e., *multicollinearity*). The reciprocal of the VIF is called the *tolerance*.
- **Partial Regression Leverage Plots.** Graphs of the dependent variable (on the y-axis) versus each independent variable (on the x-axis) from which the effects of the other independent variables in the model have been removed.

Models.

The slope of a line fit to the *leverage plot* is equal to the *unstandardized regression coefficients* for that independent variable.

Too many things to look for...

OBSERVATIONS

Finally, the observations used to create the statistical model are evaluated using five diagnostic statistics:

- 🐾 **Residuals.** *Residuals* are the differences between the observed values and the model's predictions.
- 🐾 **DFBETAs.** The changes in the regression coefficients that would result from deleting the observation.
- 🐾 **Studentized Deleted Residual.** A measure of whether an observation of the dependent variable might be overly influential.
- 🐾 **Leverage.** A measure of whether an observation for an independent variable might be overly influential.
- 🐾 **Cook's Distance.** A measure of the overall impact of an observation on the coefficients of the model.

Again, don't worry about how the diagnostic statistics work, that's what professional statisticians have to do. Just know that they exist and if you see them, you'll know a professional statistician is involved.

11 REASONS TO DOUBT A REGRESSION MODEL

Finding a model that fits a set of data is one of the most common goals in data analysis. *Least squares regression* is the most commonly used tool for achieving this goal. It's a relatively simple concept, it's easy to do, and there's a lot of readily available software to do the calculations. It's taught in most Stats 101 courses. Everybody uses it … and therein lies the problem. Even if there is no intention to mislead anyone, it does happen. So, be aware.

Here are eleven of the most common reasons to doubt a regression model.

Models.

1. NOT ENOUGH SAMPLES.

Accuracy is a critical component for evaluating a model. The *coefficient of determination* is the most often cited measure of accuracy. Now obviously, the more accurate a model is the better, so data analysts look for large values of R-squared … but there are subtle issues to consider.

R-squared is designed to estimate the maximum relationship between the dependent and independent variables based on a set of samples (cases, observations, records, or whatever). If there aren't enough samples compared to the number of independent variables in the model, the estimate of R-squared will be unstable and inflated. The effect is greatest when the R-squared value is small, the number of samples is small, and the number of independent variables is large.

More things to look for.

You can't control the magnitude of the relationship between a dependent variable and a set of independent variables, and often, you won't have total control over the number of samples and variables either. The inflation in the value of R-squared, however, can be assessed by calculating the *shrunken R-square*.

For an R-squared value above 0.8 with 30 cases per variable, there isn't much shrinkage. The original R-squared value will be a good estimate. Lower estimates of R-square, however, experience considerable shrinkage. Consequently, treat such regression models with some skepticism.

2. NO INTERCEPT.

Almost all software that performs regression analysis provides an option to not include an intercept term (i.e., the model's constant term) in the model. This sounds convenient, especially for relationships that presume a one-to-one relationship between the dependent and independent variables. However, when an intercept is excluded from the model, it's not omitted from the analysis; it is set to zero.

Models need intercepts.

Look at any regression model with "no intercept" and you'll see that the regression line goes through the origin of the axes. With the regression line nailed down on one

Models.

end at the origin, you might expect that the value of R-squared would be diminished because the line wouldn't necessarily travel through the data in a way that minimizes the differences between the data points and the regression line. Instead, R-squared is artificially inflated. This happens because when the correction provided by the intercept is removed, the total variation in the model increases. However, the ratio of the variability attributable to the model compared to the total variability also increases. Hence, there is an increase in R-squared.

I'm shrunken, like an R-square.

The solution is simple. Always have an intercept term in the model.

3. STEPWISE REGRESSION.

Stepwise regression is a data analyst's dream. Throw all the variables into a hopper, grab a cup of coffee, and let the silicon chips tell you which variables yield the best model. That irritates hard-core statisticians who don't like amateurs messing around with their numbers. You can bet, though, that at least some of them go home at night, throw all the food in their cupboard into a crock pot, and expect to get a good meal out of it. Same thing.

The cause of some statistician's consternation is that stepwise regression will select the variables that are best for the dataset, but not necessarily best for the population. Probabilities for tests of the model are optimistic because they don't account for the stepwise procedure's ability to capitalize on chance. Moreover, adding new variables will always increase R-squared, so you have to have some good ways to decide how many variables are too many.

I've got my coffee. Let's do some stepwise regression.

There are ways to assess the number of independent variables in a model. Most notably are diagnostic statistics such as AIC, BIC, and Cp. Also, the results can be verified using a different dataset, either by splitting the dataset into *training* and *testing* datasets or by collecting new samples.

Using stepwise regression alone isn't a fatal flaw, but like with guns, drugs, and fast food, you have to be careful how you use it.

Models.

4. OUTLIERS.

Outliers are a special irritant for data analysts but are also great thought provokers. Sometimes they tell you things the patterns don't. They're not really that tough to identify but they cause a variety of problems that data analysts have to deal with.

The first problem is convincing reviewers not familiar with the data that the outliers are in fact *outliers*. Second, data analysts have to convince reviewers that what they want to do with them, delete or include or whatever, is the appropriate thing to do. One way or another, though, outliers will wreak havoc with R-squared (Chapter 7).

Outliers aren't just a fanciful notion. Look back in Chapter 4 at the scatter plot showing the **Caitlin Clark effect.** They're real.

5. RELATIONSHIPS.

Least-squares regression assumes that the relationship between a dependent variable and a set of independent variables is linear, or at least, *intrinsically linear,* so it can be modified with an appropriate *transformation*. If the relationship is actually nonlinear, the R-squared for the linear model will be lower than it would be for a better fitting nonlinear model.

Relationships are a relatively simple problem to fix, or at least acknowledge, once you know what to look for. Graph your data and go from there.

6. OVERFITTING.

I may be getting overfit for the tube.

Overfitting involves building a statistical model solely by optimizing statistical parameters. It typically involves using a large number of variables and transformations of the variables. The resulting model may fit the data almost perfectly but will produce erroneous results when applied to another sample from the population.

The concern about overfitting may be somewhat overstated, like the correlation-causation admonition. Overfitting is like becoming too muscular from weight training. It doesn't happen suddenly or simply. If you know what overfitting is, you're not likely to become a victim. It's not something that happens in a keystroke. It takes a lot of work fine-tuning variables and what not. It's also usually easy to identify overfitting in other people's models. Simply look for a conglomeration of manual numerical adjustments, mathematical functions, and variable combinations.

Models.

7. MISSPECIFICATION.

Misspecification involves including terms in a model that make the model look great statistically even though the model is problematic. Often, misspecification involves

One of us is not like the other.

placing the same or a very similar variable on both sides of the equation. This can happen when both dependent and independent variables in a model are indexed to the same baseline or standard.

One example is the use of Gross Domestic Product to index variables in econometric studies. Index the dependent variable or the independent variables but not both. Another form

of misspecification involves creating a prediction model having independent variables that are more difficult to generate than the dependent variable. Also, if you need to forecast something a month in advance, don't use predictors that are measured at intervals longer than a month.

8. MULTICOLLINEARITY.

Multicollinearity occurs when a model has two or more independent variables that are highly correlated with each other. The consequences are that the model will look fine, but predictions from the model will be erratic.

If you ever tried to use independent variables that sum to a constant, you've seen multicollinearity in action. In the case of perfect correlations, such as these, statistical software will crash

Sometimes a strong correlation isn't a good thing.

because it won't be able to perform the matrix mathemagics of regression. Most instances of multicollinearity involve weaker correlations that allow statistical software to function, yet the predictions of the model will still be unreliable.

Multicollinearity occurs often in fields of study in which many variables are measured in the process of model building. Diagnosis of the problem is simple; look for large correlations between the independent variables.

Models.

9. UNEQUAL VARIANCES.

Regression, and practically all parametric statistics, requires that the variances in the model residuals be equal at every value of the dependent variable. This assumption is called *equal variances*, *homogeneity of variances*, or coolest of all, *homoscedasticity*. Violate the assumption and you have *heteroscedasticity*.

Heteroscedasticity is assessed much more commonly in analysis-of-variance models than in regression models. This is because the dependent variable in ANOVA is measured on a categorical scale while the dependent variable in regression is measured on a continuous scale.

I've got my own bin.

The solution to this is fairly simple—use *binning*. Break the dependent variable scale into intervals, like in a histogram, and calculate the variance for each interval. The variances don't have to be precisely equal, but variances different by a factor of five or more are problematic. Unequal variances will wreak havoc on any tests or confidence limits calculated for model predictions. It can often be fixed using a transformation.

10. AUTOCORRELATION.

Autocorrelation involves a variable being correlated with itself. It is the correlation between data points with previously generated data points (termed a *lag*). Usually, autocorrelation involves time-series data or spatial data, but it can also involve the order in which data are collected, called *serial correlation*. The terms *autocorrelation* and *serial correlation* are sometimes used interchangeably, but they're really not.

If the residuals of a model are autocorrelated, it's likely that the variances will also be unequal. That means, again, that tests or confidence limits calculated from variances should be suspect.

11. WEIGHTING.

Statistical software for regression allows you to weight the data points. You might want to do this for several reasons. *Weighting* is used to make data points that are more reliable or relevant more important in model building. It's also used when each data point represents more than one value. The issue with weighting is that it will change the *degrees of freedom*, and hence, the results of statistical tests. Usually this is OK, a necessary change to accommodate the realities of the model. However, if you ever come upon a weighted least squares regression model in which the weightings are arbitrary, don't believe the test results.

Models.

TROUBLESHOOTING REGRESSION MODELS

There are many technologies we use in our lives without really understanding how they work, like sewing machines, microwave ovens, and the internet. It's not how to use these mechanisms; most everybody knows that. It's understanding them well enough to fix them when they break.

Regression analysis is like that too. Only with regression analysis, sometimes you can't even tell if there's something wrong without consulting an expert.

Here are some tips for troubleshooting regression models.

There are a lot of diagnostic tests you can perform on a model.

If you have created your own regression model, not counting just checking a box on a graphing program to add a trend line, you still may not know about some of the more subtle pitfalls you can encounter. That's what Table 37 summarizes.

The biggest red flag that something is amiss is TGTBT, *too good to be true*. If you encounter an R-squared value above 0.9, especially unexpectedly, there's probably something you haven't considered. It may be an *in-line outlier*, *no intercept*, or a *model misspecification* in which the same metric is hidden in variables on both sides of the equals sign.

Another red flag is inconsistency. If estimates of the model's parameters (i.e., the coefficients for the independent variables, the model's constant, and the error term) change appreciably between datasets, there's probably something wrong. Likely culprits include a small *sample size*, *multicollinearity*, or *overfitting*.

Finally, if predictions from the model are less accurate or precise than you expected, there's probably some fundamental issue involved. Possibilities include *non-representative* (biased) *sampling*, small *sample size*, and lack of *variance control*.

Table 37 provides some other hints for troubleshooting a model you developed, or at least know a lot about.

If you are skeptical about a regression model that was developed by others, your biggest challenge will be that you won't have access to all their diagnostic statistics and plots, let alone their data. You can always ask the model creator for the information you want, but they may not be amenable to releasing it. Model creation can be a messy process, so they may not have documented their actions in a format that would be useful to anyone besides themselves.

Models.

If, you're reading about a model in a book, website, or the media, you've probably got all the information you're ever going to get. You have to be a statistical detective. Table 38 provides some clues you might look for.

Data relationships are complicated so the models that describe them are going to be complicated too, even if they are simplified representations of reality. Statistics provides a variety of techniques for creating models of data relationships and scores of diagnostic statistics and graphics to assess their validity. Still, there are quite a few ways that models of data relationships can be misleading. So, don't be surprised if you see a statistical model described in a media presentation that looks a bit sketchy and there's no way for you to prove it.

Don't be discouraged if this is way more information than you expected and can even comprehend. It's like a smorgasbord, take what you want and come back later when you want more. You may run into a media story about a regression model that does something amazing and want to consider its legitimacy. This chapter will give you some ideas to think about.

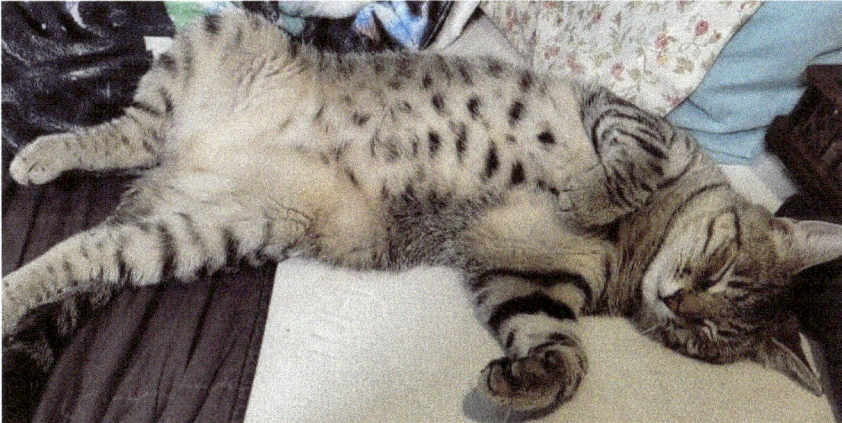

I like to nap after feasting on all this information.

Models.

Table 37. Diagnosis of your own regression models.

Your Model	Identification	Correction
Not Enough Samples	If you have fewer than 10 observations for each independent variable you want to put in a model, you don't have enough samples.	Collect more samples. 100 observations per variable is a good target to shoot for although more is usually better.
No Intercept	You'll know it if you do it.	Put in an intercept and see if the model changes.
Stepwise Regression	You'll know it if you do it.	Don't abdicate model building decisions to software alone.
Outliers	Plot the dependent variable against each independent variable. If more than about 5% of the data pairs plot noticeable apart from the rest of the data points, you may have outliers.	Conduct a test on the aberrant data points to determine if they are statistical anomalies. Use diagnostic statistics like leverage to evaluate the effects of suspected outliers. Evaluate the metadata of the samples to determine if they are representative of the population being modeled. If so, retain the outlier as an influential observation (AKA leverage point).
Non-linear relationships	Plot the dependent variable against each independent variable. Look for nonlinear patterns in the data	Find an appropriate transformation of the independent variable.
Overfitting	If you have a large number of independent variables, especially if they use a variety of transformation and don't contribute much to the accuracy and precision of the model, you may have overfit the model.	Keep the model as simple as possible. Make sure the ratio of observations to independent variables is large. Use diagnostic statistics like AIC and BIC to help select an appropriate number of variables.
Misspecification	Look for any variants of the dependent variable in the independent variables. Assess whether the model meets the objectives of the effort.	Remove any elements of the dependent variable from the independent variables. Remove at least one component of variables describing mixtures. Ensure the model meets the objectives of the effort with the desired accuracy and precision.
Multicollinearity	Calculate correlation coefficients and plot the relationships between all the independent variables in the model. Look for high correlations.	Use diagnostic statistics like VIF to evaluate the effects of suspected multicollinearity. Remove intercorrelated independent variables from the model.
Heteroscedasticity	Plot the variance at each level of an ordinal-scale dependent variable or appropriate ranges of a continuous-scale dependent variable. Look for any differences in the variances of more than about five times.	Try to find an appropriate Box-Cox transformation or consider nonparametric regression or data mining methods.
Autocorrelation	Plot the data over time, location or the order of sample collection. Calculate a Durbin–Watson statistic for serial correlation.	If the autocorrelation is related to time, develop a correlogram and a partial correlogram. If the autocorrelation is spatial, develop a variogram. If the autocorrelation is related to the order of sample collection, examine metadata to try to identify a cause.
Weighting	You'll know it if you do it.	Compare the weighted model with the corresponding unweighted model to assess the effects of weighting. Consider the validity of weighting; seek expert advice if needed.

Models.

Table 38. Diagnosis of other people's models.

Another Analyst's Model	Identification
Not Enough Samples	If the analyst reported the number of samples used, look for at least 10 observations for each independent variable in the model,
No Intercept	If the analyst reported the actual model (some don't), look for a constant term.
Stepwise Regression	Unless another approach is reported, assume the analyst used some form of stepwise regression.
Outliers	Assuming the analyst did not provide plots of the dependent variable versus the independent variables, look for R-squared values that are much higher or lower than expected.
Non-linear relationships	Assuming the analyst did not provide plots of the dependent variable versus the independent variables, look for a lower-than- expected R-squared value from a linear model. If there are non-linear terms in the model, this is probably not an issue.
Overfitting	Look for a large number of independent variables in the model, especially if they different types of transformation
Misspecification	Look for any variants of the dependent variable in the independent variables. Assess whether the model meets the objectives of the effort.
Multicollinearity	Assuming relevant plots and diagnostic statistics are not available, there may not be any way to identify multicollinearity.
Heteroscedasticity	Assuming relevant plots and diagnostic statistics are not available, there may not be any way to identify heteroscedasticity.
Autocorrelation	Assuming relevant plots and diagnostic statistics are not available, there may not be any way to identify serial correlation.
Weighting	Compare the reported number of samples to the degrees of freedom. Any differences may be attributable to weighting.

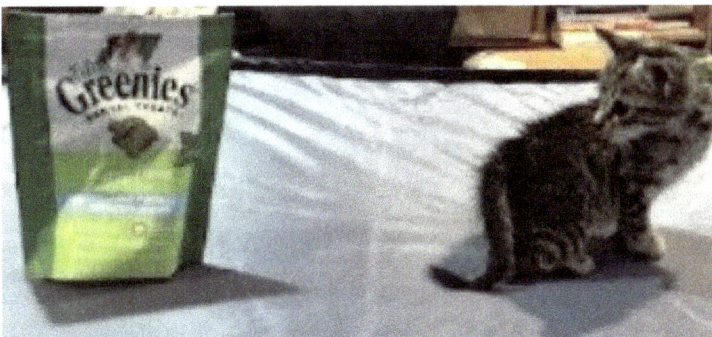

If you can't get the information you need to evaluate a model, just have a few treats and take a nap.

Models.

*I need to think a lot more about all this modeling
stuff if I'm going to be statistically literate.*

Models.

CHAPTER 9. LITERACY.
How To Evaluate Statistics In Life.

INTRODUCTION

Information is more convincing if it has statistics and graphs to lend it credibility. As a consequence, people often try to enhance their works with them. That's a good thing. Occasionally however, the statistics and graphics are misleading, maybe by accident or by design. Either way, you need to be able to recognize possible issues. You need to be statistically literate.

Statistical thinking is the **process** of how practitioners explore a phenomenon in a population. It's inductive reasoning with numbers. *Statistical literacy* is understanding, at least in general terms, the process of statistical thinking.

That's what this Chapter is about.

THE BEST SUPERPOWER OF ALL

If you're not a mutant, an extraterrestrial, an adventure seeker prone to outlandishly fortuitous accidents, or a wealthy scientific genius who can engineer all sorts of wondrous gadgets, don't despair. You can still have your own Superpower. In fact, it's the best Superpower of all. It's the power of *critical thinking*.

What's critical thinking? It's simply being able to assess the truthfulness and validity of the things people say. That may not sound as awesome as super speed or morphing into an animal, but it has distinct advantages. With critical thinking, you don't need a special costume. You don't need to hide your power or protect

No human can resist my Super Cuteness.

your identity or explain why your clothes are torn apart. It won't leave you physically exhausted and doesn't involve fisticuffs (usually).

You can use critical thinking anywhere in any situation. You can use it on your teachers and classmates, your boss and co-workers, the trolls on social media, politicians and pundits, salesmen and ministers, and everyone else who tries to get into your head. And, you can learn to think critically whether you're sixteen or sixty.

Literacy.

LEARNING TO THINK CRITICALLY

Thinking critically is a skill. Some people have it as a raw, innate talent and others have to develop it. Like in sports, you first have to learn the fundamental concepts, then consciously practice them until they become routine, and then continue to use them until you no longer even think about them. But unlike in sports, you'll be able to put the skills to use for a lifetime, not just for a few years when you're in your physical prime.

Even those with an innate ability to think critically have to practice. Whether you're a sports professional or a superhero, practice prepares you for the unpredictable occasions when you have to put your power to the test. But practicing your critical thinking skills doesn't require the grueling hours every day in the gym that athletes have to commit to. It doesn't requires formal studies or a college education, though they are beneficial. It just involves some easy mental exercises. You only have to be aware of how to develop the skill.

Here's a six-stage process that you can use to develop your own critical-thinking skills.

STAGE 1. LISTEN.

Start simply. Pay attention to the conversations you have, the internet forums you follow, the TV you watch, especially the ads, and any other communication you may read, watch, or hear. Then, decide if the communication is meant to persuade you to do or think or believe something. If it doesn't, ignore it for the time being.

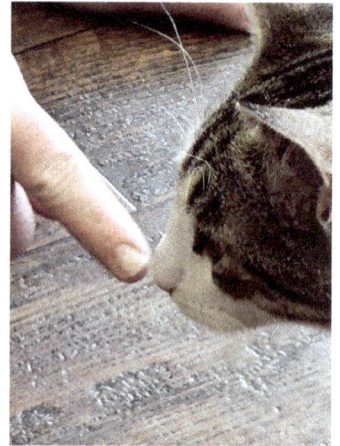

Pay attention.

Focus on those communications that want you to buy a product, or accept a belief, or support a position. You need to be able to recognize these arguments at sight, almost without thinking. Take as long as you need to get good at this. Remember, it took Spiderman more than a few tries to learn to cling to walls, but if he hadn't, he wouldn't have been able to swing from building to building.

STAGE 2. PARSE.

Once you can easily identify those attempts to persuade you, the next step is to pick apart the pieces of those arguments. Think of arguments as consisting of three parts:

- 🐾 **Premises.** The facts the argument relies on
- 🐾 **Logic.** The way the premises are manipulated

Literacy.

🐾 **Conclusion.** The result of the argument.

You need to learn to identify these pieces before you can start evaluating the validity of an argument. This process wouldn't be difficult if you only talked to other critical thinkers. Unfortunately, most people aren't critical thinkers, though many claim to be. The torrent of mental chaos you'll encounter from them is daunting. So again, start simply.

Every superhero has a weakness; mine is catniptonite.

Look up examples of formal arguments on Wikipedia and then other internet and textbook sources. When you're comfortable with picking out the parts of formal arguments, you can move on to the chaotic musings of the *idiocrasy*. In those, you'll find more of a challenge.

The parts of arguments from people who don't think critically don't always come in the customary premise-logic-conclusion order. Some arguments don't spell out all the premises. The logic behind arguments is often unstated. The conclusion is the only thing you can count on being present, but it may be the first. Sometimes, it's the only part of an argument you'll hear.

With practice, you'll get good at it. And when you do, you'll find that, even at this point, you are developing a greater awareness of critical thought than most of your cohorts.

STAGE 3. CHECK.

This is where learning to think critically gets interesting. Stage 3 involves looking at the components of arguments, which you should now be really good at picking out, and checking them for common flaws. Here are some things you should look for.

PREMISES

Are the premises actual facts or just someone's claim? The argument may or may not cite a source, but even if it does, that doesn't always mean the fact is valid.

Sometimes a source is biased or just a regurgitation of another biased source. Get into the habit of searching the internet for verification. Start with relatively unbiased sites like factcheck.org and procon.org, then start looking at other sites. You'll develop a feel for the biased sites and how they spin information.

Check if sources are known for bias at sites like allsides.com and adfontesmedia.com. Before long, you'll find you're developing a highly capable bullcrap detector that's even better than Spidey sense.

LOGIC

As soon as I grow into this helmet, I'll be able to ride on the Catcycle.

Don't worry too much at this point about whether the logic is correct. Instead, look for obviously *incorrect* logic, indicated by the presence of any of the most common fallacies.

There are dozens if not hundreds of fallacies. Don't try to learn a lot of them. Start with a few that are usually easy to spot:

Judgmental language fallacies involve making arguments using emotion-laden words. Phrases like religious fanatics, free-loading welfare recipients, lazy unemployed, and greedy bankers all impart more than just the essence of the argument. In today's contentious society, judgmental language is hard not to find.

Ad hominem is an attack on the opponent rather than the opponent's position. This happens all the time on political forums and often involves a reference to Hitler, illegal activities involving animals, or body parts.

Appeals to false authority involve basing arguments on information from someone who isn't really an expert. Think of all those celebrities hired to act in infomercials. Avoid peer pressure and the hive mind; think for yourself. Also beware of any reliance on common sense, which is more myth than magic.

After you are comfortable with these fallacies, read on in this chapter about others like *straw man*, *cherry picking*, *red herring*, *oversimplification*, *begging the question*, *slippery slope*, *equivocation*, and *quoting out of context*. Take them one step at a time.

We have more claws than Wolverine, and in a year, we'll know how to use them.

It's easy to read lists of fallacies and even memorize them but it's not so easy to recognize them in actual arguments, especially on social media. Don't get discouraged. Learning how to think critically takes time to master. It's a marathon not a sprint, and a lifelong race at that.

Literacy.

CONCLUSION

Argument conclusions can be of two types. *Deductive arguments* use premises about general information to conclude something specific. Sherlock Holmes, Doctor Who, and Patrick Jane (The Mentalist) are all masters of deduction, albeit fictional. *Inductive arguments* use premises about specific information to conclude something general. Statistics is inductive reasoning using specially-collected data. Beware of inductive arguments based on anecdotes, like Ronald Reagan's apocryphal welfare queen.

Induction often involves statistics and probability because counterexamples are such easy argument killers. *Abductive reasoning* involves finding the most likely explanation for a single observation when there isn't adequate information. Other than ignoring anecdotes, settle for being able to distinguish the two basic types of arguments, *induction* and *deduction*.

The point of this stage is for you to be able to identify the more obvious red flags. If you complete this stage, you will be far ahead of most people. Enjoy your mental prowess and use it every day.

STAGE 4. TRIAGE.

You'll find that you can spot a lot of faulty arguments just by knowing these few things to look for. In fact, you'll probably find most of the arguments you listen to are faulty in one way or another. And that's a problem. You have to be judicious with how you use your new power to think critically. You can't just engage in battle with every troll who wants to argue over a movie, an athlete, or worst of all, a politician.

Still, with great power comes great responsibility. You just have to slap down some infuriating arguments. Given that, you'll need to develop a sense for when you have to step up to expose misinformation and idiocy versus when you can roll your eyes and let it slide.

Further, you'll have to develop a sense for when you have inflicted enough damage to withdraw. Heroes don't slaughter their enemies; humiliation works just as well. You must become a guerrilla thinker. Pick your battles. Fight them in earnest. Then dissolve into the shadows. This is harder to do than it seems.

With great power comes great responsibility.

Literacy.

STAGE 5. ANALYZE.

Most disciples of critical thinking get at least as far as stage 4. Elite thinkers go far beyond that to thoroughly analyze all the components of an argument.

Analyzing arguments is challenging. It requires knowledge of a broad variety of subjects, the development of sophisticated analytical skills like statistics and logic, the availability of resources that can support your quest for the truth, and lots and lots of practice. This learning process never ends. The more you know the more sophisticated are the arguments you'll can take on.

Here are some of the things you might explore as you become an elite thinker.

I am Fireworks-Kitten, hear me roar.

Premises. Indisputable facts make good premises but not all facts are indisputable. Some premises purported to be facts are actually opinions or *factoids* (i.e., assertions that are made so commonly that they are assumed to be true). Elite thinkers also do not just consider the source of the fact because facts from even obviously unbiased, primary sources may not be entirely valid. Some facts, especially in science, evolve as topics are studied further.

Data analysis can be idiosyncratic. For example, different statisticians may come to different conclusions from the same dataset because of the way they scrub the data,

I am an elite thinker.

transform variables, and conduct analyses. One sign of a flawed data analysis is a lack of reference points like baselines and control groups. There are many other red flags that elite thinkers know to look for. In time, so can you.

Logic. There are many more fallacies that you can learn to recognize, though you'll probably have to do your own research to go beyond the easy pickings found in this chapter. But elite thinkers don't just look for errors (fallacies), they also consider the proper use of logical processes, the rules used to convert premises into conclusions.

In deductive reasoning, logical processes are called propositional logic rules. For example, using the letters P, Q, and R for premises:

Literacy.

- *Modus Ponens.* If P entails Q and P is true then Q is also true.
- *Modus Tollens.* If P entails Q and Q is false then P is also false.
- *Hypothetical Syllogism.* If P entails Q and Q entails R then P entails R.

As you might guess, there are many more logical rules. Presumably, this is what Spock spent all those years studying on Vulcan.

Inductive reasoning requires an appreciation of statistical thinking. In essence, statistical thinking posits that everything is connected, everything has inherent and extraneous variability, and extraneous variability needs to be controlled. Fatal flaws in inductive arguments usually stem from a failure to understand these concepts.

Conclusions. For any critical thinker, the validity of an argument is paramount, but for elite thinkers, the subtleties of how an argument is constructed and presented is also important. This is where having a good understanding of modes of communication, writing styles, propaganda, pragmatic context, and subliminal and nonverbal communications are essential.

GETTING STARTED

So, that is an informal path to becoming a critical thinker. There are other paths you might take, like formally studying the topic in school. Whatever approach fits best with your talent, lifestyle, and opportunities. You just have to start, and be patient that you will become more successful at it every day.

I am a Time Lord.
I can sleep 18 hours a day and still know
when to wake you up for gooshy food.

If you haven't consciously followed one of the paths, you may not be the critical thinker that you believe you are. Nothing comes without effort.

Getting started isn't difficult, it just takes practice, and then, more practice. The more you practice the more you'll learn. The more you learn the better you'll be at it. Just give it a try. Once you start experiencing the rewards, you'll understand why critical thinking is the best superpower of all.

SCIENCE VERSUS THE SCIENTIFIC METHOD

While statistics is used in many fields outside of STEM, the process of statistical analysis is closely aligned with the scientific method.

Science is our perception of how things work. The scientific method is how we determine what is the current state of science. Science is the product of the successful application of the scientific method. They are not the same. For one thing:

Literacy.

Science changes;
the scientific method doesn't.

When people say *trust the science* what they really mean to say, or should mean to say, is *trust the scientific method*. Science is constantly in a state of flux. It is never settled because there are always new things to learn.

In the 1950s, high school students were taught that there were electrons having a negative charge, protons having a positive charge, and neutrons having no charge. Earlier generations never learned about any of these in school. It was all too new and unsettled. Today, everybody learns about protons, neutrons, and electrons but they may not have heard about all the different kinds of elementary particles.

KNOWLEDGE NEVER STANDS STILL

Statistics is a science that, as you might expect, also changes frequently. There are new statistical tests and procedures introduced every year. Some catch on and some don't. There are even new disciplines in statistics that emerge.

Surveys were around for decades before computers and the internet made them ubiquitous. *Geostatistics* grew out of mining geology in the 1950s. *Sports statistics* date back to the 19th Century but gained prominence in the 1980s with *sabermetrics*. *Data mining* evolved from before WWII but emerged as a distinct field of statistics after *big data* became prevalent fifty years later. Word-frequency graphics evolved from *tag clouds* in the mid-2000s to *wordles* (not the game that first appeared in 2021) and finally *word clouds* a decade later. *Data science* became established as a career path in the 2010s. The same kind of evolution occurs in every STEM field, driven by new scientific knowledge and new technologies.

Knowledge never stands still. Even science is constantly changing.

Don't expect all the things you learn in school to stay the same for your entire life. This is true for every discipline. Even interpretations of classic works of literature, music, and art are reevaluated every year by aspiring PhD candidates.

Use school to learn how to think critically and teach yourself new things on your own. You'll need those two skills from the day you start school until long after you retire.

MAKING COOKIES

Creating science is like making cookies—you need a recipe, ingredients, and tools to combine the ingredients and bake the dough.

Literacy.

- 🐾 The **recipe** is the scientific method.
- 🐾 The **ingredients** are the knowledge of the discipline and the data from the experiments.
- 🐾 The **tools** are logic and philosophical principles.
- 🐾 The **dough** is the raw result.
- 🐾 The **cookies** are the interpreted results that have been peer-reviewed, reported in professional publications, and debated in the discipline community.

If you've ever made cookies, you know that if you use quality ingredients and follow the recipe, everything will turn out fine. It helps if you have some experience with the tools you'll use and with making cookies in general. Making science is kind of like that.

THE RECIPE

The internet has scores of websites that aim to explain the *scientific method*, often as *infographics*. Some are more detailed than others, some have steps that others don't. Even so, in real life, it's more complex than you might imagine.

Ooooh. Smells like science.

The scientific method is not a rigid formula, it's more of a guideline for what things to include in research and when to include them. The scientific method is different from "scientists' methods," which are just practices individual researchers use often because they have found them to work in the past.

For example, a researcher might limit his experiments to thirty samples because that's what he was told he should do by his thesis advisor. It's not part of the scientific method, it's just a habit the researcher adopted. It's like how a cookie maker might put his own personal stamp on his results by adding a teaspoon of nutmeg to every recipe whether it's called for or not.

Although the scientific method doesn't change, how it is implemented does. For one, how researchers design and implement an *observational* study is very different from how they design and implement an *experimental* study (see Chapter 6). Different mindsets, different populations and phenomena, and different hypotheses, but both types of study still rely on the scientific method.

Statistical studies follow the same basic steps as for the scientific method, only there is more attention paid to fundamental statistical concepts, such as *populations*, *scales of measurement*, *variance control*, and *statistical assumptions*.

Here's what the scientific method for statistical studies looks like:

Literacy.

1. Make an observation, have a thought, or get into an argument on social media.
2. Do background research. Somebody may have already invented that wheel. Remember the geologist's old admonition, "a month in the field will save you an hour in the library."
3. Define the research question to be investigated. Determine if the research will be *observational* or *experimental* as this will establish what statistical designs will be applicable. Note whether the question involves data description, comparison, or relationships as this will influence what statistical techniques will be needed.
4. Depending on the information available on the research question, either:
   a. Collect more observations anecdotally to refine the question for a preliminary study, or
   b. Design a preliminary study to answer the question and identify needs for additional data, or
   c. Design a confirmatory study to answer the question definitively.

5. Define the *phenomenon* to be investigated and the metrics that will be used to characterize the phenomenon. Identify the instruments and procedures for generating data on the metrics. Determine if the procedures and instruments will provide appropriate accuracy and precision. Identify *scales of measurement* for all metrics as this will influence what statistical techniques will be applicable.
6. Define the characteristics of the *population* to be investigated. Decide what kinds of inferences might be made to the population. Identify an appropriate *sampling scheme* for obtaining a *representative sample* from the population. Select sample collection locations, frame, or group assignments, as appropriate. Identify appropriate *variance control* approaches of *reference, replication*, and *randomization*.

The scientific method sounds simple but it's really complicated and hard to get right.

7. Develop a hypothesis that can be tested. Write *null* and *alternative hypotheses* (Chapter 6). Estimate the number of samples that will be needed for the analysis considering the resolution needed for the number of grouping variables and tests to be carried out.
8. Collect data using appropriate quality control and variance reduction procedures. This is the crux of the research. If the data collection is faulty, either because of a bad design or implementation, the research study is a

Literacy.

> *A fact is a simple statement that everyone believes. It is innocent, unless found guilty. A hypothesis is a novel suggestion that no one wants to believe. It is guilty, until found effective.*
>
> Edward Teller, 1908-2003, Hungarian-American theoretical physicist and chemical engineer.

failure. If the data analysis is problematic, it can be repeated so long as the data are good.

9. Process and analyze the data. All analyses start with *data scrubbing* and an *exploratory data analysis*. Further analyses will depend on the objective of the study—classify/identify, compare, predict/explain, or explore. Look for *violations of assumptions*.

10. Test the hypothesis and reevaluate as necessary. Make and test predictions based on the hypothesis. Draw conclusions and report findings. Document the *metadata*.

Both the scientific method and cookie making can be viewed as either *batch* (once-and-done) or *iterative* (continually repeating) processes depending on the scope of the goal. Deep scientific research usually involves many experiments based on evolving knowledge, but so too can the search for the very best recipe for peanut butter cookies. Some scientific research involves a single, straightforward experiment, just to find out something. Sometimes you make cookies just to try out a new recipe.

THE INGREDIENTS

The ingredients of the scientific method are *domain expertise* (i.e., the knowledge of the discipline) and the data from the experiments. Even before you think about collecting data from an experiment, you need to know your stuff. You can't make cookies if you don't know where the kitchen is.

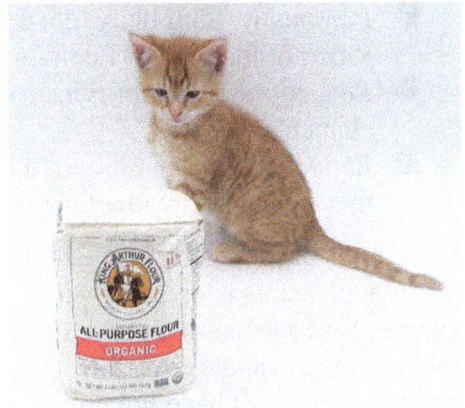

You need good ingredients to make good cookies.

You need domain expertise to create hypotheses and generate data, and you need data to test hypotheses and create results. Domain expertise is like the measuring spoons and cups that are used to get the right mix of ingredients. Data are like the ingredients. They are the evidence that will support or refute research hypotheses.

There are many ways that data go wrong (Chapter 3) just as there are many ways that baking ingredients can become stale or contaminated. When you're making cookies, it's not uncommon to substitute for an ingredient you don't have or something different you want to try. You might substitute non-gluten flour for all-purpose flour

Literacy.

or add cinnamon just because you like the taste. With data, you might correct *errors*, replace *outliers*, or add data *transformations*. You have to use the best ingredients you can.

THE TOOLS

The tools of the scientific method are the domain expertise, logical processes, and philosophical principles that are used to construct the research question, hypothesis, and experimental design. Logic is more than just the fallacies, it encompasses methods of reasoning and constructing arguments. Philosophical principles are goals or guidelines for developing a research project. Examples include:

Use all the tools you have available.

- *Empiricism*. Knowledge comes from experience and observation.
- *Rationalism*. Science must be based on facts and logical reasoning rather than on opinions, emotions, and belief.
- *Inclusiveness*. Incorporating all aspects of domain knowledge into a research question.
- *Universality*. Being true or appropriate for all situations.
- *Parsimony*. Simplicity of a research question. Also referred to as Occam's Razor or the Law of Economy.
- *Reductionism*. Simplifying a complex phenomenon into discrete, fundamental elements.
- *Refutability*. The ability of a hypothesis to be disproven. In statistical testing, this is managed with *effect size*, *confidence*, *power*, and other test details.

These tools of the scientific method aren't discussed much, but clearly, they are essential elements in creating science. Like tools used in making cookies, mixers and ovens, for instance, you don't have to know a lot about how they work if you're just licking the beaters.

> *The characteristic of scientific progress is our knowing that we did not know.*
>
> Gaston Bachelard, French philosopher

FROM DOUGH TO COOKIES

If you're making cookies, once you finish making the dough, you bake it to complete the process. If you're conducting research, once you finish analyzing the data, you document your work to complete the process.

Reporting research results is like baking cookie dough—it puts all the effort into parts that can be consumed by anyone, any time, any place.

Literacy.

One cookie for each of us.

There's no guarantee that either a research report or a cookie will be good or even "as expected." There might have been accommodations or shortcuts taken that affected the results. The research design, the recipe, may have been inferior. There may have been steps taken to optimize research results, like searching for significance (*p-hacking*). Like adding extra sugar to a cookie recipe, it seems like a good idea but others won't be able to use the original recipe and get the same results.

How results get packaged will affect how they are perceived. Cookies can be cut into shapes and decorated, then arrayed on a platter or stored in a zipper-storage bag. Research reports can be kept private or released to the public. They can be aimed at a particular audience, from non-technical to expert. They can be placed in peer-reviewed journals or reported in the main-stream media. Each type of publication appears different to the readers. There will be different types of comments, debates, and follow-up. Some people will be satisfied and some will want more.

Expectations matter. Reports written by experts that appear in prestigious publications are accepted without challenge just as cookies from professional bakers are expected to be good tasting. But these expectations are not always fulfilled. Sometimes the recipes aren't followed adequately or the ingredients are substandard. Some results are bad to begin with and some go stale over time. When that happens, just make more cookies.

I would like more, please.

What is necessary with both research and cookies is to be an unbiased, informed consumer. But, this is often not easy. As Carl Sagan once said, "We live in a society exquisitely dependent on science and technology, in which hardly anyone knows anything about science and technology." In that regard, research and baking are quite different.

Literacy.

WHEN SCIENCE GOES WRONG

Clearly, science and scientists are wrong on occasion even when they don't intend to be. That is to be expected. They are, after all, merely humans. Even if the scientific method isn't all that difficult to understand, it is incredibly difficult to put into practice, simplified flowcharts notwithstanding. As a consequence, scientific studies are too often poorly-designed, poorly-executed, misleading, or misinterpreted. Most of the time, this is inadvertent although sometimes not.

Sometimes bad science just happens.

There are many ways that the scientific method can be perverted, if not ignored altogether, to produce erroneous results. Most research characterized as *bad science* is probably the result of bias on the part of the researcher. Sometimes, it is a consequence of the topic not having a theoretical basis or being near the limits of our current understanding. And of course, in rare cases, it is intentional.

Categories of bad science go by many names, all of which are pejorative. In the following classification, category definitions vary between sources of the definition. Some topics appear as examples in more than one category.

Sometimes, referring to a study as *bad science* is a way to discredit research that challenges mainstream scientific ideas. It is *judgmental language.* Like an *ad hominem* argument, invoking terms related to bad science has been used to silence dissenters by preventing them from receiving financial support or publishing in scientific journals.

The difference between bad science and true science that strictly follows the scientific method is that true science will eventually correct illegitimate results, even if it takes decades.

That said, here are ten categories of bad science:

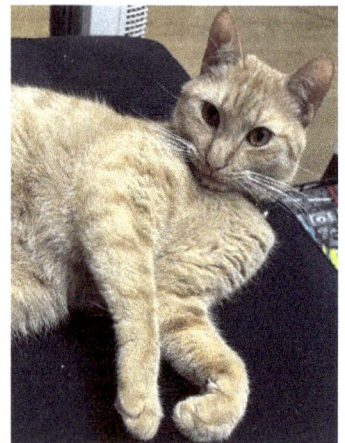

- Pathological science
- Pseudoscience
- Fringe Science
- Barely Science
- Junk Science
- Tooth-Fairy Science
- Cargo-Cult Science

Wow. That's a lot.

- Coerced Science
- Taboo Science
- Fraudulent Science.

The definitions of these forms of bad science described in the following sections have been "cleaned up" so there isn't so much overlap in the categories. For instance, some sources use *pseudoscience* for every form of *bad science*.

I created the last three categories—coerced, taboo, and fraudulent sciences—to highlight three forms of bad science that happen but don't have a cool name.

While this may seem like a fairly dismal portrayal of science, bear in mind that the vast majority of today's science is real and legitimate. But, to truly be literate, you must understand the bad as well as the good.

PATHOLOGICAL SCIENCE

Pathological science occurs when a researcher holds onto a hypothesis despite valid opposition from the scientific community. This isn't necessarily a bad thing. Most scientific hypotheses go through periods when they are ignored in favor of the hypothesis that is accepted as the status quo. It is only with persistence and further research that a hypothesis will be accepted.

Sometimes the change from a traditionally-held hypothesis to a new cutting-edge hypothesis is evolutionary and sometimes the change is revolutionary. For example, the change from the Expanding-Earth hypothesis to the Continental-Drift hypothesis was revolutionary; the change from the Continental-Drift hypothesis to Plate Tectonics was evolutionary.

I just want everyone to believe me.

The pathological part of pathological science occurs when the researcher deviates from strict adherence to the scientific method in order to favor the desired hypothesis or incorporate wishful thinking into interpretation of the data. Usually, the hypothesis is experimental in nature and is developed after some research data have been generated. The effects of the results are near the limits of detectability. Sometimes, other researchers are recruited to perpetuate the delusion.

Researchers involved in pathological science tend to have the education and experience to conduct true science so their initial results may be accepted as legitimate. Eventually, though, failure to replicate the results damages its credibility.

Cold fusion is considered by some to be an example of pathological science because all or most of the research is done by a closed group of scientists who sponsor their own conferences and publish their own journals.

PSEUDOSCIENCE

Pseudoscience involves hypotheses that cannot be validated by observation or experimentation. Thus, they are incompatible with the scientific method but still are claimed to be scientifically legitimate.

Pseudoscience often involves long-held beliefs that pre-date experiments. Consequently, it is often based on faulty premises. While less likely to be popular in the scientific community, pseudoscience may find support from the general public.

I get to have nine near death experiences.

Examples that have been characterized as pseudoscience include numerology, free energy, dowsing, Lysenkoism, graphology, body memory, human auras, crystal healing, grounding therapy, macrobiotics, homeopathy, and near-death experiences.

The term pseudoscience is often used as an inflammatory buzzword for dismissing opponents' data and results.

FRINGE SCIENCE

Fringe science refers to hypotheses within an established field of study that are highly speculative, often at the extreme boundaries of mainstream studies.

Fringe science sounds good but it's at the extremes of what is accepted.

Proponents of some fringe sciences may come from outside the mainstream of the discipline. Nevertheless, they are often important agents in bringing about changes in traditional ways of thinking about science, leading to far-reaching *paradigm shifts*.

Some concepts that were once rejected as fringe science have eventually been accepted as mainstream science. Examples include heliocentrism (sun-centered solar system), peptic ulcers being caused by *Helicobacter pylori*, and chaos theory.

The term *protoscience* refers to topics that were at one point mainstream science but fell out of favor and were replaced by more advanced formulations of similar concepts. The original hypothesis then became a pseudoscience. Examples of protosciences are astrology evolving into the science of astronomy and alchemy evolving into the science of chemistry.

Other examples that have been labeled as fringe science include feng shui, ley lines, remote viewing, hypnotherapy and psychoanalysis, subliminal messaging, and the MBTI (Myers-Briggs Type Indicator). Some areas of complementary medicine, such as mind-body techniques and energy therapies, may someday become mainstream with continuing scientific attention.

The term fringe science is considered to be pejorative by some people but it is not meant to be.

Maybe I'll be able to turn the metal in my bell into gold.

BARELY SCIENCE

Barely science might be perfectly acceptable science, except it is too underdeveloped to be released outside the scientific community. Barely science may be based on a single study, or pilot studies that lack the methodological rigor of formal studies, or studies that don't have enough samples for adequate resolution, or studies that haven't undergone formal peer review. Typical sources of barely science include researchers under pressure to demonstrate results to sponsors or announce results before competitors. Consumers see barely-science more than they know, in both movies and real life.

There's a lot of junk back here but I don't know if it's junk science.

JUNK SCIENCE

Junk science refers to research considered to be biased by legal, political, ideological, financial, or otherwise unscientific motives. The concept was popularized in the 1990s in relation to legal cases.

Forensic methods that have been criticized as junk science include polygraphy (lie detection), bloodstain-pattern analysis, speech and text patterns analysis, microscopic hair comparisons, arson burn pattern analysis, and roadside drug tests. Creation sciences, faith healing, eugenics, and conversion therapy are considered to be junk sciences.

Sometimes, characterizing research as junk science is simply a way to discredit opposing claims. This use of the term is a common ploy for devaluing studies

Literacy.

involving archeology, complementary medicine, public health, and the environment. Maligning analyses as junk science has been criticized for undermining public trust in real science.

The Tooth Fairy is legitimate but unexplainable.

TOOTH-FAIRY SCIENCE

Tooth-Fairy science is research that can be portrayed as legitimate because the data are reproducible and statistically significant but there is no understanding of why or how the phenomenon exists.

Placebos, endometriosis, yawning, out-of-place artifacts, megalithic stonework, ball lightning, and dark matter are examples. Chiropractic, acupuncture, homeopathy, therapeutic touch, and biofield tuning have also been cited as being tooth-fairy science.

CARGO-CULT SCIENCE

Cargo-cult science involves using apparatus, instrumentation, procedures, experimental designs, data, or results without understanding their purpose, function, or limitations, in an effort to confirm a hypothesis.

Cargo-cult science applies to *replication* studies. There are two forms of replication. *Direct replication*, called *repeatability*, involves reproducing the original research design and procedures as exactly as possible. *Conceptual replication*, called *reproducibility*, involves using different, hopefully improved, designs and procedures to examine the original hypothesis.

Examples of bad cargo-cult experimentation might involve replication studies that use simplified procedures, lower-grade chemical reagents, instruments not designed for study conditions, or data obtained using different populations and sampling schemes.

Attempts to *reproduce* Millikan's oil-drop experiment have been cited as examples of cargo-cult science but those studies actually used improved procedures and achieved more accurate results. Actual cargo-cult studies probably never get reported when they fail to repeat the original results.

I understand plasma globes but not ball lightning.

Literacy.

COERCED SCIENCE

Coerced science occurs when researchers are compelled by authorities to study sometimes-objectionable topics in ways that promote speed over scientific integrity in reaching a desired result. There are many notable examples.

Do what I tell you to do. Don't ask questions.

During World War II, virtually every major government pushed their scientists and engineers to achieve a variety of desired results. In the 1960s, JFK successfully pressured NASA to land a man on the Moon by the end of the decade even though the goal and schedule were daunting. In the 1980s, Reagan prioritized efforts on the Strategic Defense Initiative (SDI) even though the goal was considered to be unachievable by experts. Many governments restrict research on their country's cultural artifacts to only those individuals who agree to severe preconditions including censorship of announcements and results.

Businesses, especially in the fields of medicine and pharmaceutics, place great pressure on research staff to achieve results. Elizabeth Holmes, founder of the medical diagnostic company Theranos, is an example. Businesses are also known to conceal data that would be of great benefit to society if they were available. Examples include results of pharmaceutical studies (e.g., Tamiflu, statins) and subsurface exploration for oil and mineral resources.

Finally, academic institutions predicate tenure appointments in part on journal publications and grant awards, both of which rely on researchers finding *statistical significance* in their analyses (*p-hacking*).

TABOO SCIENCE

Taboo science refers to areas of research that are limited or even prohibited either by governments or funding organizations. Sometimes this is reasonable and good. For example, research on humans has become more restrictive after the atrocities that occurred during World War II. During the Cold War, U.S. military and intelligence agencies obstructed independent research on national security topics, such as encryption.

We don't want you to conduct that research

Some taboos, however, are promoted by special-interest groups, such as political and religious organizations. Examples of topics that are difficult for researchers to obtain funding for include: effectiveness of methods to control gun violence; ancient civilizations, archeological sites and artefacts, health benefits of cannabis and psychedelics; resurrecting extinct species; and some topics in human biology such as cloning, genetic engineering, chimeras, synthetic biology, scientific aspects of racial and gender differences, and causes and treatments for pedophilia.

FRAUDULENT SCIENCE

Fraudulent science consists of research in which data, results, or even whole studies are faked. Creation of false data or cases is called *fabrication*; misrepresentation of data or results is called *falsification. Plagiarism* and other forms of information theft, conflicts of interest, and ethical violations are also considered aspects of fraudulent science.

The goals of fraudulent science are usually for the researcher to acquire money, including funding and sponsorships, and enhance reputation and power within the profession.

Unfortunately, there are too many examples of fraudulent science. Perhaps the most notorious is the 1998 case of Andrew Wakefield, a

Fraudulent science? What's up with that?

British expert in gastroenterology, who claimed to have found a link between the MMR vaccine, autism, and inflammatory bowel disease. His paper published in The Lancet, which was retracted in 2010, is thought to have caused worldwide outbreaks of measles after a substantial decline in vaccinations. Wakefield later became a leader in the anti-vaxx movement in the U.S..

Another infamous example involves faked images in a 2006 experimental study of memory deficits in mice, which subsequently led to a decade-long unproductive diversion of funding for Alzheimer's research.

Sometimes, fraudulent actions are subtle and go unnoticed even by experts. Examples include pharmaceutical studies designed to accentuate positive effects while concealing undesirable side effects. Sometimes, well-meaning actions have unforeseen ramifications, such as when definitions of medical conditions are changed resulting in patients being treated differently. Examples include obesity, diabetes, and cardiac conditions.

From 2000 to 2020, 37,780 professional papers were retracted because of fraud (The Retraction Watch Database [internet]. New York: The Center for Scientific Integrity. 2018. ISSN: 2692-465X. Accessed 4/13/2023. Available at:

Literacy.

http://retractiondatabase.org/). Those retractions are considered to represent only a fraction of all fraudulent science.

PUTTING NUMBERS TO THE SCIENCE

Just as some science can go wrong, so too can some statistical analyses. Adding numbers to a complex operation presents even more challenges to consider. The way to handle these complications is to understand at least some of the ways that information can be made relevant or misleading. When it comes to data and statistics, the ways things can go wrong seem to be endless.

DATA CHALLENGES

Part of being statistically literate is understanding that all analyses are not equal. Some suffer from problems with data and some suffer from problems with analyses of the data. It's like with cooking, the best chef can't prepare appetizing meals with stale ingredients nor can an inept cook prepare edible meals using the finest ingredients.

Sometimes you have to go to great lengths to get the data you need.

Every dataset has more problems than non-statisticians can imagine. Even in Stats 101, instructors provide only flawless data for students to practice on, which is understandable. Data scrubbing skills come not just from training but also from extensive experience. It's not just errors and anomalies in data, it's the characteristics inherent in how the data are generated.

Data generation can be challenging because it is hard to access or because it is hard to assess.

HARD-TO-ACCESS DATA

Some data seem to resist being collected. Statisticians find it difficult to obtain enough data, even any data, on some subjects. For example:

Remote Data. Some data are difficult to access because the subjects being measured are far away or in hard-to-access areas. Examples include astronomical and oceanographic data, microscopic and subatomic data, and even data from foreign countries.

Sequestered Data. Some data are difficult to access because they are concealed or are difficult to understand. Some sequestered data may involve animals that are wary of human contact, like cryptids. Coelacanths, giant squids, platypuses, kangaroos, and gorillas were once considered to be cryptids. Even common animals like fish and small mammals avoid being captured and measured. People have secrets too, such as

Literacy.

those in closed communities. Governments classify all kinds of information, keeping it from being analyzed. Some data are encrypted or are in foreign languages hindering their availability. Some populations to be surveyed are difficult to *frame*.

Complex Data. Some data are difficult to access because the process involved in generating them is complicated or time-consuming. Scientific information, from DNA to fission-tracks, is an example. People with secrets are also complex in that they are often hard to find and hard to extract information from. Investigators involved in public health and law enforcement address such hard-to-find subjects using *snowball sampling* (also called *referral sampling* or *network sampling*) to identify new subjects from information provided by existing subjects.

Expensive Data. Some data are difficult to access in quantity because they are expensive to generate. Medical and environmental data are examples. Patients who have had even routine tests done, like colonoscopies and cardiac stress tests, are shocked to see the cost of the tests that are (hopefully) covered by insurance. Unique or technologically advanced tests may not be covered, so they are not conducted because of their costs. It can cost over $1,000 for a

I don't think I have enough money to pay for all the expensive data I need.

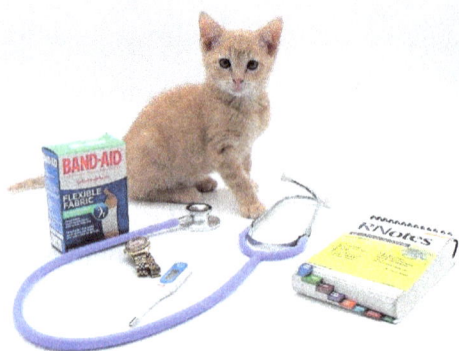

commercial laboratory to analyze an environmental sample for organic pollutants. And, that doesn't include the cost of collecting the samples, which can be tens of thousands of dollars.

I'm training to be a nurse.

Restricted Data. Some data are difficult to access because there are administrative restrictions preventing its use. This is particularly true in human resources, business operations, and law enforcement. Medical information is protected by HIPAA (i.e., the Health Insurance Portability and Accountability Act of 1996). The Dickey Amendment prevented the U.S. Center for Disease Control and Prevention (CDC) from collecting data on firearm violence from 1996 to 2020.

AWOL Data. Some data are difficult to access because they aren't numbers. Most *metadata* aren't numbers though they do provide essential information. *Missing data* can't, of course, be analyzed and may present challenges if there is a statistical

Literacy.

reason for why they are missing. (If missing values are truly random, they are said to be *missing-completely-at-random*. If other metrics in the dataset suggest why they are missing, they are said to be *missing-at-random*. However, if the reason they are missing is related to the metric they are missing from, that's bad. Those data values are said to be *missing-not-at-random*.) *Censored data* are data that are qualified as less than (<) or greater than (>) some quantification threshold, such as readings outside the range of the measuring device. They have to be analyzed using special statistical procedures. Neither missing data nor censored data can just be ignored.

HARD-TO-ASSESS DATA

Even if data exist, they can be a challenge to analyze in some instances because of the circumstances under which they were generated. These issues may be overtly stated in *metadata*, buried deep in a description of methods in a statistical report, or not mentioned at all. Examples include:

Overlapping Data. Some data are difficult to assess because one (or more) of the variables characterizing a phenomenon can have multiple responses for a single sample. This can happen in survey questions that allow respondents to "check all that apply." Another example is cause-of-death, which can have a primary cause and multiple contributing causes. In these situations, percentages do not sum to 100% and some descriptive statistics are virtually meaningless.

Different Definitions. Some data are difficult to assess because the definition of a metric characterizing a phenomenon is not consistent between data sources. For example, definitions of mass shootings might specify different minimum numbers of victims or durations, and may exclude warfare, robberies, and domestic violence. The number of employees might include all workers, exclude part-timers or contractors, or be expressed as FTEs (i.e., full time equivalents) rather than head counts.

Changing Conditions. Some data are difficult to assess because the conditions that affect the data change. In sports, for example, rules, facilities, equipment, coaching strategies, drugs, and of course, players are changing constantly. Sometimes an underlying characteristic doesn't change but the perception of the characteristic does as the understanding of it advances. Centuries ago, an individual's race and ethnicity were largely based on the simple recollections of ancestors. Today, home DNA kits can provide extensive familial information, even an

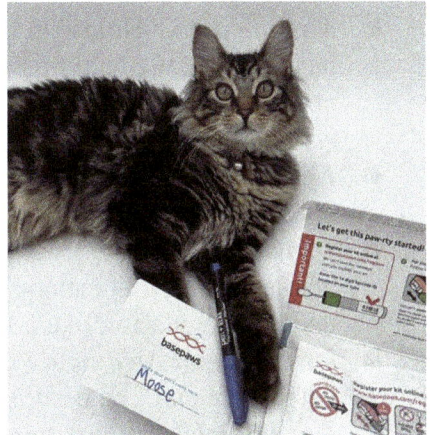

I know my ancestry.

individual's Neanderthal ancestry. Sex was taught as a binary characteristic a century ago. Today, science isn't sure it is.

Literacy.

STATISTICAL SLEIGHT-OF-HAND

Behind-the-scenes, things happen in statistical analyses that you don't hear about. They are almost always handled professionally and legitimately, but it's the reason why statistical analyses can never be reproduced exactly.

DATA SCRUBBING

Data Scrubbing involves addressing errors, anomalies, and other problems in a dataset before it is analyzed. Unlike datasets you might see in Stats 101, real-world data is packed full of all kinds of errors in both individual data points and across whole metrics. They can be easy to find or buried deep in the *metadata* and must be *repaired*, *replaced*, or *removed* before analysis. Examples include:

Data scrubbing shouldn't involve getting wet!

- **Invalid data.** Values that are originally generated incorrectly, sometimes from a problematic measurement tool.
- **Incorrectly-recorded data.** Values that are usually attributable to transcription errors, which appear randomly in a dataset.
- **Incorrectly-coded data.** Values that result when information for a nominal-scale or ordinal-scale metric is entered inconsistently, either randomly or for an entire metric.
- **Data quality exceptions.** Entries related to data verification and validation that are not meant to be analyzed, such as sample replicates.
- **Missing data.** "Holes" in the data matrix that have to be addressed on the basis of why they are missing, whether the cause is random or not (i.e., *missing-at-random*, *missing completely at random*, or *missing not at random*).
- **Extraneous data.** Non-representative values that may occur when overlapping datasets are merged.
- **Dirty data.** Individual data points that have erroneous characters as well as whole metrics that cannot be analyzed because of some inconsistency or textual irregularity, like concatenated data, aliases, and misspellings.
- **Useless data.** Any metric that has no values, values that are all the same, or values that can't be used, like *PII* or encrypted data.
- **Invalid fields.** Anomalous entries in a dataset caused by errors in electronic transmissions, missing or extra delimiters, or incorrect formatting (text versus values).
- **Out-of-spec data.** Values that don't have consistent precisions, units, or measurement scales.

Literacy.

❧ ***Out-of-bounds data.*** Values that are outside the theoretical boundaries of the measurement, for example, a pH of -50.
❧ ***Messy data.*** Metrics with outliers, censored data, excessively large ranges, or large or non-constant variances (i.e., *heteroscedasticity*).
❧ ***Corrupted data.*** Non-representative values that may occur when data are extracted/scraped, processed, reorganized, or reformatted improperly.
❧ ***Mismatched data.*** Non-representative values that may occur when multiple datasets are merged from different sources that use different data definitions with different contexts or different times, or inconsistent aliases.

And of course, there are many more, as highlighted in Table 39.

Data scrubbing can easily take 80% of the time spent on a statistical analysis and isn't nearly as much fun as actually doing the analysis. No data analyst scrubs their data in the same way or to the same extent as other data analysts. This means that it is unlikely that independent analyses of the same raw dataset will use exactly the same data.

DATA TRANSFORMATIONS

Transformations are mathematical methods of changing the scales or other aspects of variables in a statistical analysis to improve the analyses. There are many different data transformations that fall into one of four categories.

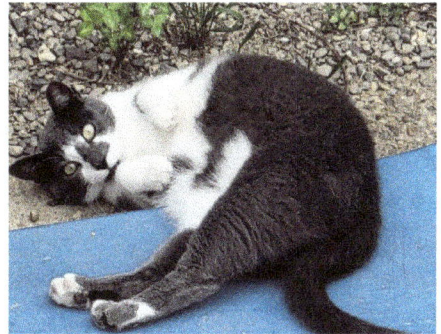

I'm transforming, from a kitten to a cat.

❧ ***Sample adjustments.*** Methods for fixing missing, erroneous, or unrepresentative data points, such as replacing missing and anomalous data, averaging duplicates, and even deleting problematic samples and variables. There is no consensus for when and how adjustments should be made so data analysts are left to make their own decisions.
❧ ***Dependent-variable transformations.*** Methods for changing the scale of the dependent variable usually to minimize the effects of violations of statistical assumptions. Some data analysts use transformations and other analysts use alternative statistical procedures (e.g., non-parametric statistics) instead.
❧ ***Independent-variable transformations.*** Methods for creating new variables from the original independent variables using mathematical functions like logarithms, square roots, z-scores, and the like. Transformed variables have better correlations with the dependent variable or they are not used. While most statisticians use transformations when appropriate, some statisticians avoid them because of a fear they will introduce bias.

Table 39. Seven categories of data errors and how to identify and fix them.

Error	Extent		Location		Identification			Recovery		
	Data Point	Data Set	Metric	Random	Visual	Statistics	Graphs	Repair	Replace	Remove
Invalid Data										
Bad generation	*	*	♦			+	+			O
Recorded wrong	*			♦		+	+			O
Bad coding	*	*	♦	♦	+	+	+		O	
Wrong thing measured		*	♦	♦	+	+				O
Data quality exceptions	*		♦	♦	+					O
Missing and Extraneous Data										
Missing-completely-at-random data	*			♦	+				O	
Missing-at-random data	*		♦		+	+	+		O	
Missing-not-at-random data	*	*	♦		+	+	+		O	
Uncollected data	*		♦	♦	+				O	
Replicates	*		Observations		+					O
QA/QC samples	*		♦		+					O
Extraneous unexplained data	*		♦	♦	+					O
Dirty Data										
Incorrect characters	*			♦	+			O		
Problematic characters	*	*		♦	+			O		
Concatenated data		*	♦		+				O	O
Aliases	*		♦	♦	+			O	O	
Useless data		*	♦		+	+	+			O
Invalid fields	*		♦	♦	+			O	O	O
Out-of-Spec Data										
Out of bounds data	*		♦		+	+	+		O	
Different precisions		*	♦		+			O		
Different units		*	♦		+			O		
Different scales		*	♦				+		O	
Large data ranges		*	♦	♦	♦	+	+		O	
Messy Data										
Outliers	*			♦		+	+		O	O
Large variances		*	♦			+			O	
Non-constant variances		*	♦			+	+		O	
Censored data	*		♦		+				O	
Corrupted Data										
Electronic glitches	*			♦	+					O
Bad reorganization	*	*	♦	♦	+			O		
Bad extraction		*	♦		+			O		
Bad processing		*	♦		+			O		
Mismatched Data										
Different sources		*	♦		+				O	
Different definitions		*	♦		+				O	
Different contexts		*	♦		+				O	
Different times		*	♦		+				O	

❧ *Supplemental variables.* New metrics created by combining information from untapped data sources. For example, original independent variables can be combined into sums or ratios. Most statisticians do not create supplemental variables for fear of *overfitting* a model.

Most statisticians use transformations whenever they are appropriate while other statisticians avoid them at all costs. As a consequence, a given set of raw data could

Literacy.

be restructured in a variety of ways making it unlikely that analyses conducted by different statisticians would be exactly the same.

DATA WEIGHTING

Weighting involves using an algorithm to assign greater (or lesser) importance to observations based on some meaningful quantity. An example is *interpolation,* in which data from sampled times or locations are used to estimate measurements for times or locations that have not been sampled. A greater weight is often assigned to measurements made closer (in space or time) to the point being interpolated. Data measured over an extended period of time are often weighted to give more recent measurements greater importance. Pollsters use weights to improve the match between the sample's demographics and the population's demographics. Weights are also used when results

I think that I'm at the perfect weight.

from multiple sources are combined, such as with independent political surveys. This is a form of *meta-analysis*.

ANALYSIS TURMOIL

As you can probably conclude from the information in Chapters 2 through 8, there are countless ways that a statistical analysis can go wrong, from picking inappropriate techniques for analyzing the data to implementing the techniques inappropriately. Even if everything is done correctly, there are innumerable options within techniques that can cause analyses conducted by different statisticians to yield different results. Here are just a few examples:

- 🐾 *Population.* A *population* and *representative samples* from it are concepts. You can't see them. You can't just recognize them by sight. Different researchers might characterize a population in different ways, and because of that, select very different samples that they believe are representative.
- 🐾 *Phenomenon.* Phenomena can be concrete or abstract. Different researchers may use different metrics to characterize them.
- 🐾 *Variance.* How researchers control variability will make a huge difference in the data themselves. How researchers analyze variability will make a difference even if they all use the same data.
- 🐾 *Assumptions.* Analysts treat violations of statistical assumptions differently. Some analysts try to correct violations of Normality and equal- variances using transformations. Some analysts immediately reject parametric approaches in favor of nonparametric approaches. Some analysts just view

Literacy.

calculated probability estimates with skepticism. And some novice analysts don't even pay any attention to violations of assumptions. Different approaches bring different results.

- ❧ *Graphs.* There are so many ways to graph data considering *foundation*, *framework*, and *facade* that it would be an unlikely coincidence if graphs prepared independently by different researchers were the same.
- ❧ *Models.* There's a big difference between putting an Excel trend line through a set of data and creating a statistical model. A trend-line requires checking a box; a statistical model requires many hours of scrubbing data, reviewing diagnostic statistics, and considering alternative models. Both have legitimate uses but results of the two won't be the same. Just checking the box is like putting a glob of ketchup on a Saltine cracker and calling it a pizza.

PUTTING WORDS TO THE NUMBERS

Even in the ideal world that statisticians dream about, where all samples are representative of their population and all statistical assumptions are perfectly met, any analysis can become distorted when it enters the world of the written word. This is virtually guaranteed when the individual making the presentation doesn't understand the nuances of the scientific method, data, and statistics.

Presentations are not always created and interpreted with complete objectivity. There may be *preconceived notions* of what is thought to be true. There may be *biases* about what the writer or reader of a presentation wants to be true. There may be *fallacies* involved that cause confusion so that a presentation is misunderstood. There may be *flaws*, both unintentional and intentional mistakes, that make the presentation invalid. There may also be *misunderstandings* that can lead individuals astray.

Fallacies, biases, flaws, and preconceived notions are like graphs in that there are too many of them to keep straight in your mind. Some are slight variations of others or alternately named. Some are specific to certain situations and some are common to all cases and all parts of a presentation.

Writing a statistical report, or even writing about someone else's statistical analysis, requires special tools.

Literacy.

Here are six categories of fallacies, biases, and preconceived notions that may influence the creation, results, or interpretation of presentations of statistical analyses:

- Misrepresentations
- Suspect Evidence
- Appeals to Others
- Bounded Thinking
- Faulty Processing
- Number Confusion.

MISREPRESENTATIONS

Every presentation starts with definitions that establish the scope and target of a research project. Sometimes researchers forget that their audience may not understand the research topic in the same way that they do. Too often, these fundamental definitions are overlooked by readers. Here are some causes of failures to characterize what a presentation applies to.

Equivocation (also called *doublespeak fallacy* and *fallacy of ambiguity*) involves using words in a different sense than the one the audience will understand and not defining the intended meaning. This is particularly troublesome in political discussions involving terms like socialism (State socialism versus democratic socialism versus social democracy) and entitlement (feeling of deservedness versus a type of mandatory spending). Equivocation occurs in statistical analyses when existing data are used without providing proper context. This can happen intentionally or unintentionally when data from different sources are merged.

Red Herrings are any aspect of a presentation that is meant solely to distract from deficiencies in the research. Red herrings may be true and irreverent, or just plain false. The *Chewbacca defense* is an example of a red herring.

I'll pick only the cherries I like.

Cherry picking (also called the *fallacy of incomplete evidence*) is the practice of using only the data or information that supports a preconceived notion. Unintentional cherry-picking is surprisingly common because the amount of information available on the internet makes it impossible to review every potential source.

Availability Bias is the result of using information that is the easiest to acquire instead of information that is more relevant to the population and the phenomenon.

Literacy.

Snow Jobs (also called *Information Bias*) involve providing overwhelming amounts of marginally-relevant evidence that audiences cannot evaluate properly to distract them from more relevant information.

Straw Men are mischaracterizations of opposing research that intentionally misrepresent the true nature of the research so that it is easier to discredit and reject.

SUSPECT EVIDENCE

Evidence consists of the fundamental facts, premises, information, and background science that lead to hypotheses, data gathering, and analysis. Making a *claim* is not the same as providing evidence. A claim involves belief; evidence requires proof.

Faulty evidence is where some research and presentations of results go astray.

There are many terms that describe types of evidence. Some characterize the forms of evidence (e.g., physical evidence as opposed to testimony evidence), some characterize the effectiveness of evidence (e.g., exculpatory evidence versus insufficient evidence), and some characterize the role of evidence (e.g., corroborating evidence versus conflicting evidence). In evaluating a statistical presentation, it is absolutely necessary to understand what the evidence is, how effective it is, and what its role is in the presentation. Figure 74 highlights thirty-one types of evidence.

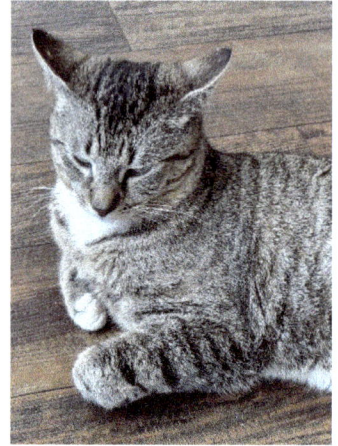

You call this evidence?

The level-of-evidence in law is determined by the burden-of-proof, which is the amount of evidence required to win a claim in court. Levels of evidence include:

- 🐾 ***Beyond a reasonable doubt.*** The highest standard of evidence, this is the main burden of proof in criminal cases.
- 🐾 ***Clear and convincing evidence.*** A higher standard than preponderance of the evidence, this is often used in family law and administrative law cases.
- 🐾 ***Preponderance of the evidence.*** The most common standard in civil cases, this requires the plaintiff to prove that the defendant is more than 50% responsible.
- 🐾 ***Probable cause.*** The standard used by police to justify a search or arrest.
- 🐾 ***Reasonable suspicion.*** The standard used by police to justify a stop or search.
- 🐾 ***Judicial notice.*** When a court assumes a fact is proven without any evidence because it is so well known.

In medical research, levels of evidence are based more on statistical rigor, including:

- 🐾 *Meta-analyses.* The highest level of evidence involving systematic reviews of studies that use randomized control trials.
- 🐾 *Randomized control trials.* Experiments in which participants are randomly assigned to different groups.
- 🐾 *Quasi-experimental studies.* Experiments in which participants are not randomly assigned to different groups.
- 🐾 *Observational (non-experimental) studies.* Experiments in which data are inherent to participants rather than being randomly-assigned.
- 🐾 *Meta-synthesis.* Experiments in which qualitative data from individual studies are combined to create new interpretations.
- 🐾 *Qualitative studies.* Experiments involving non-numerical data.
- 🐾 *Expert opinions.* Testimony and reports from expert authorities, panels, committees, and organizations, as well as literature reviews. Not based on new research.

Table 40 summarizes levels of evidence for medical research. In general:

- 🐾 Evidence from a research study is better than an expert opinion.
- 🐾 Quantitative evidence is better than qualitative evidence.
- 🐾 Evidence from studies that have analyzed data mathematically is better than evidence from studies that have analyzed data intuitively.
- 🐾 Evidence from multiple studies that reach the same conclusion is better than evidence from a single study.
- 🐾 Evidence from studies with a large sample size is better than evidence from studies with a small sample size.

These guidelines are fairly common to most fields of research.

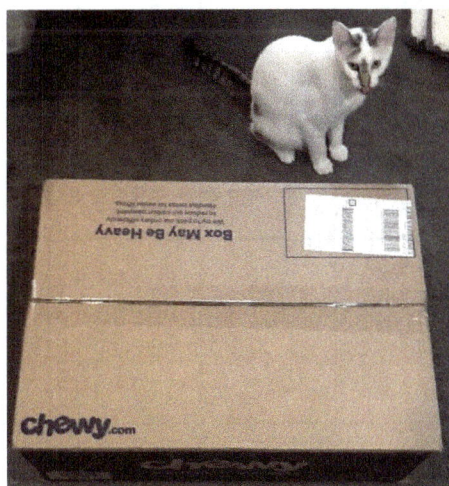

I can tell what's in the box.
Just open it already.

Literacy.

A MARCH OF EVIDENCE

31 TERMS FOR INFORMATION USED TO SUPPORT ARGUMENTS

1 ADMISSIBLE Evidence that can be presented in court because it is factual, relevant, authentic, and valuable.	**2 PRIMARY** Original objects, photographs, recordings, documents, and information that can be provided for further examination and assessment	**3 DIRECT** A fact that needs no other support	**4 DEMONSTRATIVE** Information presented through objects, pictures, models visual aids, physical evidence, and other means.	**5 PHYSICAL** Objects found at a scene having a connection to an event.	**6 PHYSICAL, CLASS** Physical evidence associated with a group, like blood type, tire or shoe tread pattern, or model of firearm.	**7 PHYSICAL, INDIVIDUAL** Physical evidence associated with a unique individual, like fingerprints or DNA
8 SECONDARY Evidence of lesser relevance or import than primary evidence such as certified copies, published reports, newspaper clippings, and sound recordings	**9 TRACE** Physical evidence deposited when two objects come in contact such as residues, oils, bodily fluids, hair, fibers, soil, wood, paint, and plant material	**10 IMPRESSION** Imprints left in deformable materials by more rigid materials. Such as footprints in wood, or impact marks in plaster walls	**11 FORENSIC** Evidence generated from procedures that are based on scientific principles and have been validated using the scientific method	**12 DIGITAL** Information like text messages, emails and GPS data from computers, mobile phones and other types of electronic devices	**13 STATISTICAL** Data or analytical results from reliable sources that establish probabilities, patterns or correlations	**14 DOCUMENTARY** Documents (typed, handwritten, or stored), photographs, and audio/video recordings considered more compelling than oral evidence
15 ANALOGICAL Analogical evidence compares similar things in order to clarify or explain what has happened	**16 ANECDOTAL** Stories that form a Connection not a strong type of evidence	**17 TEXTUAL** Support from other writing like quotations, dialog, summaries, or paraphrasing	**18 HYPOTHETICAL** Projections about past or future situations used to explain complex events or motivations considered a weak form of evidence	**19 TESTIMONY** Information provided by a witness who responds to questions usually under oath	**20 EXPERT TESTIMONY** Facts or opinions related to a case provided by an individual accepted as knowledgeable in a relevant topic.	**21 ORAL** Testimony or audio/video recordings of speech or sounds
22 CHARACTER Information that portrays a person's personality, morality, and reputation.	**23 HABIT** Evidence of a person's consistent, repeated activities and reactions during specific circumstances	**24 CIRCUMSTANTIAL** Background information that implies a connection but not decisive proof. Also called indirect evidence	**25 CORROBORATING** Evidence that supports, confirms, or authenticates other evidence	**26 CONFLICTING** Two or more pieces of evidence that suggest different, even contradictory, conclusions	**27 PRESUMPTIVE** Plausible assumptions based on supporting evidence. Also called prima facie evidence	**28 EXCULPATORY** Evidence that might justify actions, refute their occurrence, or question their relevance
29 INSUFFICIENT Evidence that is, in sum, not adequate to meet the burden of proof in a case	**30 HEARSAY** Second-hand information that cannot be verified, examined, and assessed	**31 INADMISSIBLE** Evidence that cannot be presented because it is not relevant, was not obtained properly, or is prejudicial, or is hearsay	*Evidence* is the information that establishes the credibility of an argument. It is a general term used in many situations not just legal proceedings. It may be called background or premises, or be embodied in data and metadata. There are at least 31 terms for evidence, used to characterize the form, role, and effectiveness of evidence. In evaluating a presentation, be sure you understand the evidence			

RANDOMTERRABYTES.NET/

Figure 74. Thirty-one terms referring to types of evidence.

Literacy.

Table 40. Levels of evidence used for medical research.

Level of Evidence	Examples of Levels of Evidence for Medical Research			
	British Association for Psychopharmacology, 2014, UK	National Institute for Health and	Bandelow et al., 2013, Germany, based on Scottish Intercollegiate Guidelines Network protocol	Canadian Psychiatric Association, 2006, Canada
1	*For experimental research* **Meta-analysis**—Evidence from meta-analysis of randomized double-blind placebo-controlled trials. **Placebo-controlled trials**—Evidence from at least one randomized double-blind placebo-controlled trial. *For observational research* Evidence from large representative population samples.	**High**—Further research is very unlikely to change confidence in the estimate of the effect.	++—High quality meta-analyses, systematic reviews of RCTs, or RCTs with a very low risk of bias. +—Well conducted meta-analyses, systematic reviews, or RCTs with a low risk of bias. −—Meta-analyses, systematic reviews, or RCTs with a high risk of bias.	Meta-analysis or replicated randomized controlled trial that includes a placebo condition.
2	Evidence from at least one randomized double-blind comparator-controlled trial (without placebo). Evidence from small, well designed but not necessarily representative samples.	**Moderate**—Further research is likely to have an important impact on confidence in the estimate of the effect and may change the estimate.	++—High quality systematic reviews of case control or cohort studies. High quality case control or cohort studies with a very low risk of confounding or bias and a high probability that the relationship is causal. +—Well conducted case control or cohort studies with a low risk of confounding or bias and a moderate probability that the relationship is causal. −—Case control or cohort studies with a high risk of confounding or bias and a significant risk that the relationship is not causal.	At least one randomized controlled trial with placebo or active comparison condition.
3	Evidence from non-experimental descriptive studies. Evidence from non-representative surveys, case reports.	**Low**—Further research is very likely to have an important impact on confidence in the estimate of the effect and is likely to change the estimate	Non-analytic studies, e.g., case reports, case series.	Uncontrolled trial with at least ten or more subjects
4	Evidence from expert committee reports or opinions and/or clinical experience of respected authorities. Evidence from expert committee reports or opinions and/or clinical experience of respected authorities.	**Very low**—Any estimate of effect is very uncertain.	Expert opinion.	Anecdotal reports or expert opinion.

After "Short- and Long-Term Use of Benzodiazepines in Patients with Generalized Anxiety Disorder. A Review of Guidelines [Internet]. Ottawa (ON): Canadian Agency for Drugs and Technologies in Health; 2014 Jul 28. APPENDIX 5, Summary of Guideline Evidence Levels and Strength of Recommendations. Available from: https://www.ncbi.nlm.nih.gov/books/NBK254098/"

Literacy.

In medical research, there is a preference for *experimental* studies over *observational* studies. There is also a preference for evidence created by individuals with advanced degrees in a discipline directly related to a topic and having a long history of successful research that has been published in peer-reviewed journals.

The important things to remember about evidence, whether you are evaluating a news story or a post on social media, are that the evidence must exist, must validate the assertions being made, and must be cited in a way that can be verified.

APPEALS TO OTHERS

We are all insecure, we just manifest our insecurities differently. Some people are shy and reserved. Some people are bullies and attention seekers. Some people procrastinate or agonize over even simple decisions. We all, at times, need to lean on the thoughts of others. That's when our ability to evaluate arguments and presentations is most valuable and most vulnerable.

We all often defer to the opinions of others:

- Experts and purported experts
- Well-known people and celebrities
- Mainstream and alternative news sources
- Spokespersons and paid actors
- Relatives, trusted acquaintances, classmates, and co-workers
- The general public
- Our inner selves (gut feel).

This can present misperceptions in a variety of ways:

Appeal to Authority involves relying on information from sources, both individuals and media, that are not specifically qualified on a topic. The authority may be an expert or a recognized source in a related field but not on the topic of concern. Celebrities in entertainment and sports are experts in their own career endeavors but not necessarily on health or the law.

The Bandwagon Fallacy (also called *Argument from Common Sense*, *Appeal to Common Belief*) is assuming something must be true because everybody believes the same. Most people believe explanations labelled as conspiracy theories are all false even though many of them have been proven to be true (e.g., Watergate, pedophilia in the Catholic Church and the Boy Scouts). An *Information Cascade* is where people repeat the opinions of others, especially on social media, giving the impression of a consensus.

The Dunning-Kruger Effect is when people trust their own interpretations over those of experts because they mistakenly believe their abilities are greater than they actually are. Intuition and gut feelings are also not always reliable.

Literacy.

So, be wary when a presentation relies on statements made by or about individuals rather than facts and data. Also, don't be impressed by things just because you don't understand them. In particular, don't fear *jargon*. Assess why it's there. Is it used as technical shorthand, as a means to obfuscate or intimidate, or just something copied from another source.

BOUNDED THINKING

Bounded thinking happens when the creation, result, or interpretation of a presentation is

Y'know, it's usually better not to limit how you think.

limited in some way. The most common example may be presentations indicating binary alternatives, that is, either/or situations.

A *Double-Edge Sword* is a situation in which the alternatives have both favorable and unfavorable aspects.

A *Morton's Fork* is a choice between two unfavorable alternatives, known more simply as a *dilemma* or being *between-a-rock-and-a-hard-place*.

A *Buridan's Ass* is a choice between equal alternatives, good or bad.

A *Hobson's Choice* isn't really a choice of alternatives, it is a choice of accepting an offer or not, that is *take-it-or-leave-it*.

A *false dilemma* is when only some of many choices are presented. This happens in surveys when there are too many alternatives to consider.

I have a bottom-up view of this.

A *Complex Alternative* is a situation in which accepting one choice implies denying the other choices. This is used in surveys to explore the underlying opinions of respondents. *Mutually-exclusive-alternatives* are choices in which there is no overlap of conditions.

A *Catch-22* is a situation in which there are no real alternatives. For example, you need specialized experience to get a certain job but can't get the experience without having such a job.

Perspective bias is when people have an unbalanced focus on details versus the overall situation. It may be *top-down* or *bottom-up*. It may also include ignoring atypical information (outliers).

Literacy.

Confirmation bias is the tendency of individuals to believe information that aligns with their own beliefs as opposed to contrary evidence.

Disciplinary Blinders is when an analysis is conducted or a presentation is prepared without considering information or expertise from other relevant domains. For example, there are thousands of publications involving statistical analyses but many did not

I have a top-down view of this.

have a qualified statistician on the team that prepared them. It's like when you don't have a screwdriver so you use a butter knife.

Deliberate Ignorance (also called *closed-mindedness* and the *three monkeys' fallacy*) is when people decide not to pay attention to a presentation or even any information related to it. *Plausible deniability* is when people decide not to pay attention to a presentation so that they can deny knowledge or responsibility for any negative consequences of their inaction.

Argument to ignorance is assuming something is true simply because it hasn't been proven to be false.

Excluded Outliers is where data or information that don't agree with the bulk of the data or information are disregarded. *Conspiracy theories* are novel claims that are often rejected without evidence simply because they appear to be unlikely.

A *middle-of-the-road-fallacy* (also called *marginalization of the adversary*) is where the legitimacy of information is based on it being midway between extreme alternatives. "The middle of the road is for yellow lines and dead squirrels" (armadillos in Texas and mongooses in Hawaii).

Excluded middle is where a belief that if a little of something is good than more would be better (or if less of something is good, none would be better).

Status quo (also called *argument-to-tradition* and *Procrustean fallacy*) is when individuals accept information because it is an established belief in society.

Presentism is when current beliefs and attitudes are applied to historical events and practices.

Hyperbole is where a presentation overstates (or understates) the magnitude or importance of a result.

Literacy.

FAULTY PROCESSING

Statistics isn't just mathematics, it's a way of thinking. Sometimes the numbers are right but the way they were created is misguided. Here are a few ways that it can happen.

Indulging variance is the failure to control or even consider variance. Studies that ignore variance and just address the average or most common case tell a misleading story.

Overfitting is where an excessive number of metrics are included in an analysis in order to find positive results. Predictions made from the data are accurate while predictions made from new data often are not.

Data dredging (also called *data fishing*) involves looking for any relationships in large datasets without relying on hypotheses about why the relationships might exist.

I just wanted a sip of water.

Worst-case fallacy (also called the *just-in-case fallacy*) is where reasoning is based on low-probability conditions or scenarios rather than those that are more likely. This bias is used often in deterministic data modeling to compensate for not considering variance.

Where-there's-smoke-there's-fire (also called *hasty conclusion* or *jumping to a conclusion*) is where a conclusion is reached without considering enough evidence.

Sweeping generalization (also called *stereotyping*) is where a broad assertion is applied to specific cases without support. The *pars pro toto fallacy* is where a few legitimate examples are used as evidence that all cases are valid. *Overgeneralization* (also called *hasty generalization*) is where a broad generalization is used to refute particular cases.

Oversimplification (also called *reductionism* and *sloganeering*) is where misinformation is spread by presenting overly simple answers, slogans, or memes in response to complex questions.

Non sequiturs are statements or conclusions that do not follow from the premises offered.

Literacy.

NUMBER CONFUSION

There are more ways that a statistical analysis can unintentionally and unknowingly go wrong than you can possibly imagine. The cases of intentional deceit pale by comparison and are usually easier to spot. Here are some of the many ways statistical analyses can be susceptible to fallacies, biases, preconceived notions, and other flaws.

PLANNING AND FUNDAMENTALS

If a statistical analysis is flawed, the cause probably occurred in the planning phase and involved fundamental aspects of the statistical design. Here are a few examples:

Goldilocks' quest is to have just the right number of samples, variables, tests, and other elements of a statistical analysis; not too few and not too many. Too few or too many samples leads to inappropriate resolution. Too many variables lead to confusion, overfitting, and misspecified models Too many statistical tests lead to false results purely by chance. Goldilocks' quest is a laudable goal but difficult to attain.

Phantom populations are collections of items or individuals that don't share enough relevant characteristics to be considered a statistical population. For example, comparisons between countries can suffer from disparities in cultures, governments, and laws that make inferring results to other countries invalid. A survey of "people-wearing-red-hats" might include fans of the Cincinnati Reds, Shriners, Cardinals in the Catholic Church, certain gang members, and Trump supporters, which would certainly not constitute a legitimate statistical population. Canadian researchers found one such phantom population when they tried to create a control group of men who had not been exposed to pornography.

Complex questions are those questions that require a binary response even though they do not have mutually-exclusive alternatives.

Hoyle's fallacy is the belief that a low-probability event can never happen, like winning the lottery.

The *gambler's fallacy* is the belief that the history of a random event occurring is an indication of the event occurring in the future.

I don't understand your question.

SAMPLES AND METRICS

Samples, variables, and metrics can all be inapplicable, inadequate, incorrect, or poorly selected. Examples include the following.

Literacy.

Sham samples are samples that don't adequately represent a statistical population. This is a common criticism of some election polls, especially exit polls, where matching the overall demographic of voters can be challenging.

Survivorship bias involves using data that remains after some unaccounted for filtering process. This was a key concept in armoring U.S. bombers during World War II. Survivorship bias was avoided when armor was placed on areas of planes where they had suffered the least damage on missions because planes that had more damage in those areas did not return. Likewise in World War I, the British found that head injuries increased after they introduces a new helmet. The increase was attributable to more soldiers surviving head wounds because of the helmets. Other examples include biasing future research because only articles with significance statistical testing results are published and biasing survey results when invitees decline to participate thus ensuring their opinions do not survive.

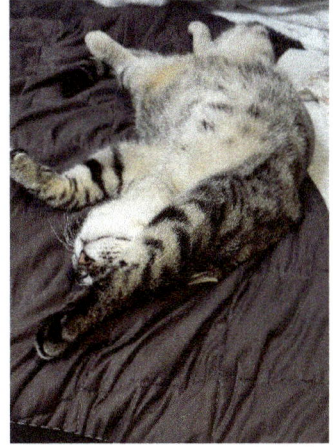

I'm a survivor ... just a little bit sleepy.

McNamara fallacy, also called the *quantitative fallacy*, is where analysis of complex situations is biased because only qualitative metrics are used while ignoring qualitative metrics. This fallacy stems from *measurability bias,* the belief that something that cannot be measured quantitatively is not worth measuring at all.

Cost bias is where information or objects that are more expensive or are more difficult to obtain are believed to be intrinsically better.

ANALYSIS

Besides all the ways that number crunching can make an analysis go bad, absent or confused statistical thinking can be just as fatal. Here are a few ways that it can happen.

Simpson's fallacy involves assuming that data relationships seen in the entire dataset are mirrored in subgroupings of the data.

False precision involves believing that data and results are more precise than they actually are. Examples include expanding chart axes to infer unsupportable resolution and treating differences in survey percentages that are less than the margin-of-error as consequential.

Significant insignificance and *insignificant significance* can occur when a statistical test isn't designed properly or the data violate statistical assumptions. Just as correlation doesn't necessarily imply causation, significance doesn't necessarily imply *meaningfulness*. Furthermore, sometimes studies do not report nonsignificant

Literacy.

results, which could be exactly what you're looking for. This is a form of survivorship bias.

Misdirected models involve researchers creating models based on biased or mistaken theories, and then using the model to explain data or observed phenomena in a way that fits the researchers preconceived notions. It is a variation of *begging the question* and a sure path to *pathological science*.

Extrapolation intoxication occurs when analyses involve areas outside the range of the available data. This can occur, for example, when predictions are made for freezing conditions based on data from ambient temperatures. Other examples include inappropriately extrapolating tests on animals to humans, showing information on maps that are not in visual range, using surveys of one demographic to infer information for a different demographic, and the like. Perhaps the only example of extrapolation that is accepted by statisticians is time-series analysis to predict the future. The issue is how far into the future is reasonable.

Post hoc fallacy is the fallacy of assuming that one event causes another simply because it occurred before, that is, A caused B because A happened before B.

RESULTS

Sometimes, everything is fine right up to the end of the analysis and reporting when fate steps in and the stats hit the fan.

Unintended consequences refers to unfavorable events that occur as the results of an action. Famous examples include the *Hawthorne effect* (people changing their behavior because they are being observed) and the *cobra effect* (a solution to a problem that makes the problem worse). There are probably unintended consequences following virtually any policy, rule, or law instituted by an organization because affected individuals will attempt to circumvent or capitalize on the conditions placed on them.

Go away. You're not the results I was looking for.

Paralysis by analysis (also called *procrastination* and the *nirvana fallacy*) is where reaching a definitive conclusion is deferred until additional data are collected or analyses are completed. It is similar to *Goldilocks Quest*.

File drawer bias is the common practice of rejecting the publication of non-significant results. It will never be known how much research effort has been wasted

Literacy.

by investigators studying hypotheses that have already been tested and rejected but not communicated to others.

Bear in mind that just because an argument is based on a fallacy doesn't necessarily mean that it is false. This is called the *fallacy fallacy*.

HOW TO EVALUATE STATISTICAL PRESENTATIONS

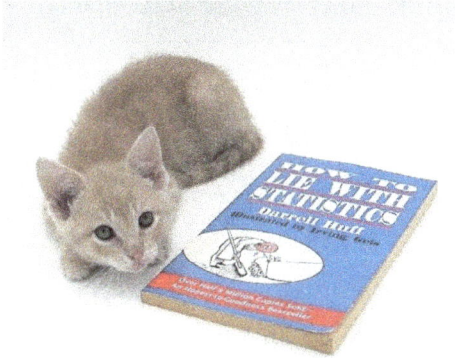

Sometimes it's hard to tell if a presentation that uses statistics is legitimate or not.

If you haven't figured it out by now, it's really hard to look at a statistical analysis conducted by someone else, where most of what they did is undisclosed, and decide if their results are legitimate.

Huff made it sound easy in **How to Lie with Statistics**, but it's not. It's even hard for professional statisticians to do. That's why peer-review processes are an integral part of professional publishing.

Even so, peer reviewers don't always catch fatal flaws in submitted manuscripts. Over 37 thousand professional papers were retracted from 2000 to 2020 and those retractions don't include unintentionally bad analyses.

So, don't feel inadequate or discouraged if you have a hard time judging the legitimacy of a statistical analysis. You are not alone.

To evaluate a statistical analysis or a media presentation about a statistical analysis, start with the easiest element to assess—the source of the presentation, the research organization who performed the analysis, and the sponsor who funded the work. Sources are the easiest to assess because information about them is almost always available on the internet. If there is no information, that's a red flag.

Assessing the way the analysis is presented comes second. It's harder to assess because of all the biases and fallacies to consider. However, there are a few red flags to look for:

- Clickbait titles and sensationalized text (i.e., *hyperbole*)

> *What comes full of virtue from the statistician's desk may find itself twisted, exaggerated, oversimplified, and distorted-through-selection by salesman, public-relations expert, journalist, or advertising copywriter.*
>
> Darrell Huff, **How to Lie with Statistics**, p. 101.

Literacy.

- Language meant to provoke fear and outrage
- Requests for likes, retweets, and donations
- Lack of context about the study relative to the claims of the presentation.

Assessing the statistical analysis itself comes third. Most of it is likely to be far too complex for any evaluation except for a few fundamental issues, like populations and graphs. Table 41 is a summary of things that might go wrong and what to look for.

Table 41. Examples of things that might go wrong in a statistical presentation.

Concern	Examples of What Might Go Wrong		What To Look For
	Unintentionally	Intentionally	
Source	Reporter unfamiliar with STEM.	Funding provided by a source known to be biased. Source hidden or masked through a third party.	Look up reputation of source on the Internet.
Data	Data not representative of population. Errors introduced when multiple data sets were merged.	Outliers treated improperly. Outright data fraud.	Statistics and graphs of sample demographics compared to descriptions of population.
Analysis	Misleading data generated by miscalibrated meter or misleading survey question.	Model misspecified or overfit.	Not much can be done without advanced expertise.
Presentation	Presentation derived from another source.	Minor finding highlighted to gain more attention in the media. Clickbait.	Look up original study. Compare presentation to others on the same topic.

STEP 1. EVALUATE THE SOURCE.

Start your evaluation of a statistical presentation with the sources. First and foremost, beware of any revelation in which someone says or writes: "I was told that …"; "some people say that …"; "I know people who think that …"; "I heard a rumor that …"; "I read somewhere that…"; "people are always telling me that …"; "wherever I go people say that …"; "it has been reported that …"; or "many sources confirm that …". These phrases are staples for politicians, fake-news outlets, and social media.

For a data presentation to be credible, it must identify its sources—the *sponsor* who funded everything, the *organizations* who oversaw the researchers, and the *researchers* who conducted the study. Sources are a good place to start because there will probably be information about all of them on the internet and you really only need to identify major red flags.

The reputations of the sponsor, the organizations, and the researchers are important but mostly for the sponsor. If the sponsor is a government agency, no problem. However, if the sponsor is a private organization, there is more to consider.

Does the sponsor have a history of funding any biased research? Might there be a conflict-of-interest with other activities of the sponsor? Do they stand to profit from the research? These could all be red flags.

Literacy.

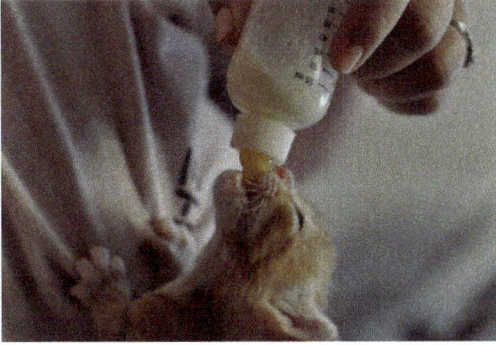

You're a great sponsor.

A positive sign would be if they have successfully funded similar research. The same considerations would apply to the organization overseeing the researchers' work.

For the researchers themselves, information on their expertise (i.e., knowledge and experience) would be more relevant, and easier to find and assess. Their expertise on a specific phenomenon, though, may be harder to assess.

Knowledge of a phenomenon comes in many flavors, most commonly defined by training. More training is better than less training but it also depends on the quality of the training and its relation to the phenomenon. An individual may be academically trained on a specific phenomenon, academically trained in a general discipline related to a phenomenon, self-trained on a phenomenon, have no training related to a phenomenon, or be indoctrinated into misconceptions about a phenomenon. Obviously, the most knowledgeable individuals would be those with the most high-quality training, even if it were self-training.

Some individuals who are experts in their research topic may have *disciplinary blinders.* They may have a team of knowledgeable staff working on the central topic of the research but no one with the same level of expertise in support roles, like statistical analysis. Of course, every research team is unique but such omissions often become apparent in time, sometimes in peer review and sometimes after publication.

Like knowledge of a phenomenon, experience with a phenomenon comes in many flavors. The experience can be directly with the specific phenomenon, general experience related to the phenomenon, or experience with related phenomena. But while training is usually a structured program requiring participation, experience can represent anything from hands-on performance of experiments down to watching other people work.

A decade of experience may represent ten years continually learning more about a phenomenon or just a year's worth of experience observing a phenomenon repeated ten times. Consequently, experience may not be so relevant and can be difficult to evaluate. Look for an individual's progression of responsibilities or at least job titles.

STEP 2. EVALUATE THE PRESENTATION.

The way a statistical analysis is presented in the media is undoubtedly a concern if you want to understand the truth. Perhaps the biggest reason why media articles

Literacy.

involving STEM can be so misleading is because they are written by non-experts to attract attention, not convey facts. They get their information from researchers, or articles the researchers have written, or articles about those articles.

It's easy for non-STEM writers to misunderstand some things, which can be just as consequential as outright lies and are probably much more common. Add to that all the fallacies, biases, and preconceived notions that are unintendedly included in the reporting and you the reader are at a disadvantage.

The research reported on in a media presentation is probably on the cutting edge of science or else it would be old-news. Mainstream science, even evolving mainstream science, will appear in professional journals, not social media. Established science appears in textbooks; nobody reads those. The research may actually be fringe-science or *pathological science* if not *barely science*. Those categories draw more public attention than even real science.

These steps are difficult but important.

To tell how credible the science may be, look to see if the media presentation is based on talking to a researcher or on a journal article. A single researcher doesn't guarantee that the information will be comprehensive and unbiased. An article in an established professional journal, on the other hand, probably would because there would be at least some review process.

If you really want to dig into some media presentation of a statistical analysis, start by tracking down the original research. If the basis of the article is an individual, they'll probably have a webpage on their research, or articles that they've written, or something else on the internet that will provide some context for their research. If the basis of the article is a STEM journal, you'll probably be able to find at least an abstract and maybe the whole article.

From those starting points, look for other information on the same topic especially by other researchers. Do they all agree? Consider how the presentation you're evaluating fits into the context of the other sources of information. The process sounds vague and unproductive but you will be surprised at what pops up.

If you're looking at a statistical presentation for your employment instead of in the media, be clear on whether it's supposed to be a *data summary* or a *data analysis*.

Summary reports contain data lists, sorts and queries, and simple descriptive statistics. Analyses describe objectives, data, hypotheses, and results. Summaries provide information; analyses provide knowledge.

Literacy.

It's like with your bank account. Sometimes you just want a quick summary of your balance. That information has to be readily available whenever you might need it and both you and the bank have to be working with exactly the same data. If you want to assess patterns in your spending, though, you have to conduct an analysis.

Say you want to figure out how much you're spending on commuting over the past five years, you'll have to compile the data and scrub out anomalies, like the cross-country driving you did on vacation, to look for patterns. Analyses involve much more than a glance, they take time, sometimes, a lot of time.

I don't think that'll be enough money.

Finally, beware of cherry-picked facts and data. Find the most sensational claims in the presentation and try to trace them back to the original analysis. Given the contexts under which the claims were made, decide if they were overstated. It's a lot of work but, if you can access the original research, it is often worth the effort.

STEP 3. EVALUATE THE RESEARCH.

Evaluate the actual statistical research last because it will be the most difficult to understand and it's the least likely to be the biggest problem (compared to the presentation). Moreover, you won't be able to tell if something is wrong unless you're a statistician and have some basic understanding of the topic of the research. Nevertheless, here are a few things to look for.

DATA

Were the data generated specifically for the research or were they obtained from another organization, such as a government agency, a college or university, or a professional data-acquisition company? Are they *primary data*, *secondary data*, or *tertiary data*?

One way to judge the quality of a dataset is to determine if it can be accessed, even if you would have to pay for it. You don't have to actually obtain and look at the data, just knowing that they are available probably means that the owner thinks they're good.

If the data were generated specifically for the research, there's not much you can do even if you could access them. *Data scrubbing* and an *exploratory data analysis*, what you would have to do to assess the data, go way beyond what you would want to do to evaluate a statistical analysis. You also probably wouldn't have information on the procedures used to collect the data, that is, the *metadata*. Furthermore, there

Literacy.

may be some domain-specific metrics that you wouldn't know enough about to evaluate. Even if you did, who would decide if your analysis is more legitimate than the original researcher's. As a consequence, evaluating a dataset just to judge if an analysis is legitimate is a lot of work for an uncertain return.

ANALYSIS

If you're evaluating a media presentation of a statistical analysis,

> It's easy to lie with statistics; it is easier to lie without them.
>
> Frederick Mosteller, 1916-2006, American mathematician and statistician.
>
> It is easy to lie with statistics. It is hard to tell the truth without statistics.
>
> Andrejs Dunkels, 1939-1998, Swedish mathematician and writer.

don't expect to find a major flaw in the statistical analysis. You'll need access to the original research report, not just an article about it, and at least some understanding of statistics. Even so, you'll probably never be able to determine if the researchers did all the right things in all the right ways.

You may, however, be able to identify simple and fundamental errors, things like *phantom populations* and *sham samples*. If you know something about the topic of the analysis, consider the metrics too. Are they reasonable? Are the assumptions underlying the analysis reasonable? Be wary of any mentions of *causation* and *significance*.

One place you might find some issues in a statistical analysis is in the statistical graphics. Were the graphics originally prepared by a statistician or were they created later for a media presentation? There are scores of ways that data presentations can unintentionally or intentionally mislead the audience. Look carefully at the *axes* and be sure the data are presented in context and are not cherry-picked. Follow the guidelines of *foundation, framework*, and *facade*.

STEP 4. MAKE YOUR JUDGMENT.

If you find something questionable about the analysis, consider its importance. A minor or presumed flaw won't necessarily mean that the whole presentation is invalid (i.e., the *fallacy fallacy*). Keep your emotions in check and consider the bigger picture. If you still have reservations about the presentation, think about motivations. Who stands to gain from what the article is reporting? Follow the money.

Finally, **know when to quit**. The analysis and the presentation may both be legitimate or there may not be enough information to tell. Be satisfied that you've accomplished your aim of being statistically literate.

Literacy.

Don't be discouraged if all the aspects of statistical thinking, statistical literacy, and critical thinking make your head explode. It's a learning endeavor that will span your lifetime. Take it one step and one experience at a time. As you gain knowledge and confidence, you'll understand why critical thinking and statistical literacy are the greatest superpowers of all.

WHERE TO GO FROM HERE

If you enjoyed reading **Stats with Kittens** you might want to go further in becoming statistically literate. If so, your next step would be to take an introductory course in statistics, Stats 101. That's where you'll learn about the theoretical background of the statistics summarized in **Stats with Kittens** and how to perform many different types of analysis. The fundamental concepts, though, are the same—populations, phenomena, data, probability, uncertainty and so on.

Fundamental elements of introductory applied statistics are shown in Figure 75.

What should my next step be?

After completing Stats 101, you can decide if you want to take more courses in statistics. They will be useful in any career you decide to pursue. You may even consider pursuing a career as a statistician, data scientist, or any of the dozens of other job titles that use statistics in their work. If you want to get on with a career you've already chosen but want to use the statistics you learned in Stats 101, consider reading **Stats with Cats**.

Stats with Cats: The Domesticated Guide to Statistics, Models, Graphs, and Other Breeds of Data Analysis, 2nd edition provides a roadmap for getting from a problem or question to a data-driven solution. For people who have completed Stats 101 and are faced with conducting a statistical analysis on their own, **Stats with Cats** describes how to use statistics on real projects, from goal setting and planning through selection of analysis methods and presentation of results. It shows how to avoid and diffuse potential disasters—in the data, the analysis, the project, and the project participants—before running afoul of them. The book even describes how to critique other people's statistical reports.

For experienced professionals, **Stats with Cats** provides a variety of tips and tricks to help you analyze your data from identifying and correcting data problems to avoiding fatal flaws in a data analysis. For experienced data analysts, statisticians, and data miners, **Stats with Cats** provides advice on planning and managing an analysis, dealing with people and problems, and communicating results.

Literacy.

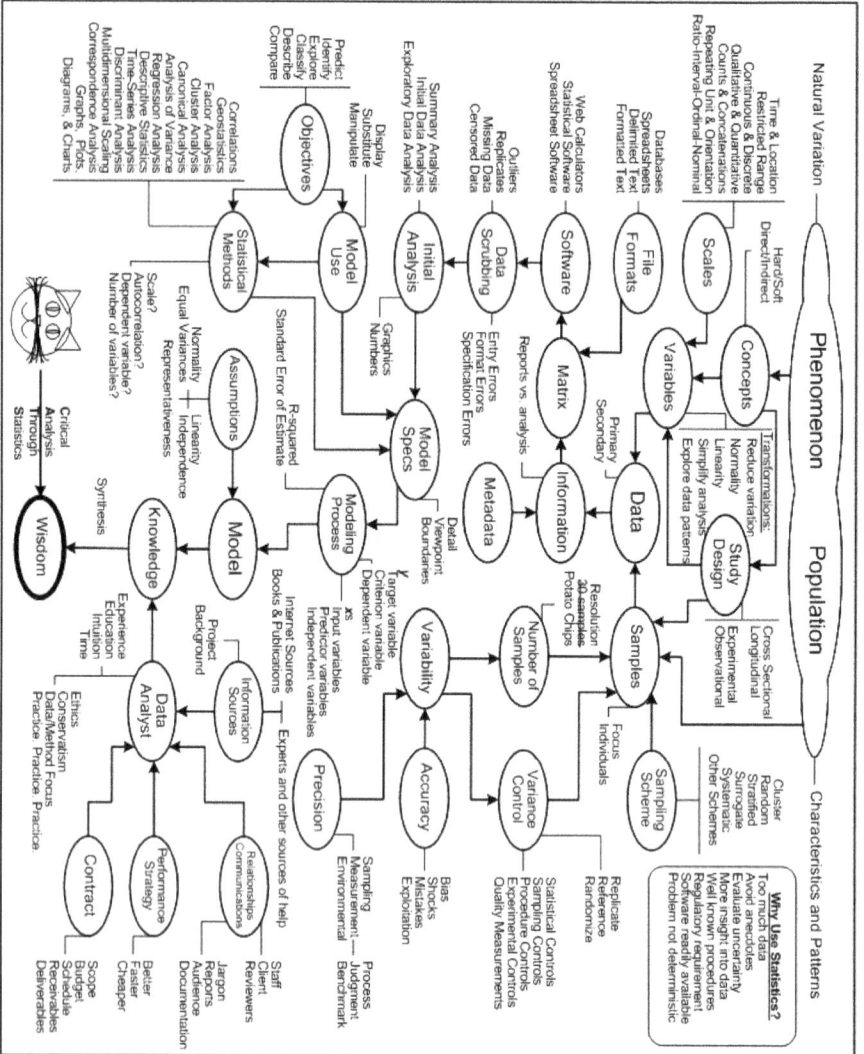

Figure 75. Fundamental elements of applied statistics (From C. Kufs. 2011. Stats with Cats. p.333).

Stats with Cats contains 31 chapters divided into 7 parts, covering topics that will help you decide how to conduct your own statistical analysis.

Part I. The Lost Treasures of Stats 101. Reviews some of the basic jargon and concepts you'll need to know to do your own analysis. Part I will refresh your memory about the basics you learned in Stats 101 that you'll need to analyze your own data.

There are a lot of pieces to the statistical puzzle.

Part II. Frisky Business. Helps you decide if you should do the data analysis yourself or get someone to do it for you. It describes how to set up a data analysis project and what software and information sources you might use. It also describes problems that you may encounter on projects involving data.

Part III. Is that a Dataset in Your Pocket? Provides a strategy for deciding what variables to measure and how to measure them, how to select samples including guidance on how to decide how many you'll need, how to recognize and control sources of variability, and how to format data that software can analyze.

Part IV. Statistical Foreplay. Describes common types of data errors, how to find and correct them, and what you can do about duplicate data, missing data, censored data, and outliers. It also describes ways you can augment your dataset to make your analysis more thorough.

Part V. Getting Serious with Data. Describes what to calculate, what to plot, and what to look for when you first explore your data.

Part VI. A Model for Modeling. Describes the process you go through to create a statistical model including statistical analysis techniques you probably didn't hear much about in Stats 101, why even the most credible models can fail, and what you might do about it.

Part VII. Saving the World One Analysis at a Time. Describes how to write data analysis reports, how to critique a statistical analysis even if you don't know a lot about statistics, and provides some suggestions for how you can apply the things you learn about statistics in your own lives.

And, of course, **Stats with Cats** includes enough pictures of cats so your experience will be stress-free if not enjoyable.

Literacy.

I can't wait for Stats with Cats.

Literacy.

GLOSSARY.

Abductive reasoning—A type of reasoning that involves concluding the most likely explanation for an observation. It's also known as *inference to the best explanation*, *educated guess*, or *best guess*. It is different from deduction, which begins from a general rule based on data; abduction begins from past knowledge.

Accuracy—How close measurements are to their true value. The ability of a measurement device to produce a true value.

Acquiescence bias—Survey responses that tend to accentuate positive opinions about a topic.

Ad hominem—A logical fallacy in which the person making an argument is criticized instead of the argument itself.

Additive model—A statistical model in which the terms of the model are combined by addition, that is the sum of constant coefficients multiplied by variable values.

Glossary.

Adjusted correlation coefficient—A Pearson correlation coefficient that measures the strength of a data relationship after correcting for the number of variables and the number of data points. Also called the *shrunken correlation coefficient* because it is always the same or smaller than the unadjusted correlation.

Aggregators—Pundits that calculate statistics from sets of election polls in order to minimize the effects of inconsistent polls.

Aim—The reason a graph is created and what it is expected to show, such as data properties or relationships.

Algorithm—Step-by-step procedure for performing some operation, often by computer.

Alphanumeric—Consisting of letters and numbers.

Alternative hypothesis—The hypothesis that is adopted when the null hypothesis is rejected on the basis of a statistical test.

Analogy—One of Hill's criteria for causality. A relationship is more likely to be causal if there are proven relationships between similar causes and effects.

Analysis bias—Systematic inaccuracies attributable to how data are processed and analyzed, including data scrubbing and the selection and implementation of analysis techniques.

Analytical models—Mathematical equations derived from scientific laws that produce exact solutions that apply everywhere.

ANCOVA—Analysis of Covariance, a statistical technique for removing the effects of one or more continuous variables to help detect differences between averages of a continuous dependent variable in groups defined by one or more qualitative independent variables.

Anecdotal evidence—Evidence based on unverified personal observations that were collected in a non-statistical manner.

Anecdote—A personal story or observation.

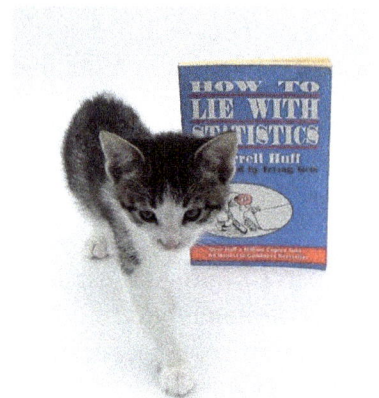

Glossary.

Anonymous—Polls in which the real-world identities of respondents are not known. Polls in which the real-world identities of respondents are known but not released are called confidential.

ANOVA—ANalysis Of Variance, a statistical technique for detecting differences between averages of a continuous dependent variable in groups defined by one or more qualitative independent variables. There are many variations of the basic ANOVA method including, one-way, two-way, and n-way designs, fixed-effects, random effects, and mixed effects designs, factorial designs, and repeated measures designs.

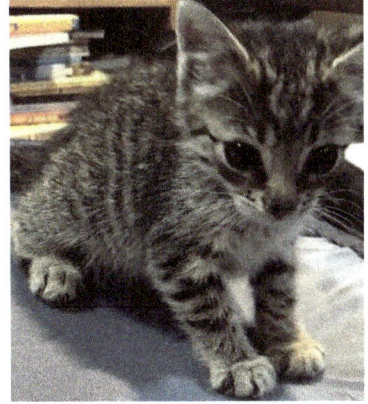

Appeal to common belief—The fallacy of basing an assertion on what many people believe without proof. It is also called *bandwagon fallacy*, *consensus fallacy*, *argumentum ad populum*, and many other names. It's a very popular fallacy.

Appeal to false authority—A fallacy in which the opinion of a person who is not a legitimate expert is used to support an argument.

Appeal to ignorance—The fallacy of rejecting an assertion because it has not been proven to be false.

Applied statisticians—Individuals who use statistical procedures created by theoretical statisticians to analyze real-world data. Applied statistics involves using statistics to analyze real-world data.

Arc diagram—charts that show organizational connections in one-dimension using arcs. They have no scale but are dawn to show direction, importance, or frequency.

Area charts—Charts in which the y-axis represents quantity, the x-axis represents time or another metric, and the data are represented by areas. They are like bar charts in which the bars are stretched to fill the entire plot area.

Area sample—Samples limited to a specifically-defined location or segment of a population.

Area survey—Surveys in which the population is defined by a physical or alternately defined area rather than

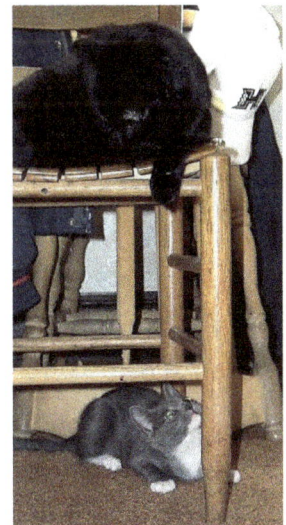

demographics based on probabilities. A sample from such a bounded area is called an area sample.

ARIMA—AutoRegressive, Integrated, Moving Averages model, a statistical technique for forecasting time-series. Also referred to as Box-Jenkins modeling or time-domain analysis.

Arithmetic mean—A measure of central tendency calculated by summing the data and dividing by the number of data points.

Array—Matrix.

Artificial intelligence—The concept of using machines to replace humans goes back thousands of years to the use of mechanical devices to regulate irrigation and natural water flows, keep time, create music or sounds, provide security, and perform other functions in the place of humans. Since the 1950s, the concept evolved from single-function machines to digital technologies and algorithms for mimicking capabilities of the human mind, including learning and problem solving.

Association relationship—A data relationship in which two events or conditions occur together without any direct interaction. There are reasons that associations occur unlike spurious relationships which occur randomly.

Association rules—A statistical modeling method to explore relationships between qualitative variables. Results often take the form "if you did this, you might also want to do these things, too."

Assumptions—Conditions required for the accuracy of statistical analyses that are difficult to verify or meet exactly so they are assumed. Key assumptions of statistical tests and models include sample representativeness, model additivity and linearity, and independence, Normality, and equal variances (homoscedasticity) of errors.

Attribute—Characteristic of an individual sample. Another term used for variables.

Attrition—When individuals on a survey panel drop out or are removed by the panel manager.

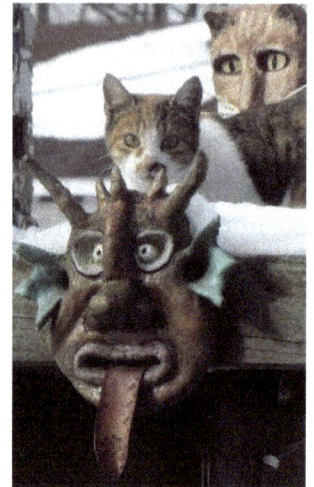

Glossary.

Audience—The individuals who are expected to read a statistical report or other document. The contents, formatting, and writing style of a report should be geared to support audience understanding.

Autocorrelation plot—A line plot of the order of occurrence of data points.

Autocorrelation—Synonymous with serial correlation if the data points are collected at constant time intervals.

Autoregressive—Prediction of a variable from other measurements on the same subject, usually over time.

Availability bias—Misleading samples, data, or information selected primarily because they are easy to obtain. *Availability sampling* occurs when planned sampling points are ignored in favor of more accessible points. This happens in most areas of data collection from environmental sampling to surveys. Availability bias also occurs when data are too difficult or expensive to obtain.

Average—Arithmetic mean. One measure of central tendency of a data distribution defined as the sum of all measurements divided by the number of measurements.

Axis lines—The sides of a graph that represent the scale values of data being plotted, usually referred to as *axes*. An axis may be divided by short lines called *tick marks*. Tick marks may be extended across a plot area to facilitate determining a data point's value on an axis. These are called *gridlines* because they are usually drawn for both axes thus forming a grid. *Reference Lines* may also be drawn perpendicular to an axis as a benchmark for data patterns.

Bad science—Any scientific endeavor that does not follow the scientific method and other scientific principles.

Balanced—An ANOVA design in which the number of samples is the same, or nearly the same, in all groups or blocks of the design.

Balloon charts—Balloon charts are much like bubble charts in that they show qualitatively a third data dimension using the size (diameter or area) of the balloon or bubble. Balloon charts often use spheres rather than circles with a line extending from the balloon to the x-axis.

Glossary.

Bandwagon Fallacy—Assuming something must be true because everybody believes it is. Also called argument from common sense and appeal to common belief. Similarly, an information cascade is where people repeat the opinions of others, especially on social media, giving the impression of a consensus.

Bar charts—Bar charts (or bar graphs) have an ordinal or continuous-scale variable on the vertical axis and a nominal or ordinal-scale variable on the horizontal axis. The quantitative axis usually represents frequency. Data are represented by bars.

Barely science—A type of *bad science* that might be perfectly acceptable science except that it is too underdeveloped to be released outside the scientific community.

Batch process—A process that is conducted from a set beginning to its completion. In data analysis, that involves completing the steps of data collection, scrubbing, analysis, and reporting without repeating any of the steps. Statistical analyses are usually batch processes. In contrast, continuous processes involve steps that are repeated with no real end. In data analysis, continuous processes might involve constantly collecting new data and updating reports. Statistical reporting, such as with dashboards, is usually a continuous process.

Bayesian probability—Methods based on Bayes Law that seek to refine probability estimates by successively incorporating new information. It is an alternative to frequentist probability.

Beating the data until it confesses—From a quote by Ronald H. Coase, a British Economist, "if you torture the data long enough, it will confess to anything." It is sometimes used as an admonition against capitalizing on chance relationships by data dredging or data fishing.

Begging the question—Begging the question involves assuming the conclusion of an argument as a premise. It is essentially circular reasoning.

Bell curve—The Normal distribution. Normal distributions don't always resemble the profile of a bell. They may be flatter or more peaked. It's just a convenient comparison that is made in introductory statistics.

Glossary.

Benchmark—The accepted standard against which a data value is made, a reference point.

Best guess—Same as *educated guess* and *abductive reasoning*.

Beta coefficients—Regression coefficients that are standardized so that they are comparable despite the original units of the variables. Also called *standardized regression coefficients*.

Beta—The probability of making a Type II error in a statistical comparison, (i.e., accepting a false null hypothesis). Power is 1-beta.

Between a rock and a hard place—A choice between two undesirable alternatives. Also called a dilemma.

Between-groups variance—Variation attributable to the treatment in an ANOVA.

Beyond a reasonable doubt—Evidence that must convince a jury that there is no other explanation that makes sense. The jury must be virtually certain of the defendant's guilt.

Bias—Systematic deviation, whether intentional or not.

Big data—Datasets so large (terabytes and larger) that they require special techniques to manage and analyze, often characterized as having the properties of volume, velocity, variety, veracity, value, validity, volatility, variability, visualization, and vulnerability.

Big picture—An overall view of a situation, also called top-down view. It is the opposite of a bottom-up view that focuses on details, also called in the weeds. In statistics, the big picture involves overall descriptive statistics and dashboards, tests of model effectiveness, and report summaries and conclusions. The bottom-up view focuses on data and anomalies, subgroup descriptive statistics, tests of model components, analytical methods, and metadata.

Binomial—Binomial can refer to several different things in statistics. A binomial variable or sample refers to measurements with only two alternatives, like coin flips or yes-no responses to survey questions. A binomial test assesses whether a

Glossary.

proportion of a binary variable is equal to some value. The binomial distribution is a discrete probability distribution of the number of successes in a sequence of independent experiments.

Bins—Bins are the intervals of values that make up a histogram for a continuous-scale metric.

Birthday paradox—The counterintuitive fact that only 23 people are needed for there to be a 50% probability that at least two members of a group will share the same birthday.

Biserial correlation coefficient—Measures the strength of a data relationship where one variable is measured on a continuous scale and the other variable is measured on a binary scale.

Bivariate plot—Two variable plots, usually referring to scatter plots.

Blanks—Samples collected to assess variance in studies involving chemical analyses.

Blinding—Withholding information about a study's details from study participants to protect against bias. Single blinding involves not informing study subjects which treatment group they have been assigned to. This is often accomplished through the use of placebos. Double blinding involves also not informing the people conducting the study, both study leaders and those who interact with the subjects, which treatment group the subjects have been assigned to. Blind samples are samples whose origins are not identified to laboratories.

Block clustering—A data mining technique that simultaneously clusters both variables and observations in a dataset, also called bi-clustering, co-clustering, and two-mode clustering.

Block diagrams—A three-dimensional drawing projected onto a two-dimensional surface. Block diagrams may also refer to an illustration of a system in which parts or functions are represented by blocks connected by lines that show the relationships.

Block-design ANOVA—The process of grouping samples in ANOVA to try to control extraneous variability. The nature of the blocks is not important so long as

they control variability in the main effects. Also called a block effect. Continuous-scale variables used for the same purpose are called covariates.

Bootstrapping—A procedure for creating a new dataset by resampling existing data values with replacement. Jackknifing involves resampling without replacement.

Borders—The lines enclosing a graph and its plot area. Some graph creators use them and some don't.

Bottom-up—A viewpoint that focuses on details, also called in the weeds. In statistics, the bottom-up view focuses on data and anomalies, subgroup descriptive statistics, tests of model components, analytical methods, and metadata.

Boundary windows—Boundary windows address the question "should I give somebody my advice?" The axes represent expertise and receptiveness.

Bounded thinking—Bounded thinking happens when the creation, result, or interpretation of a presentation is limited in some way by the creator's limited perspective. Ways that presentations may be limited include: false binary (either/or) alternatives, unbalanced focus on details versus the overall situation, disciplinary blinders, confirmation bias, closed-mindedness, accepting wide-held information without evidence, and ignoring outliers.

Box-whisker diagram—A statistical graphic consisting of a rectangle (the box) with lines extending from the two opposite ends (the whiskers). Also called box plots. Box plots depict several different statistical measures of central tendency and spread of a dataset. The box usually represents the interquartile range, which is the center 50 percent of the dataset. The small square in the center of the box represents the median. Sometimes a line is used instead of a square. The whiskers represent the upper and lower ends of the data distribution ending with the minimum and maximum. There are quite a few variations of box plots.

Brainstorm diagrams—A diagram used to show relationships between associated ideas, words, images and concepts, also known as a mind-map.

Glossary.

Branching—Alternative sets of follow-up survey questions asked on the basis of a respondent's answer to a previous question

Bubble diagram—A two-variable scatter plot in which a third variable is represented by the size of the data points. Bubbles are usually represented as circles whose sizes may be defined by either their diameters or their areas.

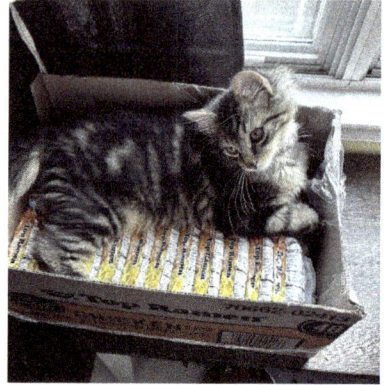

Bulletin boards systems—BBSs were computer servers that allowed users to upload and download software and data, play online games, and message other users. They were popular in the 1980s but declined in the mid-1990s with the emergence of internet web browsers.

Burden of proof—The standard that a party seeking to prove a fact in court must satisfy.

Buridan's ass—A dilemma in which a person cannot choose between two equally attractive and attainable alternatives.

C&RT—Classification and Regression Trees, also called CART, is a nonparametric, nonlinear method of predicting continuous dependent variables or classifying categorical dependent variables using a hierarchical, binary decision structure.

Calendars—Displays of data by date, including days that have no data associated with them. Also called calendar plot graphs.

Candlestick charts—A chart similar in appearance to a box plot that is used to describe price movements of a security, derivative, or currency, usually over short periods of time. Candlesticks show open and close prices in the thick body, and high and low prices in the lines.

Canonical analysis—A statistical method for examining relationships between multiple dependent variables and multiple independent variables (or any two sets of variables). Also called canonical correlation analysis.

Canonical correlation coefficient—A Pearson correlation coefficient that measures the strength of a data relationship involving two sets of more than one variable.

Capitalizing on chance—The result of analyzing a large number of variables or performing multiple statistical tests on the same data leading to statistically significant results that are really attributable to random fluctuations in the data.

Cargo-cult science—A type of *bad science* which involves researchers using apparatus, instrumentation, procedures, experimental designs, data, or results without understanding their purpose, function, or limitations, in an effort to confirm a hypothesis.

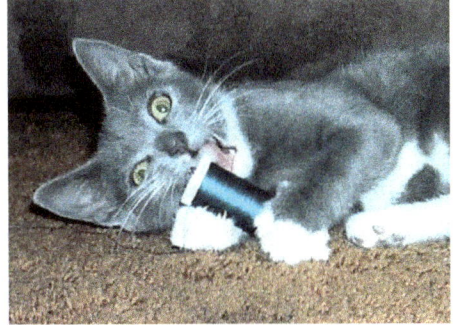

Case—Individual objects on which information is collected in a statistical study and represented as rows in a matrix or spreadsheet. Also called record, survey respondent, patient, or other description of the entity on which measurements are made.

Catch-22—A paradox in which the solution to a problem is impeded by the problem itself so that there are no viable alternatives, a no-win situation. From the 1961 novel and 1971 movie of the same name.

CATS—Critical Analysis Through Statistics. Also refers to the Rulers of the Universe who visit Earth disguised as small furry creatures with whiskers and claws.

Causal relationship—A data relationship in which one event or condition causes or triggers another event or condition.

Cause-and-effect diagram—Cause-and-effect diagrams are used to present ideas for what conditions or events might be responsible for an observed effect. They are commonly used in quality control activities, especially in the statistical field called Six Sigma. They are also called fishbone diagrams or Ishikawa diagrams.

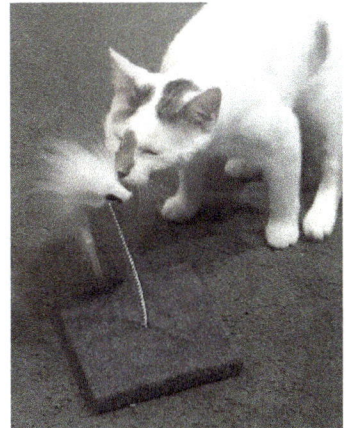

Cell—The area of a spreadsheet created by the intersection of a row and a column.

Censoring—Data that is hidden because of limitations of a scale, measurement device, or other aspect of the data collection process.

Census—Collection of data from all members of a population.

Central limit theorem—Means for a variable based on independent samples from a population will be Normally-distributed and approach the true mean of the

Glossary.

population regardless of the population's actual distribution as the sample size becomes large.

Central tendency—The middle of a dataset characterized by the mean, median, or a related statistic.

CHAID—Chi-squared Automatic Interaction Detector, similar to C&RT except that the classification decisions are not binary.

Chain-referral sampling—A sampling strategy that involves seeking referrals from sampled entities to other entities having some similar characteristics in order to access hidden or elusive populations. It is also called snowball sampling.

Characteristic of the population—The portion of a data value that is the same between a sample and the population. This part of a data value forms the data patterns present in the population revealed in a statistical analysis.

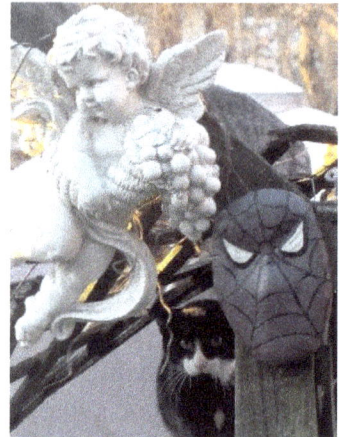

Chart junk—Tufte's term for elements of a statistical graphic that do not add information value to the data being presented.

Charts—Visual representations of data, often used synonymously with plots, graphs, and diagrams. Charts tend to involve lines and areas more than individual points.

Chemometrics—Analysis of data from studies involving chemistry.

Chernoff faces—A multivariable icon graphic in which variable values are represented by facial characteristics of a cartoon figure.

Cherry picking—Using only the most advantageous facts, data, or arguments while ignoring conflicting information.

Chewy—Refers to the online retailer of pet food and related products. It was created as a public corporation in 2019 and is headquartered in Plantation, Florida.

Chi-squared test of Normality—A statistical comparison of the difference between a histogram of the data and a histogram of the theoretical distribution. It is sensitive to the classes used for the histograms and requires at least five samples in each class.

 Glossary.

Choice factor—The percentage of a poll's first response option (p) times the percentage for the second response option (q, or 1-p) that is used in the calculation of the margin-of-error, MoE.

Chord diagram—A diagram consisting of a circle with several nodes represented on the outer part of the circle and arcs or chords drawn inside the circle to illustrate connections between the nodes.

Circular mean—An average for angles and cyclic quantities such as compass directions.

Classical probability—The interpretation of probability in terms of the frequency in an event in an infinite number of trials. Also called frequentist probability.

Classification and Regression Trees—CART. Decision-tree algorithms for creating models for classification when the dependent variable is categorical, or for regression when the dependent variable is continuous

Classification models—Models used to predict or explain group membership.

Clear and convincing evidence—A legal standard of proof requiring that the evidence must be more likely to be true than false, falling between the preponderance of the evidence and beyond a reasonable doubt standards.

Closed-ended questions—Survey questions that have a finite number of pre-established choices from which the respondent has to select.

Closed-mindedness—The reluctance of an individual to consider new ideas, perspectives, evidence, or information that challenges their beliefs or opinions.

Cluster Analysis—A statistical technique for sorting samples into groups based on continuous-scale variables.

Clusters—Groupings of data in statistical plots, usually circular in shape.

Cobra effect—A solution to a problem that makes the problem worse, also called unintended consequences. The term was based on an apocryphal story of the British government in colonial India placing a bounty on

cobras as a means of reducing the snake population only to have people breed the snakes for profit.

Codes—Values, usually numeric, used to represent the levels of a variable measured on an ordinal or nominal scale.

Coefficient of determination—The proportion of variance in the dependent variable that is explained by the independent variables. Also called R-square, R-squared, or R^2.

Coefficient of variation—The standard deviation divided by the mean. Similar to the relative standard deviation, which is the standard deviation divided by the absolute value of the mean.

Coefficient—A number, usually a constant, which is multiplied by a variable, such as in a regression equation.

Coerced science—A type of bad science in which researchers are compelled by authorities to study objectionable topics in ways that promote speed in reaching a desired result over scientific integrity.

Coherence—One of Hill's criteria for causality. A relationship is more likely to be causal if it is compatible with related facts and theories.

Coincidences—Things that happen that appear to be meaningfully related but aren't, perhaps because of serendipity (being in the right place at the right time) or happenstance (it was destiny).

Column charts—Bar charts with an ordinal or continuous-scale variable on the horizontal axis and a nominal or ordinal-scale variable on the vertical axis.

Combination diagram—Two or more charts or chart elements placed together in the same graphic to accommodate scaling issues or provide more information.

Combinations—A selection of a subset of unique items in groups of a given size taken from a larger group.

Combo charts—Two or more charts or chart elements placed together in the same graphic. Also called combination diagrams.

Common-cause data relationships—An event or condition **C** causes or influences events or conditions **A** and **B**.

Common-cause variability—In six-sigma analyses, predictable variations that can be observed in a process.

Comparison—Same as a statistical test.

Complex alternative—A dilemma in which accepting one choice implies denying the other choices.

Complex data relationships—Many events or conditions contribute to the cause or influence of another event.

Concatenating datasets—Concatenating datasets usually involves adding observations rather than variables to an existing dataset.

Concept diagrams—Diagrams for showing ideas, hierarchies, processes, and other non-numeric information. Most of the diagrams are drawn manually or from templates. There are no specific data formats. The information can range from verifiable knowledge to informal opinions

Conceptual models—Models used to convey mental images of mechanisms, processes, or other phenomena that exist or will be created. Conceptual models may take the forms of blueprints, flow diagrams, geologic fence diagrams, anatomical diagrams, and textual descriptions.

Conceptual replication—Replication of a scientific study by using different designs and procedures to examine the original hypothesis, also called reproducibility.

Conclusion—The third part of an argument after premise and logic.

Conditional probability—The probability of an event happening based on the existence of a previous event.

Confidence interval, personal—The time interval between when you graduate and when you get onto the next thing (e.g., job, more school) and realize you don't really know anything.

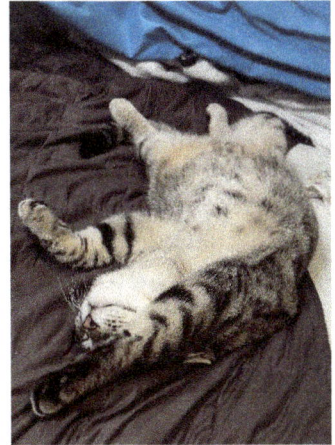

Glossary.

Confidence interval, statistics—The mean of a statistic plus and minus the variation in the statistic.

Confidence—Probability of not making a Type I (false positive) error.

Confidential—Polls in which the real-world identities of respondents are known but not released. Polls in which the real-world identities of respondents are not known are called anonymous. Confidential but not anonymous refers to data collected from human subjects where the analyst knows the identity of the subjects but keeps that information from being divulged.

Confirmation bias—The tendency to believe information and arguments that agree with prior beliefs.

Conflicting evidence—Lines of evidence from different sources that can't be reconciled and is unclear about which is more important.

Conformity bias—The tendency of survey respondents to answer questions as they believe researchers want them to rather than expressing their own opinions.

Confounding—When a variable or uncontrolled effect causes a statistical model to be misleading.

Connector lines—Lines drawn on statistical graphics to facilitate reader recognition of data commonalities.

Consecutive sampling—The strategy of identifying characteristics or conditions that are important to a study and sampling entities until a predetermined number having that characteristic or condition is reached. Also called quota sampling and total enumerative sampling.

Consistency—One of Hill's criteria for causality. A relationship is more likely to be causal if it can be replicated.

Constant—A value in a model that does not change.

Constrained questions—Questions in which a large number of possible responses is inappropriately limited causing misleading results.

Glossary.

Context diagrams—Diagrams that provide background information related to some idea or event to support discussions of the idea or event.

Context points—Points or lines used to show data for some benchmark condition.

Contextual differences—Inconsistencies between datasets being compared or merged that are attributable to differences in data definitions, differences in the conditions under which the data were generated, differences in business rules and data administration policies, or differences related to the passage of time.

Contingency table—A matrix table showing the number of observations in combinations of two or more categorical variables, used to analyze the relationship between categorical variables. Also called a cross-tabulation or cross-tab.

Continuous distributions—A theoretical distribution that describes the probabilities of a continuous random variable's possible values. The Normal distribution is an example of a continuous distribution.

Continuous scales— Scales that define a mathematical progression involving fractional levels, represented by numbers having values after a decimal point. Interval and ratio scales are types of continuous scales.

Contouring—Creating lines (called contours, contour lines or isopleths) representing equal values of a spatially-dependent variable (called a regionalized variable) on a contour map or plot. Also called gridding because a grid of interpolated values is used to construct the lines.

Control chart—A time-series graph showing changes in a process-related variable over time. There are many different types of control charts. All characteristically have a central line for the average, an upper line for the upper control limit, and a lower line for the lower control limit. There are several standard ways to interpret control charts to determine if processes are out of control.

Control group—The group of subjects in a statistical design that does not have an experimental treatment applied to it.

 Glossary.

Controlled surveys—Another name for probability surveys in statistics. This is different from survey control in land surveying.

Convenience sampling—Samples chosen entirely because of their accessibility. It is a type of non-probability sampling. Also called availability sampling.

Coordinates—Values that define an object's position in space or time.

Correction—A systematic deviation, bias, intentionally used to adjust for a different unintended bias.

Correlation coefficient—Measures the strength of a data relationship. There are many types of correlation coefficients, the most common of which is the Pearson correlation coefficient.

Correlation does not imply causation—The 19th century admonition not to infer that a large correlation coefficient means that one variable causes the other. A modern interpretation of the phrase is correlation does not ALWAYS imply causation. Furthermore, causation does not ALWAYS imply correlation.

Corroborating evidence—Evidence that strengthens or confirms existing evidence.

Corrupted data—Data that are rendered invalid when they are extracted/scraped, processed, reorganized, or reformatted improperly.

Cost bias—The belief that information or objects that are more expensive or more difficult to obtain are intrinsically better.

Count—The number of items in a dataset or group. Also referred to as frequency, sample size, or number of samples.

Covariance—A measure of how much two variables vary relative to each other.

Covariates—Continuous-scale variables used to control extraneous variance, usually in ANCOVA.

Criterion variable—The variable used to measure the outcome of a relationship. Also called a dependent variable.

Critical thinking—Assessing the truthfulness and validity of assertions based on evidence and logic.

Cross-validation—Methods used to verify statistical models, especially those used for prediction. Commonly used methods involve resampling or splitting the dataset to create separate datasets for training (creating the model) and testing (verifying the performance of the model).

Cumulative probability plot—A plot of the proportion of data in a dataset on the vertical axis versus the values of the variable on the horizontal axis. Probability plots show a comparison of the sample data to a Normal distribution (which plots as a straight line).

Curves—Data relationships that do not follow straight lines. *Curvilinear relationships* may be either *intrinsically linear* or *intrinsically nonlinear*.

Cycles—A data pattern in which points increase to a maximum then decrease to a minimum. The pattern then repeats.

Dashboard—A visual summary of important descriptive statistics concerning some business operation. Data dashboards are often formatted with easy-to-understand conventions, like gauges, bars, and text alerts like dashboards in vehicles. Dashboards represent a snapshot in time but are updated on a routine, sometimes frequent or even continuous, schedule

Data comparisons—Another term for statistical tests.

Data density— Defined by Tufte as the number of entries in a data matrix divided by the area of the data graphic.

Data dozen—Twelve types of data used in statistical and other methods of analysis, including: automatic and manual measurements; records, reports, and testimony by individuals and organizations; images; transformations; and metadata.

Data dredging—Conducting extraordinary analyses to try to find significant relationships, sometimes resulting in capitalizing on chance. Also called data fishing.

Data icons—Symbols used in place of simple data points to attract attention or convey additional information.

Glossary.

Data mining—The use of statistical, mathematical, and computer algorithms to recognize patterns in data, usually for the purpose of prediction. Commonly used data mining techniques include regression, time series analysis, classification, clustering, dimensionality reduction, decision trees, association rules, text mining, and neural networks. The term does not just mean analyzing any dataset although many traditional data analysis techniques are used. Data mining is usually applied to very large and complex datasets to explore obscure phenomena and challenging problems. The term has unfortunately been given a negative connotation because data mining techniques have been used unethically to explore private information of individuals.

Data relationship—One of many ways in which data from two variables are related statistically.

Data science—The term dates back to the 1960s, but since the 2010s, has referred specifically to a field of practice that combines statistics, computer programming, and domain expertise.

Data scrubbing—Removing nonrepresentative or inappropriate data from a dataset prior to analysis. Sometimes referred to as data cleansing.

Data series—Data from a variable (metric) contained in a column of a data matrix.

Data storytelling—The art of revealing findings about a dataset in ways that engage the audience rather than just presenting results in a more conventional manner. Storytelling usually focuses on the Big Picture while grounding the story with important details involving graphs, descriptive statistics, and the results of statistical analyses. Stepwise storytelling involves sequentially presenting simple elements each of which illustrate a single finding. All-in-one storytelling involves combining several elements to highlight the interrelationships of the findings.

Data summary—A straightforward presentation of statistical results usually involving data lists, sorts and queries, and simple descriptive statistics. In contrast, data analyses describe objectives, data, hypotheses, and results. Summaries provide information; analyses provide knowledge.

Database, relational —A data management tool that stores data in tables formed by rows and columns, akin to a spreadsheet, only there may be many tables that are

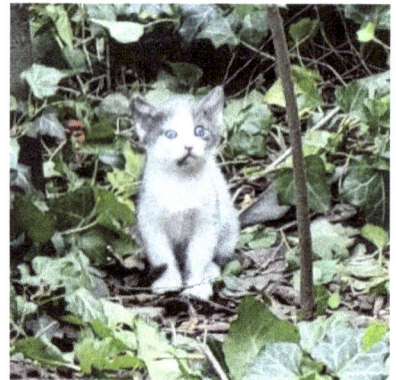

related to each other by some key variable. Relational databases can store and sort data more efficiently than spreadsheets.

Data-ink ratio—A concept defined by Edward Tufte as the non-erasable core of a graphic divided by the total ink used to print the graphic. It refers to the proportion of elements in a statistical graphic that represent data compared to all the elements in the graphic including axes and embellishments. An example of a graphic having a high data-ink ratio is a sparkline, which contains only a line connecting data values (i.e., no axes, labels, or point identifiers).

Data—Numbers and characters that convey information about a phenomenon based on measurements made on samples.

Dataset, data set—A repository for data. For statistical analyses, datasets take the form of a matrix consisting of variables in columns, samples in rows, and data in cells.

Datum—Singular form of data.

Deciles—Ten equal divisions of a dataset based on the frequency of values of a variable.

Decision trees—A flowchart-like structure in which each node represents a decision about an attribute and each branch represents the outcome of the decision.

Decomposition models—Models that separate data into component groups, often used in time-series modeling to separate trends, cycles, seasonality, shifts, and random variation.

Deduction—Using general rules to make suppositions about individual cases.

Definition window—A text-based matrix plot used to clarify a definition by dividing aspects of the definition into discrete alternatives.

Degrees of freedom—The number of independent pieces of information that must be known in order to estimate the value of a parameter. Usually calculated by the number of data points minus 1. More sophisticated designs may also subtract the number of independent variables.

Deliberate ignorance—The fallacy of ignoring readily available information that does not support a preconceived notion.

Demographics—Characteristics of individual samples used to ensure representativeness of a population or to stratify the sample.

Density curve—A simplified graphical representation of a distribution of continuous data as a smooth curve.

Dependent events—Events in which the probabilities of occurrence are not independent.

Dependent variable—The variable being tested in a scientific experiment to determine its relationship to independent variables. Also called an outcome variable or criterion variable.

Descriptive statistics—Numbers used to characterize datasets, samples, or populations according to frequency of occurrence, central tendency, dispersion, and distribution shape.

Design effects—A measure of a survey sample's representativeness to the population, calculated from the variance of the survey results divided by the variance using a simple random sample It is affected by unequal selection probabilities and weighting.

Design of experiments—The process of creating ANOVA designs to analyze the effects of multiple grouping variables on a dependent variable. DoE involves the specification of different types of factorial designs.

Deterministic—A type of model called exact because it does not have an error term. Deterministic models produce a single result for a given set of inputs compared to statistical models that include an error term and produce ranges of results.

Deterministic-empirical models—Presume that a specific mathematical relationship exists between two or more variables that allow them to be modeled without uncertainty under a given set of conditions.

Detrended probability plots—Graphs of the differences between observed and expected values (probability plot) with the linear trend removed.

Glossary.

Deviation plot—A graph that compares data values to a baseline value, also called a deviation-from-baseline plot.

Deviations—Differences between observed values and values predicted from a statistical model. Also called residuals, errors, noise, dispersion, spread, and differences.

Diagrams— Visual representations of data, often used synonymously with plots, graphs, and charts. Diagrams tend to be more artistic and fill the entire data space.

Dichotomous question—A type of close-ended survey question with two choices, usually "either/or" or "yes/no."

Differences—The term differences is a repurposed English word that has several meanings in statistics. It is often used as another term for model residuals. It can refer to means being compared in a statistical test. In time-series modeling, a difference is a data value minus a prior data value. In many statistical formulas, it refers to the subtraction of two components. And of course to add to the confusion, differences can refer to any distinctions between two ideas or objects.

Dilemma—A difficult choice between two or more alternatives. There are many categories of dilemmas including Morton's Fork, Buridan's Ass, Hobson's Choice, and false dilemma.

Direct data relationships—Simple relationships in which event or condition **A** precedes event or condition **B**.

Direct information—Measurements specifically about a phenomenon as opposed to measurements about factors associated with the phenomenon.

Direct replication—Replication of a scientific study by reproducing the original research design and procedures as exactly as possible, also called repeatability.

Directional test—Statistical tests in which only one side of the data distribution is of interest. Hypotheses for directional tests are defined using inequalities.

Dirty data—Data points that have erroneous characters as well as entire metrics that cannot be

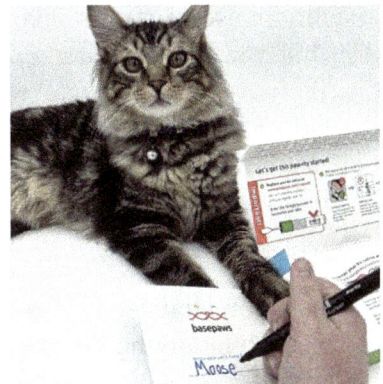

Glossary.

analyzed because of some inconsistency or textual irregularity. Use of "NA" or "ND" for data values or inequalities with censored data values will prevent software from performing statistical calculations.

Disciplinary blinders—When an analysis is conducted or a presentation is prepared without considering information or expertise from relevant domains outside of the background of the analyst.

Discrete distributions—Theoretical distributions that model the likelihood of metrics having non-continuous scales. Examples include: Binomial, Bernoulli, Hypergeometric, Multinomial, Geometric, and Poisson.

Discrete scales—Scales of measurement that have finite numbers of values. Nominal and ordinal scales are discrete scales.

Discriminant analysis—Method of analysis for models having a qualitative dependent variable.

Disjoint events—Events that are independent.

Dispersion—Variability, inexactness in a calculated statistic.

Dissent bias—The opposite of acquiescence bias, in which survey participants respond in negative ways or against what they believe the interviewer desires.

Distribution factor—The square of the two-sided t-value based on the number of survey respondents and the desired confidence level, used for calculating a survey's margin-of-error.

Distribution model—A mathematical equation of a curve that is used to represent the frequencies of items in a population. The most commonly used distribution model is the Normal distribution (AKA bell curve, Gaussian model). Other distribution models include: t, F, Chi^2, Lognormal, Weibull, Rayleigh, triangular, rectangular, logistic, Laplace, gamma, Cauchy, exponential, and beta for continuous data; and binomial, multinomial, Poisson, geometric, and Bernoulli for discrete data.

Distribution shape—The form of a data frequency distribution involving symmetry, the number and position of modes, and other characteristics.

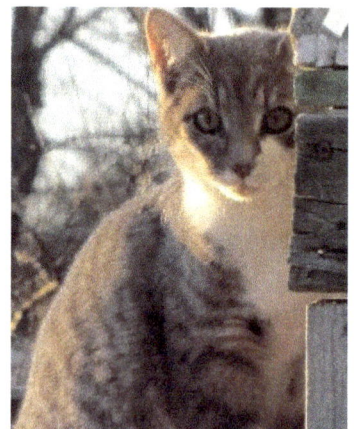

Glossary.

Distribution-free statistics—Nonparametric statistics. Statistics that do not rely on a theoretical model of probability.

Divider lines, shapes—Chart embellishments used to facilitate audience understanding of data relationship patterns.

DOE—Design of Experiments, an area of statistics involving optimization of ANOVA designs.

Domain expertise—The knowledge of a discipline that is needed to interpret the results of a statistical analysis.

Donut plot—A pie chart with a hole in the middle to allow additional labeling.

Dot plot—Like a histogram only using dots instead of bars.

Double sampling—A sampling strategy that involves using an easy-to-select sample or metric as a substitute for a difficult-to-select sample or metric. Also called surrogate sampling and proxy sampling.

Double-barreled question—Survey questions that ask about more than one idea in a single question but only allow respondents a single response.

Double-edge sword—A dilemma in which the alternatives have both favorable and unfavorable aspects.

Drawing—Most commonly, a free-hand depiction of information. May refer to a game of chance in which all the chances are unique (i.e., there are no duplicates) and there is only one winner for the game. May also refer to the process of selecting individual cards from a deck of cards.

Drop-out rate—The proportion of survey participants who do not complete a survey after starting it.

Dunning-Kruger effect—When people trust their own interpretations over those of experts because they mistakenly believe their abilities are greater than they actually are.

Duplicate—A second sample collected to replicate an original sample in order to assess variability.

Dynamic Model—A data model that includes a term related to time so that temporal changes can be analyzed.

Econometrics—Statistics involving data on economics.

EDA—Exploratory Data Analysis.

Educated guess—Another term for abductive reasoning.

Effect size—The difference in group means that can be distinguished by a statistical test.

Effective sample size— An estimate of the sample size that would be required to achieve the same precision as would be provided by a simple random sample. The effective sample size is always the same size or larger than the actual sample size.

Eisenhower window—A matrix diagram used to establish priorities for competing issues or tasks based on their anticipated urgency and importance, attributed to former U.S. President Dwight Eisenhower.

Elongated—Refers to data clusters that are longer than they are wide. Some statistical clustering algorithms are less sensitive to elongated data clusters than to circular data clusters.

Embellishments—Artistic (non-data) lines, shapes, graphics, and text used to engage the audience and help them interpret data relationships.

Empirical models—Mathematical models that are based on observations rather than scientific principles.

Environmental variability—Differences between a sample and the population that are attributable to extraneous factors. Minimizing environmental variance is difficult because there are so many causes and because the causes are often impossible to anticipate or control. Environmental variability often looks like natural variability because the causes of the variability are unknown.

Eponym—Jargon named after a person. An eponym can also refer to the actual person.

Equal variances—The assumption that the variance of model residuals is constant in all groups (or across all values for continuous variables). The assumption entails that the same is true for the dependent variable itself. Also referred to as homoscedasticity or homoskedasticity.

Equation—A mathematical statement indicating that the values of two mathematical expressions are equal or otherwise related in some way.

HOMO CE A TIC

Equivocation—The fallacy of using a word or phrase with multiple meanings without distinguishing which meaning is meant.

Error term—The element in a statistical model that represents variability in model predictions.

Error variance—The variance of the residuals from a statistical model.

Errors—Differences between observed values and values predicted from a statistical model. Also called residuals, deviations, noise, dispersion, scatter, spread, and differences.

Estimates—A calculated value for a population parameter made on the basis of a sample of the population.

Event—The outcome of an experiment to which a probability is assigned.

Everything is uncertain—In statistics, all measurements and model results are considered to be inexact, that is, to have variation associated with them. In contrast, deterministic relationships are considered to be exact, without uncertainty.

Evidence—Information supporting the validity of an argument.

Exact—Models that do not contain error terms.

Excluded middle—The fallacious belief that if a little of something is good than more would be better (or if less of something is good, none would be better).

Glossary.

Excluded outliers—The fallacy of ignoring data or information that doesn't agree with the bulk of the data or information.

Exculpatory evidence—Information that disputes an argument or absolves alleged guilt of a defendant.

Exit polls—Surveys of voters as they leave their polling venue to determine how they voted.

Experimental study—A scientific investigation in which the researcher controls the conditions (called effects) that the subjects are exposed to. In contrast, an observational study is a scientific investigation in which the inherent characteristics of subjects are studied.

Experiment—One of Hill's criteria for causality. A relationship is more likely to be causal if it can be verified experimentally.

Expert opinions—Testimony provided by individuals judged to have qualifications on a specific topic well beyond the average person.

Expertise—Advanced qualifications possessed by an individual on the basis of education, training, and experience related to a specific topic.

Explanation models—Models that aim to explain associations within or between sets of variables on the basis of known information about the topic and patterns identified in the data.

Explanatory variable—Variables used in an explanation model, which are usually quite different from variables used in a prediction model.

Exploded—Refers to sections of charts, particularly pie charts, that are offset and enlarged to highlight some aspect of the data. Statisticians do not consider exploding chart sections to be a good practice. It is sometimes used in media presentations to attract audience attention.

Exploitation—A bias applied intentionally to only selected data.

Exploratory data analysis—An initial examination of a dataset to identify important variables and influential observations, and explore data relationships and multivariable patterns. EDA is primarily graphical with some computational analyses. EDA makes no presumptions about data structure and uses the data to establish it. EDA is considered a necessary prelude to more sophisticated modeling. It tends to be more thorough than either a summary analysis (SA) or an initial data analysis (IDA) although these terms mean different things to different statisticians.

Extraneous data—Unwanted data observations that occasionally appear in datasets. They may be replicates, QA/QC samples, or unexplained values such as from merges of overlapping datasets.

Extrapolation intoxication—Conducting statistical analyses outside the range of the available data. Examples include showing information on maps beyond the physical extent of the data, inferring survey results to untested demographics, and making predictions for extreme conditions based on ambient conditions. The only example of extrapolation that is accepted by statisticians is time-series analysis to predict the future, although the pertinent issue is how far into the future is reasonable.

Extrapolation—Prediction of values outside the span of the data used to create the prediction model.

Extreme bias—Cases in which a survey participant selects responses more extreme than their belief in order to exaggerate their own belief.

Facade—Embellishments that are added to either convey additional information, draw audience attention, or make the graph easier to read.

Factoid—Assertions that are made so commonly that they are assumed to be true.

Factor analysis—A statistical data reduction technique in which the original variables are recombined into a smaller number of factors, which can then be used as new, more efficient variables in other statistical procedures.

Factorial designs—Types of ANOVA designs used to simultaneously assess the effects of individual factors (main effects) as well as the interactions between them.

Glossary.

Factorial—In mathematics, the product of an integer and all the integers below it. In statistics, a type of design for investigating main and interaction effects between two or more independent variables.

Factors—The term is used in several ways in statistics. In factor analysis, variables are recombined mathematically into a smaller number of metrics called factors. In ANOVA, main effects are sometimes referred to as factors.

Fail to reject—In a statistical test, if the null hypothesis of no difference cannot be rejected, the test is called non-significant or not significant. Researchers hate failing to reject because it impairs their ability to publish in journals that will only publish significant test results.

Fake news—False or misleading information presented as legitimate. Although the term is over a century old, it has become pervasive with the spread of social media.

Fallacy fallacy—The belief that just because an argument contains a fallacy, the entire argument is invalid.

Fallacy—Incorrect reasoning in an argument, used intentionally or unintentionally, to convince an audience about some assertion.

False dilemma—When there are so many choices to consider that only some of the many choices are presented giving the impression that they are comprehensive.

False negative—When a statistical test is judged to be nonsignificant (false) when it is actually true, also called a Type II error.

False positive—When a statistical test is judged to be significant (true) when it is actually false, also called a Type I error.

False precision—A fallacy committed by believing or causing an audience to believe that data and results have more precision than they actually do.

Fatal flaw—A fundamental weakness that ultimately leads to failure.

F-Distribution—A family of univariate, continuous distributions used to represent ratios of the variances of two normally distributed populations, used in ANOVA and ANCOVA.

Feedback data relationships—Event or condition **A** and **B** are linked in a loop. **A** causes or influences **B**, which then causes or influences **A**, and so on.

File drawer bias—The tendency of researchers and publishers to not report disappointing findings.

Fishbone diagram—A cause-and-effect diagram having the appearance of a fish skeleton. Also called an Ishikawa diagram.

Fixed-effects ANOVA—An ANOVA in which only specific levels of an effect are being compared. Most ANOVA are fixed-effect designs.

Floating axis—Graph axes that are not fixed because there is no static origin, most notably seen in waterfall charts.

Flood probability—The likelihood that the flow of a river (discharge) of a given volume or larger occurring in any year is equal to 1/year-flood. For example, the probability of a 100-year flood on a river is 1/100 (0.01) every year.

Flow chart—A diagram of the steps in a process, in which the steps are represented by boxes and the relationships between the steps are represented by arrows.

Flow lines—Chart embellishments, particularly on scatter plots, that are used to indicate how a particular observation had different values at different times or under different conditions.

Focus—What gets plotted on a statistical graphic: all or most of the individual data points for one metric; groups of data points plotted as means or medians for one metric; or only a few data points plotted as icons showing several metrics.

Forecasting—Prediction of values (forecasts) from a time-series model.

Glossary.

Foundation—The fundamental type of graph selected on the basis of the aim and focus of the graph and the characteristics of the variables to be plotted.

Frame—The list of survey invitees or process for identifying survey invitees that will represent the population based on the sample.

Framework—The specifications of a graph: the plot area; the variables (data series); the points to be plotted, and the axes.

Frankendata—Data collected by different researchers, at different times, for different purposes. Data hash.

Fraudulent science—A type of bad science in which data, results, or even whole studies are faked.

Freeper—An individual who responds repeatedly to open-invitation polls to make a certain opinion appear to be more popular.

Frequency distribution—A graphical or tabular representation of the number of observations for each possible value of a variable.

Frequency—Number of occurrences. Count. Sample size.

Frequentist probability—The interpretation of probability in terms of the number of a certain event in an infinite number of trials, independent of an observer. In contrast, Bayesian probability is based on an observer's expectation based on prior knowledge. For frequentists, the probability of an event is constant. For Bayesians, probability changes as more is known about the event.

Fringe science—A type of bad science in which hypotheses within an established field of study are highly speculative, often at the extreme boundaries of mainstream studies.

F-test—A statistical test comparing the variances of two samples.

Gamma distribution—A statistical frequency distribution used to predict the wait time until a future event occurs.

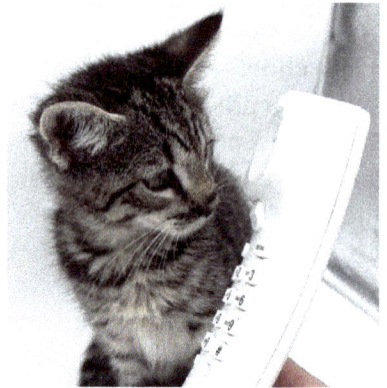

Gantt charts—A bar chart showing a project's schedule as horizontal bars representing project tasks on a timeline. The bars represent the duration of tasks and the dates they are planned to be conducted.

Gaussian distribution—Another name for the Normal distribution.

Geometric mean—A measure of central tendency calculated by multiplying all data values together and then taking the kth root where k is the number of data points.

Geostatistics—An approach to analyzing location-dependent data, involving variogramming and kriging. Geostatistics originated in the mining industry.

GIS—Geographic Information System, used in the analysis and presentation of location dependent data.

Going to the polls—Participating in a political election.

Goldilocks' quest—The effort to find the perfect number of samples, variables, tests, and other elements of a statistical analysis. Goldilocks' quest is a laudable goal but virtually impossible to attain.

Gradient—One of Hill's criteria for causality. A relationship is more likely to be causal if a greater exposure to the suspected cause leads to a greater effect.

Graph components—Foundation, Framework, and Facade.

Graphs— Visual representations of data, often used synonymously with plots, charts, and diagrams. Graphs tend to be more mathematically complex than charts and plots.

Grid sampling—A form of systematic sampling in which a grid is placed on an area being investigated and samples are selected in each grid cell. If the samples are selected at the same relative location in each cell, the scheme is called a systematic-grid sample. If the samples are selected randomly within each cell, the scheme is called a systematic-random sample. If the grid size is designed to identify entities or properties based on a probability of their occurrence, the scheme is called search sampling.

Glossary.

Gridding—Interpolation of spatially-dependent data values.

Gridlines—Lines added to graph axes at major scale intervals that extend across the plot area and are used to facilitate recognition of data point values.

Grouping variable—A metric that represents a sample's membership in one of several independent categories.

Happenstance—The belief that an event occurred because it was fated to happen (it was destiny).

Hard information—Measurements of tangible, observable objects or processes as opposed to measurements of intangible beliefs or opinions.

Harmonic mean—A measure of central tendency calculated by averaging the reciprocals of the data points and taking the reciprocal of the average.

Hasty generalization—The fallacy in which a conclusion is drawn based on insufficient or unrepresentative evidence.

Hawthorne effect—When individuals change their behavior when they become aware that they are being observed.

Heatmaps—Representations of data quantities in which values are represented as colors.

Heteroscedasticity—Unequal variances between data groups. Heteroskedasticity.

Hidden trends—Data patterns that are not apparent in graphs of all data but become noticeable when data are plotted by groups or by another factor. Simpson's paradox is an example of hidden trends.

Hierarchical clustering—A method of associating data into homogeneous groups on the basis of descriptive variables. Different groupings are created as more data points are assigned to groups, forming a hierarchical structure.

Hierarchical samples—Samples that can be combined into larger units or split into smaller units for analysis. For example, a study of educational performance could focus on individual students, classes, schools, school districts, counties, states, or nations.

Hill—Sir Austin Bradford Hill (1897-1991) was an English epidemiologist and statistician who proposed nine criteria for evaluating causal relationships in medicine.

Histograms—Vertical bar charts in which the horizontal axis represents data ranges (called bins) and the vertical axis depicts the number of samples in each data range. A histogram for a Normally distributed dataset has the largest number of samples at the center of the dataset with progressively fewer samples toward the tails of the distribution.

Hobson's choice—Dilemmas in which there is really only one choice even though it appears that there are more than one.

Homoscedasticity—Equal variances between data groups. Also spelled homoskedasticity

Horse-race polls—Surveys of voter preferences for specific candidates in an election.

Hot-button issues—Topics that elicit strong emotions in people.

Hoyle's fallacy—The incorrect assumption that the only way a certain outcome could have occurred is through a specific, complex, or unlikely process or event. Also called a junkyard tornado.

Huff—Darrell Huff (1913-2001) was an American writer best known as the author of **How to Lie with Statistics** (1954)

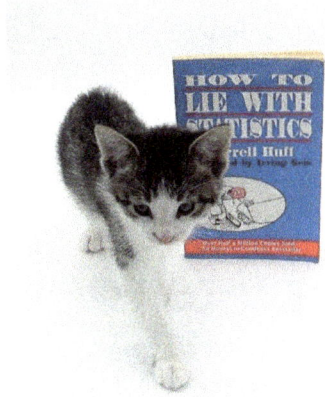

Hyperbole—Extreme exaggeration or overstatement.

Hypothesis tests—Statistical tests, formal procedures for comparing population parameters to benchmarks or other populations based on samples from the population.

Hypothetical syllogism—Arguments that deduce conclusions from premises and conditional (if-then) statements.

Glossary.

Iceberg diagram—Diagrams that explore the idea that there are some things about a concept that are overt, well understood, believable, or known to most people, there are other things about a concept that are hidden, misunderstood, unconvincing, or unknown, and there are many things in between. The graphic uses an iceberg as a visual metaphor because 90% of a floating iceberg extends below the water's surface. Creating an iceberg diagram involves starting with a drawing of an iceberg adding text to levels of the iceberg to represent what's overt versus what is hidden.

Icon plots—Scatter plots in which data points are represented by multivariate icons.

Icons—Small graphics that depict relative values of two or more variables, including star charts, ray charts, sparklines, and Chernoff faces. Icons are often plotted on larger graphs to depict additional data dimensions.

IDA—Initial Data Analysis.

ID—Identification, identity, or identify.

Important-before; important-again—A belief that if something was important in the past it will continue to be important in the future.

Imputation—The replacement of missing data points with surrogates, such as mean values, so that the records can be included in an analysis.

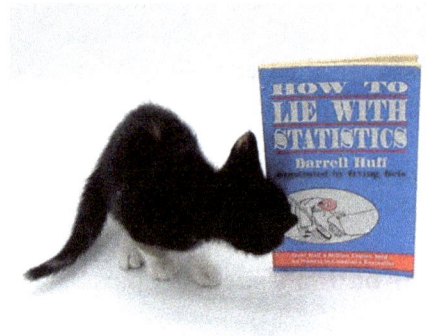

Independence—Statistical independence refers to an assumption made by most statistical modeling techniques. The assumption is that the residuals (errors) from the model are independent (uncorrelated with) each other. This is about the same as assuming that the data values themselves are independent of each other. When data are correlated, such as with spatial and time-series data, special techniques have to be used to account for the correlation.

Independent events—Events in which the probabilities of occurrence are not associated.

Independent variables— Variables that are analyzed to predict or explore relationships with a dependent variable. Also called a predictor variable or explanatory variable.

Glossary.

Indexing—Transforming variables scales to more representative or manageable values by dividing measurements by a value that facilitates comparisons.

Individual—A word used to refer to an individual sample (e.g., observation, subject, record, case, individual, or what the entity is, such as patient, respondent, or student).

Induction—Using individual cases to reach general conclusions.

Indulging variance—Failing to adequately control or even consider extraneous variance in data or models.

Inefficient metrics—Variables used in an analysis that are not appropriate for investigating the phenomenon in question. The issue is with the metric not the phenomenon. As a consequence, the strength of a relationship will be smaller than expected.

Inference to the best explanation—Another term for abductive reasoning.

Inference—The process of using samples that are representative of a population to make general statements about the population. Statistical inference usually involves testing or interval estimation.

Influence relationship— Instances in which a condition or event changes the manifestation of an existing condition or event but does not cause it to happen. Influences can be direct or mediated by a separate condition or event. Influences may exist at any time before or after the influenced condition or event. Influences may be unidirectional or bidirectional.

Influential observation—A datum that contributes substantially to an overall data relationship but is not determined to be an outlier by a statistical test.

Infographics—Combinations of visualizations, text, images, and artwork designed to inform or persuade general audiences. They are specific, elaborate, attractive, and self-contained. Every infographic is unique and must be designed from scratch for visual appeal and overall reader comprehension. There is no software for automatically producing infographics the way there is for visualizations. While visualizations explore, infographics explain.

Glossary.

Information cascade—When opinions and information are repeated by users of social media giving the impression that the information is more important or more popular than it actually is.

Information windows—A type of matrix plot used for organizing information for planning, scheduling, communications, and relationships, in which the cells contain data, tables, graphs, or text. Examples include Johari windows, Eisenhower windows, Rumsfeld windows, and boundary windows.

Information—Data with its associated metadata.

Inherent characteristics—Qualities that are a natural part of a subject and unlikely to change. A key facet of observational studies.

Initial data analysis—An initial examination of a dataset to develop an overall sense of the dataset, identify patterns or data anomalies, and evaluate fundamental statistical requirements of the data. IDA is primarily computational with some graphical analyses. IDA assumes some knowledge of the underlying data structure. Conducting either an IDA or an EDA is a prerequisite to conducting almost any type of statistical analysis. The terms SA, IDA, and EDA mean different things to different statisticians.

In-line outlier—An anomalous data point that follows a dominant data trend but greatly inflates the trend's r-square value. It may be misleading if the outlier is not actually representative of the trend.

Inset plot—A small, separate statistical graphic placed within a larger, main statistical graphic for the purpose of accommodating scale differences or providing additional information.

Insignificant significance—Occurs when a statistical test fails to reject a null hypothesis because it is poorly designed or the data violate statistical assumptions.

Insufficient evidence—When the evidence presented in an argument is not adequate to prove a fact or meet a burden of proof.

Interaction effects—Effects in a statistical design that consider the influence of two or more independent variables beyond their individual effects.

Glossary.

Internet of things—Physical devices, vehicles, appliances, and other objects that have sensors which enable them to exchange data over an intranet or the internet. The term was first coined in 1999.

Internet Relay Chat, IRC—Provides facilities for instant messaging and other communications. It began in the 1990s and is still being used.

Interpolation—Prediction of values within the span of the data used to create the prediction model. Interpolation can be either inexact (sample values are assumed to have some variability associated with them) or exact (sample values are assumed to have no variance). There are a variety of algorithms for interpolation, including: nearest neighbor, inverse distance, spline, and moving average.

Interquartile range—A measure of statistical *dispersion* calculated by subtracting the value of the 25th percentile for a variable from the value of the 75th percentile.

Intersection of events—Events or outcomes that are elements of two or more groups.

Interval scales—Interval data is like a ratio scale in that it has a consistent order and a consistent interval between values. But, the scales lack a fixed zero point. Examples include, standardized test scores, time, temperature, elevation, location coordinates, sound levels, pH, and mineral hardness

Intervention analysis—A method of analyzing time series for a shock or intervention that may have changed the nature of the series.

Interview methods—Methods used to provide questions to individuals in a survey, including in-person, telephone-voice, texts, recorded message, surface mail, email, and websites.

Intrinsically-linear, intrinsically-nonlinear—A model that has an additive linear relationship between the dependent and independent variables is called intrinsically-linear. A model that has an additive curvilinear relationship between the dependent and independent variables that can be made linear through the use of

transformations is also called intrinsically-linear. A model that has an additive or

 Glossary.

multiplicative curvilinear relationship between the dependent and independent variables that cannot be made linear through the use of transformations is called intrinsically-nonlinear. Intrinsically-linear models have regression coefficients that are constants. Intrinsically-nonlinear models have regression coefficients that are mathematical functions of independent variables, hence why they cannot be transformed to linearity.

Invalid data—Information that has been generated or recorded incorrectly, for reasons such as incorrect formatting and scale recoding. Invalid data may involve individual data points or all the measurements for a specific metric.

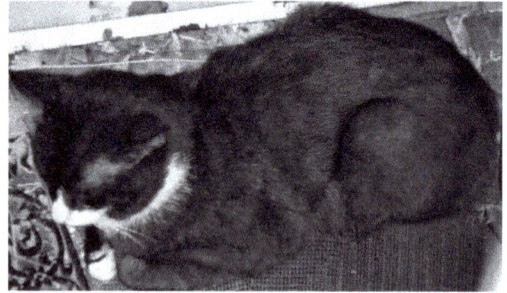

Inverse data relationships—The absence of event or condition **A** causes or influences event or condition **B**, or the presence of **A** minimizes **B**.

Inverse-distance interpolation—Based on the assumption that things that are close to one another are more alike than those that are farther apart. Inverse-distance algorithms give more weight to observations close to a prediction location and less weight to observations farther away as a function of distance.

Inverse-distance weighting—Estimating values based on the distance of the value to the closest actual data points. Estimated values are essentially weighted harmonic means.

Invitee—An individual who is asked to participate in a survey.

Ishikawa diagrams—Another name for a fishbone diagram.

Isopleths—Lines on a contour map that represent constant values.

Jackknifing—A procedure for creating a new dataset by resampling existing data values without replacement. Bootstrapping involves resampling with replacement.

Jargon—Special words or expressions with meanings specific to a profession. In statistics, jargon involves: concepts named after the person who discovered it; unique words created to convey a special meaning; and common words and phrases with alternative meanings.

Glossary.

Johari window—A method for exploring personal relationships in which subjects pick adjectives they feel describe their own personality while peers pick adjectives that describe the subject. These adjectives are then placed in the appropriate pane of a window diagram.

Joint probability—The likelihood of two or more events occurring together.

Judgment sampling—A non-probability sampling scheme in which samples are selected based on what a researcher believes would be most appropriate for a study. Judgment samples are also known as purposive samples.

Judgmental language—Words or phrases that express a biased personal opinion.

Judgments—Decisions made by data-generation technicians to create a data value.

Judicial notice—When a fact is considered to be true without formal evidence in order to save time because it is well-known and generally accepted.

Jumping to a conclusion—Reaching a conclusion without considering enough evidence. Also known as hasty conclusion, hasty generalization, and where-there's-smoke-there's-fire.

Junk science—A type of bad science in which research is considered to be biased by legal, political, ideological, financial, or otherwise unscientific motives.

Just-in-case fallacy—Reasoning that is based on low-probability conditions or scenarios rather than those that are more likely. This bias is used often in deterministic data modeling to compensate for not considering variance. Also called a worst-case fallacy.

Kendall tau correlation coefficient—Measures the strength of a data relationship where one variable is measured on an ordinal scale and the other variable is measured on an interval or ordinal scale.

Key—A variable (column) in datasets to be combined that contains information that uniquely distinguishes each observation (row).

 Glossary.

Kitten—A juvenile cat less than one year old known for their genetic predisposition to extreme cuteness and their affinity for statistics and other forms of data analysis.

Kiviat chart—Another name for a radar chart

K-means clustering—A classification method that groups observations by calculating mathematical distances of data points from user-supplied or randomly-selected centroids.

Knowledge compression—David McCandless' assertion that statistical graphics make data easier to understand because visual information is easier to understand than text. Thus, a graph can contain more, easily understood information than the same amount of text.

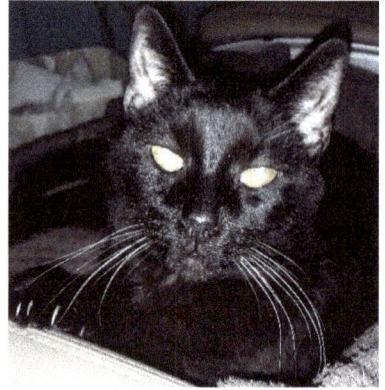

Knowledge—In statistics, knowledge is information created through the analysis of data. Wisdom is the understanding of the fundamental principles governing knowledge.

Kolmogorov-Smirnov (K-S) test—A statistical test of Normality that compares the maximum vertical distance between a probability plot of the data and a probability plot of the Normal distribution or other theoretical distribution being considered. The test tends to overemphasize deviations near the center of the distribution and the population distribution parameters must be known for the test to be exact. The Anderson-Darling test is a modification of the K-S test in which more emphasis is placed on deviations in the tails of the distribution. The Lillifors test is another modification of the K-S test which is used when the mean and standard deviation of the hypothesized Normal distribution are not known and must be estimated from the sample data.

Kriging—Interpolation method similar to inverse-distance interpolation that incorporates the correlation structure of the data into the weightings. Part of geostatistical interpolation.

Kurtosis—A measure of the relative proportion of area in the tails of a distribution relative to the center of the distribution. A negative kurtosis indicates that there is a greater frequency of values in the center of the sample distribution and the data tails are short compared to a Normal distribution. Such sample distributions are said to be platykurtic. A positive kurtosis indicates that there is a greater frequency of values in the tails of the sample distribution with a relatively tall, thin peak at the center. This form of sample distribution is said to be leptokurtic. Some older references calculate kurtosis so that a Normal distribution has a value of three instead of zero.

Labels—Considered an embellishment on a graph. Some people find them beneficial and some find them distracting.

Lags—As used in time series analysis, a first-order lag is the value of a variable before the current value. A second-order lag is the value of a variable two time periods before the current value. In general, an n-order lag is the value of a variable n time periods before the current value. The duration between measurements of the time-series variable must be constant in order to compute lags.

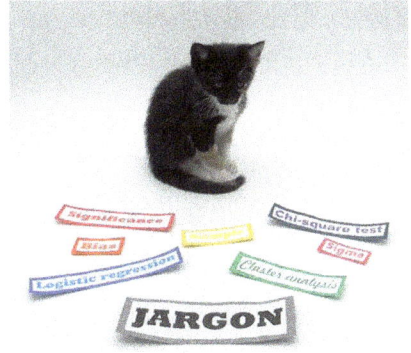

Law of large numbers—Estimates of a population parameter will become more accurate and precise as the number of samples used to calculate the parameter increases.

Law—In science, laws describe what events or results happen under certain conditions, but not why they happen. Theories provide explanations for why events or results happen.

Leading questions—Survey questions that imply or express the opinion of the researcher instead of being unbiased.

Leads—A variable in a time-series model that shifts toward future times. In contrast, a lag in a time-series model shifts toward past times.

Lean—Lean Six Sigma is a process improvement method that combines Lean manufacturing principles, which focus on eliminating waste, with Six Sigma techniques aimed at reducing process variation and improving quality.

Least squares regression—The most commonly used method of finding a best-fit line to a set of x-y data, described by equations with the form $y = ax + b$. This method finds the line that reduces the sum of the squares of the residuals (errors, differences) from the line to each data point.

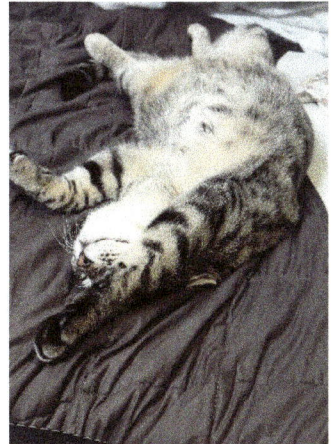

Glossary.

Left skewed—A data frequency distribution in which the tail on the left side of the distribution is longer than the tail on the right.

Left variables—In canonical analysis, left variables refer to the set of variables to the left of the equals sign. There are also right variables, referring to the set of variables to the right of the equals sign. There are no dependent/independent relationships in canonical analysis.

Legend—A box on a graph that shows information relevant to items (lines, points, data sources) on the graph.

Leptokurtic—A frequency distribution having a larger value for kurtosis than the Normal distribution, indicating that there are more data values near the mean than in a Normal distribution.

Levels—The different possible designations for values of a variable measured on an ordinal or nominal scale.

Leverage plot—A scatter plot for one term (variable) in a model showing the relationships between each data point and the model parameter for the term. The coefficient of the model term is the slope of the best-fit line to the data in a leverage plot. Also known as an added variable plot or partial regression residual plot.

Lies, damn lies, and statistics—A 19th century complaint about untrustworthy statistics. The author is unknown but Mark Twain claimed it wasn't him.

Likely voters—Individuals judged by a pollster to be likely to vote in an election. Likely voters are usually identified on the basis of participation in prior elections but every pollster has their own models for the determination.

Likert-scale question—A type of close-ended survey question like a single-choice question but where the choices represent an ordered spectrum of choices.

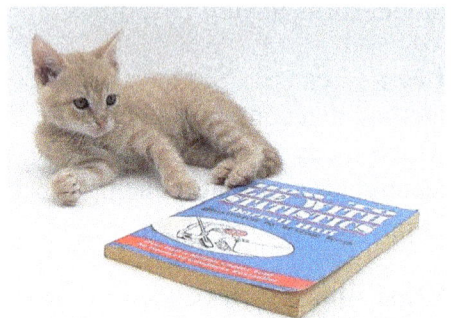

Limit-of-detection—A value on a measurement scale below or above which data values cannot be determined. Left-censored or low-censored values are below the LoD. Right censored or high censored values are above the LoD. Examples include very low concentrations of chemicals in environmental samples (left-censored) and inventories where there are too many items to count (right censored).

Line plots—An x-y (bivariate) plot in which the variable represented on the x-axis is measured on an ordinal-scale.

Linear regression—A statistical method for fitting a trend line through a set of data such that the variance is a minimum. Simple regression involves one dependent variable and one independent variable. Multiple regression involves one dependent variable and more than one independent variable. Stepwise regression involves the sequential selection of variables in a multiple regression model such that the variables accounting for the most variance are selected first.

Linear scale—An arithmetic scale representing an additive (as opposed to logarithmic) progression of values used for an axis on a graph . May be based on Steven's ordinal, interval, or ratio scales.

Linearity—Refers to the assumption that the relationship between data for dependent and independent variables in a model form a straight line on a scatter plot. Sometimes also incorrectly referred to as additivity.

Loaded questions— Survey questions meant to induce respondents to answer in a certain way.

Logic—The process of deriving conclusions from premises in legitimate arguments.

Logistic regression—A regression technique used when the range of the dependent variable is 0 to 1. Logistic regression, or logit regression, is most often used for models involving the prediction of a probability. The logistic trend line is an S-curve in which the tails of the trend flatten as they approach 0 and 1.

Lognormal distribution—A continuous probability distribution of a variable whose logarithm is normally distributed. It is used to model right-skewed data.

Longitudinal study—Research that involves observing the same individuals or variables over a long period of time in order to assess the natural progression of a phenomenon.

Long-term variability—A way to categorize variance, usually days in length and longer, often attributable to natural changes and changes in the study environment. For

comparison, short-term (immediate) variability is more often attributable to data-generation instruments and procedures.

Lower confidence limit—The smaller value in a calculated confidence interval.

Machine learning—The process by which data is analyzed using statistical algorithms without relying on explicit instructions from the researcher. Examples of machine learning techniques include naive bayes, decision trees, random forests, support vector machines, K-nearest neighbors, ensembles involving bagging, boosting, voting, and stacking, neural networks, and K-means.

Main effect—The influence of a nominal- or ordinal-scale independent variable on a continuous-scale dependent variable. Main effects ANOVAs compare one or more grouping variables independently. Comparisons of combinations of grouping variables are called interaction effects.

MANOVA—Multivariate ANalysis Of Variance, a statistical technique for detecting differences between averages of more than one continuous dependent variable in groups defined by one or more qualitative variables.

Marginalization of the adversary—Another term for a middle-of-the-road-fallacy in which it is believed that information midway between extreme alternatives must be legitimate.

Margin-of-error, population—The MoE in a survey that considers the size of the population.

Margin-of-error, question—The MoE in a survey that considers the size of the difference in two responses, **p** and **q**.

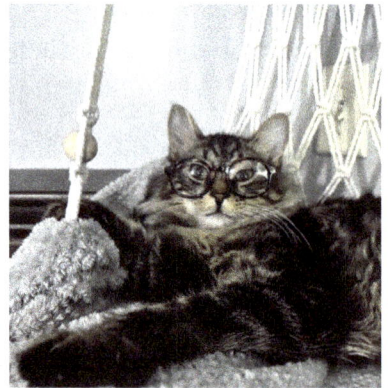

Margin-of-error—MoE is the amount of random sampling error in a survey's results, estimated from $1/\sqrt{}$number of survey responses.

MAR—Missing At Random means that there is a pattern to the occurrence of the missing values related to some other variable in the model.

Mathematical model—An equation constructed of mathematical expressions relating independent variables to a dependent variable.

 Glossary.

Matrix charts—A diagram in the form of a table or matrix in which the cells contain text, graphics, or other charts. A calendar is a common form of matrix chart.

Matrix question—A type of close-ended survey question that allows two aspects of a topic to be assessed at the same time.

Matrix—Mathematical array of numbers consisting of columns and rows.

Maximum-ordinate—The vertical axis showing data frequency in a histogram.

Maximum—The highest data value.

MCAR—Missing Completely At Random, means that there is no pattern to the occurrence of missing values.

McNamara fallacy—McNamara fallacy is where inadequate metrics are used to analyze complex situations because qualitative factors are ignored (measurability bias).

MDS—MultiDimensional Scaling is a data-reduction, variable rearrangement technique like Factor Analysis, the difference being that MDS isn't restricted to using the correlation/covariance matrix as its measure of similarity.

Mean absolute deviation— The arithmetic mean of the absolute (unsigned) differences between data values and the mean or median of the data. Sometimes used to characterize data dispersion.

Mean number of events—The parameter for the Poisson distribution.

Mean—An estimator of the center of a distribution of data measured on a continuous scale. A mean is usually an arithmetic average but may also be calculated from logarithms (geometric mean), reciprocals (harmonic mean), or quartiles (trimean).

Meaningful units—The number of units in the data range that visually provide precise, clear-cut, reliable information.

Meaningfulness—The concept that an effect size in a statistical test may or may not be large enough to be relevant in the real world. Sometimes referred to as practical significance.

Glossary.

Means plots—A graph that shows the mean and standard deviation of data by groups or changes over time.

Measurability bias—The belief that something that cannot be measured reliably is not worth serious consideration.

Measure—An attribute of an individual sample. Also called a metric or a variable.

Measurement components—Three factors involved in creating a data measurement, namely Benchmark, Process, and Judgment.

Measurement scales—The ways that the values in a set of numbers are related to each other.

Measurement variability—The differences between a sample and the population attributable to how data were measured or otherwise generated.

Measures—Metrics. Variables used to characterize a phenomenon.

Median—An estimator of the center of a distribution of data.

Mediated data relationships—Event or condition **A** causes or influences event or condition **C** and **C** causes or influences event or condition **B** so that it appears that **A** causes or influences **B**.

Mekko chart, Marimekko chart—Other terms for mosaic charts and percent-stacked-bar charts

Merging—The process of adding variables from a dataset to existing observations in a different dataset using a key variable contained in both datasets.

Messy data—Data with outliers, censored data, errors, and missing entries that have to be scrubbed before analysis.

Meta-analyses—Evaluation of multiple independent studies of the same subject in order to identify consistent findings.

Metadata—Information about how a data value was generated.

Glossary.

Meta-synthesis—Evaluation of qualitative data from multiple studies to create new interpretations.

Metrics—Measures. Variables used to characterize a phenomenon.

Middle-of-the-road-fallacy—The belief that information midway between extreme alternatives is legitimate without evidence. Also called marginalization of the adversary fallacy).

Mind map—A type of concept diagram in which a non-linear structure is used to represent hierarchical aspects of ideas. Often used in brainstorming.

Minimum—The lowest data value.

Misdirected models—Models based on biased or mistaken theories that are used to explain observed phenomena in a way that fits the researchers preconceived notions.

Missing data—Data values that are not available for some observations of a variable in a dataset. The data may be missing-at-random, missing-completely-at-random, or missing-not-at-random.

Misspecification—Including terms in a statistical model that are inappropriate for theoretical or numeric reasons.

Mistakes—Random data errors usually attributable to special causes that affect only a few data points.

Mixed-effects ANOVA—An ANOVA design that includes both fixed and random effects.

Mixture diagrams—Graphs such as bar charts that used mainly to display the proportions or amounts of data groupings.

MNAR—Missing Not At Random, also called non-ignorable missing data, means that the pattern of the missing values in a variable is related to the non-missing values in the same variable.

Model misspecification—Including terms in a model that improve its statistical performance even though the model is problematic. Misspecification often involves using dependent and independent variables that are based on the same standard and using metrics in a prediction model that are excessively difficult or time-consuming-to measure.

Glossary.

Model variance—The part of the total variance in a dependent variable in ANOVA or regression models that is explained by the independent variables. It is also known as between-group variance. The remainder of the total variance is considered to be error variance, also known as within-group variance.

Model—A representation of some phenomenon, typically used in place of the phenomenon to manipulate a process or display a result. Models can be physical, written as descriptive text, drawn, or consist of mathematical equations or computer programming.

Modes, distribution—Separate concentrations of data points in a frequency distribution. Theoretical distributions, like the Normal distribution, are usually unimodal. Describing the central tendency of multimodal data distributions is a challenge.

Mode—The value that appears most commonly in a dataset.

Modus Ponens, Modus Tollens—Commonly used logical rules for creating conclusions from premises. Modus ponens takes the form: if **P**, then **Q**; **P**; therefore, **Q**. Modus tollens takes the form: if **P**, then **Q**; not **Q**; therefore, not **P**.

Monte Carlo simulation—An approach to evaluating models using datasets randomly selected from a known theoretical distribution.

Monte Hall problem—A counter-intuitive probability paradox in which a participant must decide what to do when new information is provided.

Moose—The kitten who had his DNA tested in Chapter 2.

Morton's fork— A choice between two unfavorable alternatives, known more simply as a dilemma or being between a rock and a hard place.

Mosaic charts—Charts having rectangular arrangements of cells that show information in defined rows and columns. The cells may be different sizes and colors to convey information, like frequencies. Examples include heatmaps and treemaps.

Motion charts—Charts that show how data points change over time or conditions using motion. Creating a motion chart involves creating a series of charts and compiling them into an animated gif or a video for presentation.

Moving Average Interpolation—Assigning values to grid nodes by averaging the data within the grid node's search ellipse.

Moving average smoothing—Replacing data points with averages (weighted or not) of prior time-series or other sequential data.

Moving-average model—A type of time-series model in which a set number of prior observations are averaged to produce the next observation. The averaging often uses weights to give more recent observations greater importance.

Multicollinearity—High correlations between predictor variables in a regression model, which can result in unstable results for the correlated predictors.

Multi-level pie charts—Another name for sunburst diagrams, ring charts, and radial treemaps.

Multiple correlation coefficient—A Pearson correlation coefficient that measures the strength of a data relationship between one dependent variable and two or more independent variables.

Multiple regression—Linear regression with more than one independent variable.

Multiple-choice question—A type of close-ended survey question that is like a single-choice question except that the respondent can select more than one of the responses.

Multiplicative model—A model in which the terms are multiplied together rather than being added.

Multivariable plots—Statistical graphics that display more than one variable.

Multivariable—More than one variable.

Multivariate analysis of variance—Analysis of variance with more than one dependent variable.

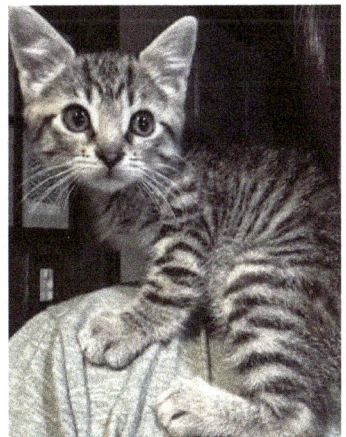

Glossary.

Multivariate—More than one dependent variable.

Multi-way ANOVA—An ANOVA design in which more than one effect is tested. Simple ANOVAs are called one-way. Multi-way ANOVAs could be two-way, three-way, and so on depending on the number of effects being tested.

Mutually-exclusive-alternatives—Choices in which there is no overlap of conditions.

Mutually-exclusive—Two or more conditions, events, survey response choices, or other related elements that can't occur or be true at the same time.

N and n—N is the number of individuals in a population; n is the number of individuals in a sample.

Natural variability—The part of a data value that is attributable to inherent uncertainty. In a completely deterministic world, there would be no natural variability.

Negative-skewed—A data frequency distribution in which the tail on the left side of the distribution is longer than the tail on the right. Also called left skewed.

Network diagrams—Charts that show hierarchies, organizational connections, and interconnections between objects, in two dimensions. They have no scale although they can be drawn to show direction, importance, or frequency.

Network sampling—An approach to identifying individuals who are hard to find and hard to extract information from by using information provided by existing subjects. Also referred to as snowball sampling or referral sampling.

Neural networks—A class of nonlinear statistical modeling methods that identify, segment, and model patterns in data and then combine the models into a single network of models used for prediction.

Neutral bias—When participants in a survey select a middle-of-the-road response or an "other" response because they have no deeply-held opinion or don't want to stand out.

Nirvana fallacy—When reaching a definitive decision or conclusion is deferred until additional data are collected or analyzed. Also called paralysis by analysis and procrastination.

No difference—Refers to non-significant results for a statistical test.

No intercept—An option in creating a regression model that forces the line through the origin. This approach will artificially inflate the value of R-square.

Nodes—Intersection points in spatial sampling schemes and statistical graphics where different items are connected.

Nominal scales— Scales that define items having no mathematical relationship to each other, such as personal and place names, IDs, product brands, and sets of associated attributes.

Non sequiturs—Conclusions that do not follow from the premises.

Non-directional test—Statistical tests that involve both tails (ends) of a data distribution.

Non-disjoint—Events or outcomes that overlap, occurring at the same time. Examples of non-disjoint outcomes include a student getting a grade of B in two different courses.

Nonlinear regression—Statistical methods for fitting nonlinear trends, such as curves, waves, and steps, through a set of data such that the variance is a minimum.

Nonparametric statistics—Descriptive and inferential statistics and procedures that do not use theoretical population frequency distribution models.

Non-participation rate—The proportion of survey invitees who choose not to participate.

Non-probability sampling—Selecting a sample from a population based on subjective judgment, convenience, or other intentional criteria rather than random choice. As a consequence, available samples in a population do not have an equal

and known probability of being selected so there is no way to quantify variability and make generalizations about the population.

Non-representative—Samples collected from a population that do not share the same pertinent characteristics of the population.

Non-response bias—Inaccuracies that occur in a survey when invitees decline to participate or when participants respond to some questions.

Nonsignificant—Failing to reject the null hypothesis when the test statistic is less than the standard value. Also said to be not significant.

Normal distribution—A continuous, unimodal, symmetrical theoretical probability distribution used commonly in statistical testing because of its association with the Central Limit Theorem. It is also called the Gaussian distribution and the bell curve.

Normal—In **Stats with Kittens** and **Stats with Cats**, the term "Normal" when capitalized refers to the distribution model, assumptions about the distribution model, or tests of assumptions about the distribution model. The word "normal" is also used as a common noun or adjective meaning typical, a mathematical term referring to a type of data standardization, and a chemistry term referring to a unit of concentration. Experience has shown that these differences in meaning cause copy editors to suffer brain spasms.

Normality—Refers to the property of datasets that have a frequency distribution that mimics a Normal distribution. The Normal distribution model refers to a specific mathematical equation.

Normalized—Changing the scale of a variable by subtracting the minimum value from each observation and then dividing by the data range. This rescales the range to 0 to 1.

Normally distributed—Data that follow a Normal (Gaussian or bell-shaped) frequency distribution.

Nugget—The spatial variance of co-located samples.

Glossary.

Null hypothesis—The hypothesis in a statistical test that is presumed to be true. The test is then conducted to see if the hypothesis can be rejected. Usually, the null hypothesis involves there being no difference or no change in some values.

Numerical models—Numerical models are mathematical equations that have a time parameter. Numerical models are solved repeatedly, usually on a grid, to obtain solutions over time. This is sometimes called a dynamic model (as opposed to a static model) because it describes time-varying relationships.

Objectives—The objective of a statistical analysis might be characterization, identification, classification, detection, prediction, or explanation.

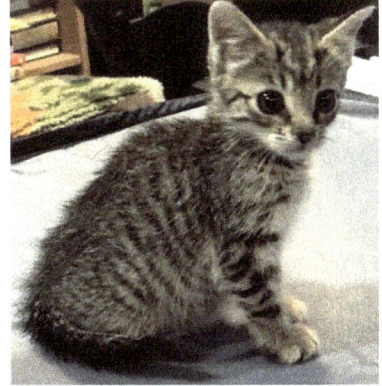

Observation—A term used for sample, subject, record, case, individual, patient, respondent, student or entity on which measurements are made.

Observational study—A statistical study in which subjects inherently possess a characteristic rather than being assigned it randomly as in an experimental study.

Odds ratio—The odds of event **A** taking place in the presence of event **B** divided by the odds of event **A** in the absence of event **B**. Two events are independent if the odds ratio is equal to 1, positively correlated if greater than 1, and negatively correlated if less than 1.

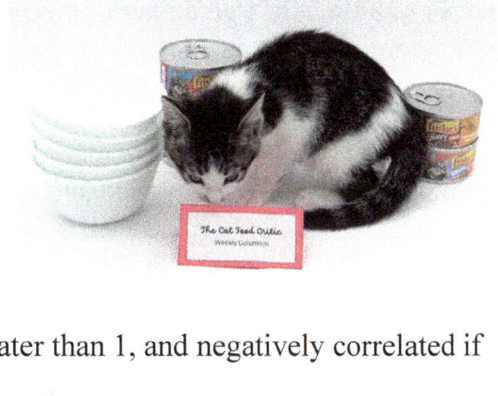

Odds—The number of favorable events divided by the number of unfavorable events. In contrast, probability is the number of favorable events divided by the total number of events. This is the same as saying that odds are the probability that the event will occur divided by the probability that the event will not occur. Odds range from 0 to infinity; probabilities range from 0 to 1. To convert from odds to a probability, divide the odds by one plus the odds. Odds are traditionally used in gambling.

Once in a blue moon—Something that happens very rarely. It refers to the appearance of a second full moon within a calendar month, which happens about every thirty-two months.

Glossary.

Once-and-done—An action that does not have to be repeated once it is completed, like graduating from high school.

One-tailed—A statistical test that considers only one extreme end of the population frequency distribution. Also referred to as directional or one-sided.

One-way ANOVA—A method to compare the means of three or more groups defined by one independent variable.

Open surveys—Surveys in which anyone can participate without regard to demographics or other framing filters. Same as an open-invitation poll.

Open-ended question—A survey question with no predetermined categories of responses, allowing respondents to provide any information they want.

Open-high-low-close charts—Used to display the opening, high, low, and closing prices of a stock in a given period of time. Similar to and sometimes used synonymously with candlestick charts.

Open-invitation poll—Surveys in which anyone can participate without regard to demographics or other framing filters. Same as an open survey.

Oracle—Classically, an oracle is an expert, prophet, or prognosticator who predicts the future. In statistics, it is a source of probabilities that combines the use of mathematical models with the oracle's judgment. Probabilities in sports and investing come from oracles, for example.

Order bias—A tendency for a survey participant to respond on the basis of the position of a question or response choice rather than a personal preference.

Ordinal scales— Scales that define measurement levels having some mathematical progression or order, represented by integers, usually positive.

Orientation statistics—Statistics for cyclic data represented on a circle or sphere.

Outcome variable—Another term used for dependent variable.

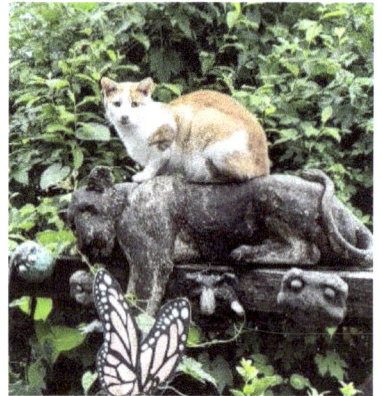

Glossary.

Outlier—A data point that is uncharacteristic of the rest of other data points in a dataset. Outliers may provide information on special conditions that affect a phenomenon or occur for no known reason.

Out-of-spec data—Data that do not have consistent precisions, units, or measurement scales.

Overfitting—The inclusion of many variables in a model, especially relative to the number of samples and the simplicity of the data pattern. Overfit models usually have high R-square values but perform poorly on new samples from the same population.

Overlapping Data—Variables that are difficult to assess because they have multiple data for a single observation. This causes percentages to not sum to 100%, thus confounding some descriptive statistics. Examples include survey questions that allow check-all-that-apply and causes-of-death that have a primary cause and multiple contributing causes.

Oversample—Collecting more samples of a scarce demographic so that additional analyses can be conducted on the group.

p Value—The probability value usually associated with a statistical test.

Panel—A pre-approved list of potential survey participants, managed by survey companies, that includes demographic information to facilitate assembling a frame for a survey.

Panes—The cells of a window diagram.

Paradigm shifts—A fundamental change in approach or underlying assumptions that happens when traditional methods become outmoded.

Paralysis by analysis—When a decision is not made because the information available is overly complex or contradictory and the decision has unacceptable consequences. This leads to overthinking and procrastination.

Parameter—A number that defines some characteristic of a population or a model. For example, the mean and variance of a Normally distributed population are the population's parameters. They are the terms in the equation of the model.

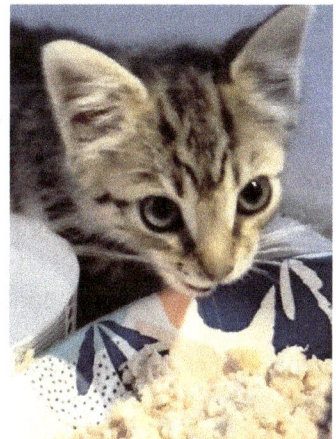

Glossary.

Parametric statistics—Statistics that use theoretical probability distributions as models for the frequency distributions of data.

Pareto chart—A specialized bar chart that displays categories in descending order of frequency, used to identify the relative importance of factors in a dataset.

Pars pro toto fallacy—Where a few legitimate examples are used as evidence that all cases are valid.

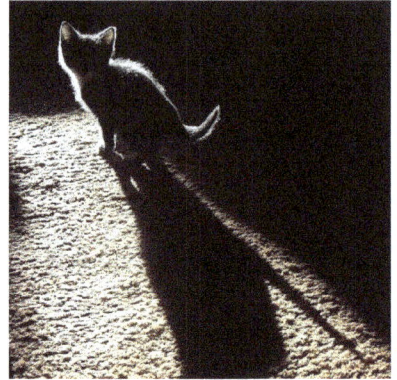

Partial correlation coefficient—A Pearson correlation coefficient that measures the strength of a data relationship between one dependent variable and one or more independent variables with the effects of other independent variables held constant.

Partial regression leverage plot—A scatter plot for one term in a model showing the relationships between each data point and the model parameter for the term. The coefficient of the model term is the slope of the best-fit line to the data in a leverage plot. Also known as an added variable plot or a partial regression residual plot.

Participant—A survey invitee who completes the survey.

Path analysis—A statistical technique for evaluating causal relationships among variables.

Pathological science—A type of bad science in which a researcher deviates from strict adherence to the scientific method in order to favor a desired hypothesis.

Pattern—A characteristic shape that data may display, particularly on scatter plots. Five commonly observed patterns are shocks, steps, shifts. cycles, and clusters.

PCA—Principal Components Analysis. A special case of factor analysis in which the original variables are recombined into a smaller number of uncorrelated factors, which are often then used in regression analysis to evaluate the relative importance of the factors to the dependent variable.

Peakedness—The height and width of a mode in a data distribution, quantified by the kurtosis.

Pearson product correlation coefficient—Measures of the strength of a data relationship when two variables are measured on continuous scales.

Percentiles—The percent of a data distribution that is equal to or below a specified value, on a scale of 1 to 100.

Percent-stacked-bar charts—Another name for mosaic charts, Mekko charts, and Marimekko charts.

Performance window—A matrix table with two rows and two columns that summarize the number of correct classifications or predictions versus the number of incorrect classifications or predictions. It is also called a confusion matrix, an error matrix, or a matching matrix.

Periodograms—A graph that shows the relative importance of possible periods (or frequencies) of a time series.

Permutation—A selection of a subset of unique items in groups of a given size and in a given order taken from a larger group.

Personally identifiable information—Data that can be combined to identify a specific individual. PII is used mainly within the U.S.; Personal Data is the approximate equivalent of PII in Europe. Examples of PII include: name; address; date of birth; email, telephone numbers, personal IDs (social security number, passport number, driver's license number, bank account numbers, credit card numbers); computer-related details (account IDs and passwords, device IDs, IP addresses, GPS tracking, cell phone records, social media accounts); biometric data (finger and palm prints, retinal scans, DNA profile); and even answers to password-security questions. This information can be augmented with gender, race, age range, former address, education, and work history.

Perspective bias—A bias caused by an individual's understanding or point of view that is mistaken, one-sided, or narrow-minded.

p-Hacking—Manipulation of data or statistical testing options or procedures in order to obtain a statistically significant result.

Phantom population—A population that is assumed to exist but in fact does not.

Phenomenon—The event, condition, or characteristic that is being studied in a statistical analysis. Characteristics of a phenomenon are defined by metrics, measures, measurements, attributes, and most often, variables.

Phi correlation coefficient—Measures the strength of a data relationship where both variables are measured on binary scales.

Physical evidence—Tangible objects, like documents, pictures, or videos, that can support an argument or presentation.

Pick charts—A 2x2 matrix table used in Lean Six Sigma (LSS) to categorize actions and ideas based on payoff and difficulty.

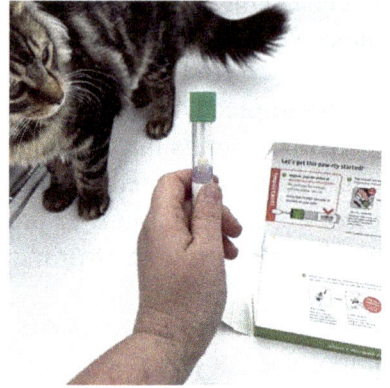

Pie charts—A type of mixture chart that shows the proportions as segments of a circle. There is no axis for scale; proportions are implied by areas of the circle, like slices of a pie. Pie charts are the most commonly used graphic to display proportions of data groups.

PII—Personally Identifiable Information. Information that can be used to determine the identity of a specific participant in a study.

Pilot study—A small-scale study conducted before large-scale research to assess the practicality, accuracy, and precision of the data generation process and to improve the statistical design of the research.

Pips—The small dots on each side of standard dice indicating a number from one to six. The number of pips on opposite sides of dice sum to 7.

Placebos—A sham treatment used to blind subjects to the details of an experiment. Placebos are used commonly in randomized clinical trials to test the efficacy of medical treatments.

Platykurtic—A frequency distribution having a smaller value for kurtosis than the Normal distribution, indicating that there are fewer data values near the mean than in a Normal distribution. Histograms of platykurtic distributions have a flattened appearance.

Plausibility—One of Hill's criteria for causality. A relationship is more likely to be causal if there is a plausible mechanism between the cause and the effect.

Plausible deniability—Assertions of individuals and organizations that they have no knowledge or responsibility for negative actions. Their claims rest on a lack of evidence that can confirm their participation.

Glossary.

Plot area—The portion of a graph bounded by the axes.

Plots— Visual representations of data, often used synonymously with charts, graphs, and diagrams. Plots tend to place more emphasis on individual data points.

Point identifiers—Shapes or icons used to represent data on statistical graphics.

Point-biserial correlation coefficient—Measures the strength of a data relationship where one variable is measured on a continuous scale and the other variable is measured on a binary scale.

Poisson distribution—A discrete probability distribution that is often used to describe the probability of an event happening a certain number of times within a given interval of time.

Polar chart—Charts for circular data that uses polar coordinates rather than rectangular coordinates. Values of data points are represented by the distance from the center point. Radial charts are a kind of polar chart.

Poll—Formerly referred to a simple, one-question, informal interview that was often conducted in person. In comparison, surveys, were formal, elaborate, longer, data-gathering efforts involving many participants sampled from a population, and conducted with as much statistical rigor as possible. Today, poll and survey are used synonymously.

Polling denialism—The notion that statistical surveys, especially in horse-race polls, have all become illegitimate and shouldn't be believed or even participated in. Polling is blamed for the way that media sources misinterpret polls.

Pollster—An individual or business that conducts statistical surveys.

Polychoric correlation coefficient—Measures the strength of a data relationship when both variables are measured on an ordinal scale.

Polygon plot—A plot that shows values for several metrics, usually for a single data point. The plots are usually small (because they are created for multiple data points) and may be used as point identifiers in icon plots.

Glossary.

Polynomial regression—Regression in which one or more of the independent variables in the model are polynomials.

Polyserial correlation coefficient—Measures the strength of a data relationship where one variable is measured on an ordinal scale and the other variable is measured on an interval scale.

Pooled variance—The sample variance of two or more groups calculated when the variances of individual groups are unequal.

Population correction—An adjustment made in calculating a survey's margin-of-error to account for how much of a population is being sampled. Bigger populations have smaller MoEs.

Population pyramid—A version of a funnel chart used to display and compare demographic frequencies, They are sometimes used for comparing two binary groups represented by back-to-back histograms.

Population—A collection of individuals or items that have some common set of characteristics. Samples of the individuals or items used in a statistical analysis must be representative of the entire population.

Positive-skewed—Data frequency distributions having a long or fat tail on the right, represented by a skewness value over 1. Also called right-skewed.

Post hoc fallacy—The assumption that one event caused another event solely because the first event came before the second event. Also called post hoc ergo propter hoc fallacy.

Post-mortem—A process conducted at the conclusion of a project to determine which parts of the project were successful or unsuccessful. Also referred to as an after-action review.

Power analysis—An evaluation of the validity and reasonableness of statistical tests involving number of samples, effect size (resolution) and the probabilities of Type I and Type II errors. Confidence is equal to (1 - probability of Type I error). Power is equal to (1 - probability of Type II error).

Power—The likelihood of a hypothesis test detecting a true effect when there is one. Power equals 1 minus beta, the probability of a Type II error, i.e., not rejecting a false null hypothesis.

Glossary.

Practical significance—A determination of whether the difference detected in a hypothesis test is large enough to be relevant in the situation being examined. Also referred to as meaningfulness.

Precision—How close measurements made on the same entity are to each other. Precision is the reciprocal of the variance.

Preconceived notions—An idea, opinion, or judgment that is influenced by personal biases, past experiences, or societal norms without consideration of actual evidence.

Prediction error—The cumulative difference between a model's predictions and the actual data. Prediction error characterizes the precision of the whole model while residuals characterize the differences for individual data points.

Prediction intervals—The interval that provides a specified confidence that an individual prediction from a statistical model will be included within its limits.

Prediction—The estimation of new, unmeasured values of a dependent variable in a statistical model. If the prediction involves extrapolation into the future, the term forecast is used instead of prediction.

Predictor variable—An independent variable in a statistical prediction model.

Premises—The introductory facts in an argument based on logic.

Preponderance of the evidence—Enough information supporting an argument that the claim is more likely to be true than not.

Presentism—The interpretation of past events in terms of present-day attitudes and concepts.

Prevalence—The percentage of the population with a disease or condition, used to calculate the Bayesian probability of the results of a medical test.

Primary source—An individual or organization that creates original information, called primary data.

Principal Components Analysis—A statistical method for combining independent variables into new variables, called components, so that the new components retain as much of the original information as possible. The goal of PCA is to reduce the total number of independent variables.

Glossary.

Probability distribution—A mathematical function that describes the likelihood of possible values of a variable occurring.

Probability models—May refer to theoretical frequency distributions or to models based on the intersection of multiple disjoint events.

Probability of detection—The ability of a medical test to detect a condition. Also called the test's sensitivity or True Positive Rate (TPR).

Probability of nondetection—The ability of a medical test to detect the absence of a condition. Also called the test's specificity or its True Negative Rate (TNR).

Probability of precipitation—The percent probability that it will rain (or snow) in some part of a specific location during a specific time period. It is determined by multiplying the forecaster's confidence that rain will fall in the area and the percentage of the area where 0.01-inches of rain is expected to occur. Weather forecasters are examples of oracles.

Probability of success—One of the two parameters of the Binomial distribution, the other being the number of independent tests.

Probability plot—A plot of the frequencies of data versus a theoretical frequency distribution. If the points form a straight line with no major aberrations, the data is likely to follow the distribution.

Probability sampling—A method of selecting samples from a population so that each member of the population has an equal, known, non-zero chance of being selected. This means either the selection process or the population itself is random. Common strategies used for probability sampling are random sampling; stratified sampling; cluster sampling; and systematic sampling.

Probability survey—A survey in which the participants are selected to match certain target percentages based on the characteristics of the population being studied.

Probability with replacement—A probability calculation in which items that are selected are returned to the sample pool after each draw. The total number of available items in the sample pool remains the same and the events are considered to be independent.

Probability without replacement—A probability calculation in which items that are selected are not returned to the sample pool after each draw. The total number of available items in the sample pool decreases by one with each selection and the events are considered to be dependent.

Probability—A way to quantify how likely an event is to occur considering all possible outcomes. Probability ranges from 0 (virtual impossibility) to 1 (virtual certainty). It can be expressed as a fraction, decimal, or percentage. Frequentist probability is based on the number of events while Bayesian probability is based on prior probabilities.

Probable cause—A legal standard meaning that there is a reasonable belief that a crime has been committed or that a particular person is responsible for a crime.

Process—Repeatable activities that are conducted as part of generating a data value.

Procrustean fallacy—When information is considered to be legitimate because it is an established belief in society. Also called argument-to-tradition, forced conformity, and status quo fallacy.

Profile lines—Data icons in which multivariate data is represented by a line, also called sparklines.

Property Charts—Charts that show some characteristic of a dataset, usually measures of central tendency, dispersion, or frequency. The most common property charts are bar charts, box plots, and maps.

Proportional area charts—Charts that rely on visual perceptions of areas rather than scales. They are like pie charts except for being square or rectangular instead of circular.

Propositional logic rules—In deductive reasoning, logical processes are called propositional logic rules. Examples include modus ponens and modus tollens.

Protoscience—Topics that were once mainstream science but fell out of favor and were replaced by more advanced formulations of similar concepts. The original hypothesis then becomes a pseudoscience. Examples include astrology and alchemy.

Glossary.

Proxy sampling—Using an easy-to-select sample or metric as a substitute for a difficult-to-select sample or metric. Also called surrogate sampling and double sampling.

Pseudoscience —A type of bad science in which hypotheses that cannot be validated by observation or experimentation are still claimed to be scientifically legitimate.

Purposive sampling—Selecting specific entities because of some characteristics they have regardless of their frequency of occurrence in a population. The entities may have expert or specific knowledge of a study's topic, represent typical or extreme cases, unique or critical cases, or minimum or maximum values. This method is often used in qualitative research when there is a need to focus on a specific type of entity, experience, or characteristic. Also called judgment sampling, purposeful sampling, and judgmental sampling.

Push polls—Polls that try to force participants to acknowledge information the pollster wants to convey. The information doesn't have to be true, and in politics, often isn't.

p-Values—The calculated probability that the results of a statistical test could have occurred by chance.

Pythagorean means—The arithmetic mean, the geometric mean, and the harmonic mean.

Qualitative studies—Studies based on non-numeric (qualitative) data. It tends to be exploratory and use small sample sizes.

Qualitative—Scales consisting of any combinations of numbers and letters that indicate classes or groups.

Quality control samples—Samples collected for the sole purpose of determining if the process of sample collection, preservation, transport, analysis, and reporting is sound. These samples are not included in a statistical analysis but are evaluated separately. They are metadata.

Quality—The acronym QA/QC refers to quality assurance and quality control. Quality assurance refers to activities taken to ensure that the data will be of a level of quality appropriate for the intended use. QA components might include staffing, training, standard operating procedures, audits, and documentation. Quality control refers to the tests and other activities taken to ensure that the quality specifications are fulfilled. In brief, QA focuses on the data generation process; QC focuses on the resulting data.

Quantile-quantile plot—A scatter plot of two sets of quantiles, one from a theoretical distribution and one from a data distribution. If the points approximate a straight line with no major deviations, it is assumed that the data can be modeled with the theoretical distribution. Also called a QQ plot. Residuals from a model can be plotted on a QQ plot to assess the Normality assumption.

Quantiles—Values that divide sorted data into equal parts. Commonly used quantiles include quartiles (four parts), deciles (ten parts), and percentiles (100 parts).

Quantitative—Scales consisting entirely of numbers representing a natural progression.

Quartiles—Three values that divide a dataset into four parts, at 25% of the data, 50% of the data, and 75% of the data.

Quasi-experimental studies—Like an experimental study except that subjects are not assigned to groups randomly.

Question skipping—Occurs when a survey participant does not respond to a question. Skipping complicates statistical analysis because it creates missing data that have to be evaluated. Questions may be skipped when a survey does not provide appropriate choices, when a participant has time constraints, is uncommitted, or is uncomfortable with selecting an unpopular choice. The most problematic situation is when the skipping is a deliberate act to skew the survey results.

Question-order bias—A tendency for a survey participant to answer a question differently because of how they answered a previous question.

Quintiles—Five groups that a dataset can be divided into based on 20%, 40%, 60%, and 80% of the data.

Quota sampling—The strategy of identifying characteristics or conditions that are important to a study, and then sampling entities until a predetermined number having that characteristic or condition is reached. Also called consecutive sampling and total enumerative sampling.

Quoting out of context—The fallacy in which a quote is removed from its surrounding text so it distorts the quote's intended meaning.

R or r—Upper case **R** is the multiple correlation coefficient, representing one dependent variable and more than one independent variable. Lower case **r** is the simple correlation coefficient, representing two variables.

Radar plots—Multivariate, circular plots consisting of axes separated at equal angles, one axis for each variable. Data need to be measured in the same units on the same quantitative scale.

Radial bar charts—A bar chart arrayed in a circle.

Radial-treemap—A chart that displays a tree (hierarchical) structure using a two-dimensional circle, with the structure expanding outwards from the center.

Random assignment—The process of allocating subjects to experimental groups such that each subject has an equal chance of being in any group.

Random effects ANOVA—An ANOVA in which generalizations are to be made to additional untested levels. Random-effect ANOVA designs are specialized and sophisticated, and thus, uncommon.

Random Forest—A classification technique in which separate decision trees are combined into a single tree.

Random sampling—Selecting a sample from a population such that each member of the population has an equal chance of being chosen.

Random TerraBytes—Charlie Kufs' blog about miscellaneous topics at randomterrabytes.net. No, it's not spelled wrong. Charlie had a company in the late 1990s that specialized in the analysis of environmental data. Terra (instead of tera) represented the environmental connection. He started Random TerraBytes so he would have a place to put blogs about everything other than statistics, which he published at statswithcats.net. The figures in Chapter 4 about sociological generations, candy bars, UFOs, and the New York Giants all came from Random TerraBytes.

Glossary.

Randomization—The act of randomly assigning subjects to treatments in experimental studies. Sometimes also used to refer to the random selection of samples from a population.

Randomized control trials—An experiment in which subjects are randomly assigned to groups, usually a control group which receives no intervention and treatment groups which do receive interventions. Often referred to as RCT.

Random—The lack of a pattern, order, or predictability in information, created by chance rather than according to a plan.

Range—The maximum minus the minimum. Also refers to the distance at which spatially correlated variables become independent (uncorrelated) in geostatistics.

Rank-biserial correlation coefficient—Measures the strength of a data relationship where one variable is measured on an ordinal scale and the other variable is measured on a binary scale.

Ranking Question—A type of close-ended survey question in which respondents are supposed to place an order on a list of unrelated items.

Rare event—An event that has a very low probability of occurrence, sometimes said to be once-in-a-blue-moon.

Rating Question—A type of close-ended survey question in which respondents are supposed to assign a relative score to unrelated items.

Ratio scale—Scales that define a mathematical progression involving fractional levels, represented by numbers having values after a decimal point, and having a unique and constant zero point. Ratios calculated using variables measured on ratio scales are meaningful whereas ratios calculated using variables measured on other scales are not.

Ratio—One number divided by another.

Raw coefficients—Unstandardized regression coefficients, the a_0 through a_n terms in a multiple regression model.

Reasonable suspicion—An objective belief that something is true that would be shared by most observers, especially related to criminal activity.

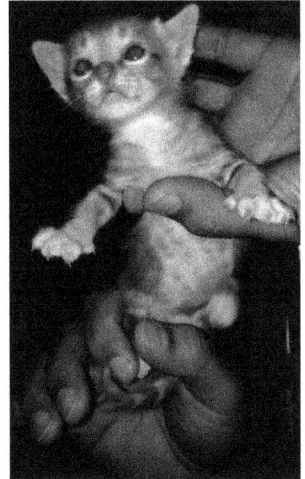

Glossary.

Reciprocal—The reciprocal of a value, **x**, is **1/x**.

Recoding—Changing the codes used to represent levels of a variable so they are more appropriate for an analysis.

Record—Record has several meanings in statistics. It may refer to observations in a dataset. It may refer to evidence about the past or previous conduct or performance of a person or organization. And, it may refer to durable forms of storage (e.g., computer files, reports, documents, notes, transcript) for text or other information..

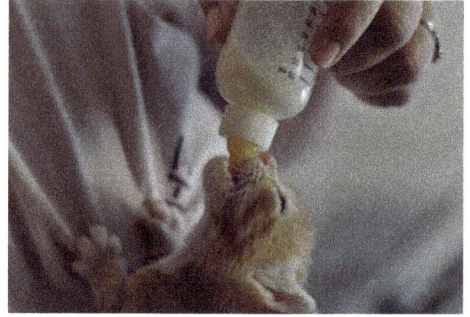

Recurrence intervals—The average time interval between the occurrence of two stream discharge events of a given or greater size, are calculated using the formula **RI=(n+1)/m**, where **m** is the discharge and **n** is the years in the data record. Also called return periods.

Red herring—Information that is intended to be misleading or distracting.

Reductionism—The idea that complex systems can be understood by analyzing their individual, simpler parts.

Reference panel—A set of DNA sequences from individuals who are representative of a particular population or ancestry that are used to classify DNA from individuals with unknown heritage.

Reference—One of the three fundamental Rs of variance control—Reference, Replication, and Randomization. Reference refers to there being some ideal, background, baseline, standard, benchmark, or at least, generally-accepted norm that can be compared to all similar data operations or results.

Referral sampling—An approach to identifying individuals who are hard to find and hard to extract information from by using information provided by existing subjects. Also referred to as network sampling and snowball sampling.

Regionalized variable—A spatially-correlated variable.

Regression coefficient (raw)—Values determined in regression modeling that are multiplied by the independent variables to estimate the dependent variable. They are constants in intrinsically linear models and mathematical functions of independent variables in intrinsically non-linear models.

Regression—Statistical processes for estimating the relationships between a dependent variable and one or more independent variables. Linear regression computes the parameters of a line that minimizes the sum of squared differences between the data and the line.

Reject, fail to reject—The two possible decisions that can be made as the result of a statistical test of hypotheses.

Relationship plots—Graphs that aim to display data patterns—shocks, steps, shifts, cycles, and clusters—or trends—linear, curvilinear, nonlinear, or complex trends.

Relationship—There are four ways that data may be related to each other. A cause is a condition or event that directly triggers, initiates, makes happen, or brings into being another effect, condition or event. An influence is a condition or event that changes the manifestation of an existing condition or event but does not cause it to happen. Associations are conditions or events that appear to change in a related ways. Spurious relationships appear to be causes, influences, or associations but are not, usually because they are based on invalid evidence or because they represent relationships that aren't understood.

Relative risk—The ratio of the probability of an event occurring in a group susceptible to a risk versus the probability of the event occurring in a group not susceptible to the risk.

Relative standard deviation—The standard deviation divided by the absolute value of mean. Similar to the coefficient of variation, which is the standard deviation divided by the mean.

Remote data—Data that are difficult to access because the subjects being measured are far away or in hard-to-access areas. Examples include astronomical and oceanographic data, microscopic and subatomic data, and even data from foreign countries.

Repeatability—The ability to benchmark and process portions of a measurement system to produce consistent results. Repeatability does not consider variability attributable to the person making the measurement.

Repeated measures ANOVA—An ANOVA design in which the data consist of multiple observations of the same subjects.

Glossary.

Replicate samples—Multiple samples of the same entity used to determine if the process of sample collection, preservation, transport, analysis, and reporting is sound. These samples are not included in a statistical analysis but are evaluated separately. May be called duplicates or triplicates depending on how many samples are collected.

Replication—Replication is used in a variety of ways to assess or control variability. Samples or variables can be replicated to test for consistency in data collection. More often, replication refers to the practice of repeating an entire study, particularly in many of the sciences, to verify previously determined results.

Reporting bias—The selective reporting or omission of information based on personal beliefs or on the outcome of the research. A common form of reporting bias is not reporting non-significant results (file drawer bias).

Representative sample—A sample of a population that accurately reflects the important characteristics of the whole population.

Representativeness—The property of a sample related to how similar the sample is to its parent population on the characteristics being studied.

Reproducibility—The ability of the measurement system and the people making the measurements to produce consistent results. By comparing reproducibility to repeatability, the effects of the judgments made by the people making the measurements can be assessed.

Repurposed words—English words and phrases that are given alternative statistical meanings (e.g., mean, confidence).

Resampling statistics—A variety of methods for using subsets of a dataset to refine or verify a model or statistic. Bootstrapping involves taking many random samples from a dataset to create more robust estimates of distribution statistics. Jackknifing involves recomputing statistics for a dataset by systematically leaving out one data value at a time and then using the jackknifed statistics to evaluate the variability and bias of the statistic. Cross-validation involves leaving a subset of the data out of an analysis and using it to verify the model created with the remaining data.

Glossary.

Rescaling—Changing the scale of a variable by adding or removing some information.

Research hypothesis—Statistical hypotheses constructed for testing elements of a research question. The null hypothesis posits that there is no effect discernable from a variable. The alternative hypothesis posits that the variable is associated with an effect.

Research question— The question that a research project sets out to answer. From a research question, statistical hypotheses are constructed for testing.

Residuals—Differences between observed values and values predicted from a statistical model.

Resolution—The ability of a statistical test to detect meaningful differences.

Respondent-driven sampling—An approach to identifying individuals who are hard to find and hard to extract information from by using information provided by existing subjects. Also referred to as network sampling and snowball sampling.

Response bias—A tendency for survey participants to respond in an insincere or false manner because of some external trigger, such as the way a survey question is worded or asked by an interviewer.

Response rate—The number of completed (and valid) survey responses divided by the number of individuals invited to participate.

Response variable—Another term for dependent variable.

Response-order bias—A tendency for a survey participant to answer a question on the basis of the position of the response on the list of choices.

Restricted Data—Data that are difficult to access because there are administrative restrictions preventing its use.

Return periods—The average time interval between the occurrence of two discharge events of a given or greater size. Return periods are calculated using the formula **RI=(n+1)/m**, where **m** is the discharge and **n** is the years in the data record. Also called recurrence intervals.

Glossary.

Ridge regression—A form of linear regression in which a small amount of bias is added during the modeling calculations resulting in regression coefficients that are slightly biased but much less variable. Ridge regression is used to address the problem of multicollinearity in the independent variables.

Right skewed—Data frequency distributions having a long or fat tail in the positive direction, represented by a skewness value over 1. Also called positive-skewed.

Right variables—In canonical analysis, there is no distinction between dependent and independent variables. Sets of variables are referred to simply as left variables or right variables.

Ring chart—An organization chart presented in a radial format. Each ring of the chart represents a different organizational level. The center of the circle represents the top or root of the hierarchy and the levels progress outwards. The sizes of the levels can be made proportional to data values so the chart can function like a pie chart. Also known as sunburst diagram, multi-level pie chart, and radial treemap.

Robopolls—Polls conducted by telephone using an automated system.

Rose diagrams—A histogram in which the bins represent intervals on a compass and is presented radially on a two-dimensional circle.

R-square or **R^2**—Coefficient of Determination. Interpreted as the proportion of variance in the dependent variable that is explained by the independent variables.

Rule of complementary events—The sum of the probability that an event will occur and the probability that the event will not occur will always be 1.

Rule of general addition—The probability of either of two events occurring is equal to the sum of the individual probabilities of each event minus the probability of both events occurring simultaneously.

Rule of special addition—For mutually exclusive events, the probability of both events occurring simultaneously is zero. The probability of either event occurring individually is equal to the sum of each event occurring individually.

Rumsfeld window—A 2x2 matrix, concept plot used to categorize ideas or events as things we know we know, things we don't know we know, things we know we don't know, and things we don't know we don't know.

Sabermetrics—Analysis of data from baseball.

Sample size factor—The number of people surveyed in the calculation of a survey's margin-of-error. The more people that are surveyed, the smaller the MoE.

Sample size—The number of samples in a statistical analysis.

Sample—A portion of a population. May refer to a single individual sample or a collection of many individual samples.

Sampling design—A strategy for collecting samples in a manner that will best represent the population. Commonly used designs for probability sampling include: stratified, random, systematic, surrogate, and cluster.

Sampling variability—The part of a data value that is the difference between a sample and the population that is attributable to how uncharacteristic (non-representative) the sample is of the population.

Sankey diagrams—A diagram used to show how one set of items transitions into another set of items. The items being connected are called nodes and the connections are called links. The sizes of the nodes and links reflect the quantity of the items.

Scale—Scales describe the relationships between numbers used to measure some attribute of an object. Scales can be qualitative or quantitative. They can be continuous or discrete. They can be classified as ratio, interval, ordinal, or nominal. They can represent counts, orientations, times and locations. These distinctions are important because they influence how the measurements can be analyzed.

Scatter plots—A statistical graphic that aims to display data values, usually for two quantitative variables, called two-dimensional (2D) scatter plots. Three-dimensional (3D) scatter plots are feasible, appearing as cubes instead of rectangles, but are more difficult to interpret because of the added dimension.

Scientific method—Science is our perception of how things work. The scientific method is how we determine what is the current state of our science. Science is the product of the successful application of the scientific method. Science changes; the scientific method doesn't.

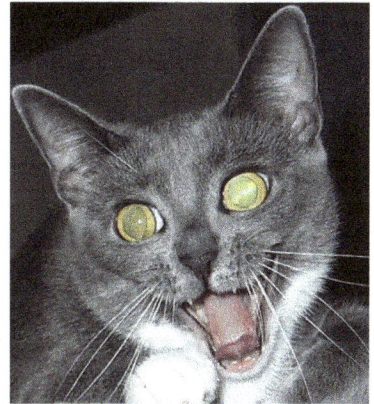

Glossary.

Scraping—Techniques used for extracting data from websites.

Scrubbing—Editing a dataset by correcting, replacing, or removing: invalid data; incorrectly-recorded data; incorrectly-coded data; data quality information; missing data; extraneous data; dirty data; useless data; invalid fields; out-of-spec data; out-of-bounds data; messy data; corrupted data; and mismatched data.

Search sampling—Grid sampling in which the size of the grid spacing is designed to identify objects of a specified size, shape, and probability of detection. The scheme is also called spatial probability sampling.

Secondary source—A secondary source recycles primary data (or other secondary data) by reformatting or analysis to create new information called secondary data.

Segmentation models—Statistical models used to assign data points into groups having common characteristics defined by variables. Also called classification models.

Selection bias—The systematic error introduced when the selection of entities from a population is not random.

Self sampling—When individuals without invitations participate in open surveys without any review by the pollster. Self-sampling is mostly limited to surveys conducted on social media for entertainment. Also called volunteer sampling.

Semantic-differential question—A type of close-ended survey question. They are like rating scale questions, in which the choices represent a spectrum of preferences, attitudes, or other characteristics, between two extremes.

Semi-structured data—Data that have design characteristics that facilitate reformatting but still need some preprocessing before analysis. For comparison, structured data are already formatted in the form of a matrix and unstructured data need extensive preprocessing before analysis.

Semivariogramming—Analyzing the relationship between spatial variance and the distance between samples. Same as variogramming.

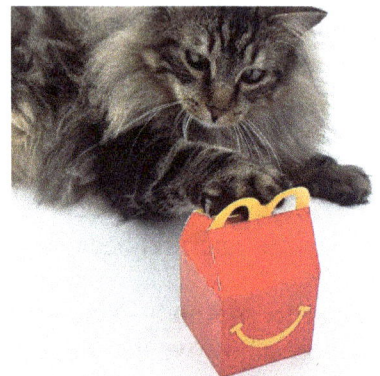

Glossary.

Sensitivity—The ability of a diagnostic test to detect a condition. Also called True Positive Rate (TPR) or probability of detection.

Sequestered data—Data that are difficult to access because they are concealed or are hard to understand. Examples include classified or untranslated documents, trade secrets, cryptids, recluses, and closed communities.

Serendipity—Being in the right place at the right time.

Serial correlation—The correlation between data points with the previously listed data points, termed a lag. Serial correlations can be calculated for any number of lags, although usually only the first few are important. If the data points are collected at a constant time interval, the term autocorrelation is more typically used.

Sham samples—Samples that are not entirely representative of the population they were supposedly drawn from.

Shapiro-Wilk test—A statistical comparison of a sample distribution to a Normal distribution. The value of the Shapiro-Wilk test is like a correlation coefficient between a probability plot of the data and a probability plot of the Normal distribution.

Shifts—Increases and/or decreases in the trends of bivariate data plots. Shifts are longer than steps and don't necessarily progress in the same direction.

Shocks—Uncontrollable short-duration conditions or events that can influence a single data point, a cluster of data points, or even most of the data in a dataset.

Short-term variability—Immediately occurring variability often attributable to data-generation instruments and procedures. In contrast, long-term variability is more likely to be related to natural changes and changes in the study environment.

Shrunken correlation coefficient—A Pearson correlation coefficient that measures the strength of a data relationship after correcting for the number of variables and the number of data points. Also called the adjusted correlation coefficient.

Side-by-side bar chart—A bar chart showing groups of individual bars placed together along a horizontal or vertical axis. Each bar represents a specific category, making it easy to see similarities, differences, and trends at a glance.

Significance level—The acceptable type I error rate (i.e., the probability of rejecting a true null hypothesis) set by the researcher before conducting a statistical test. Also

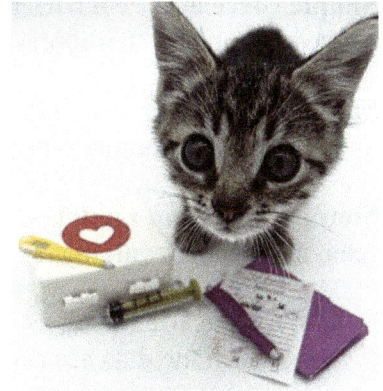

Glossary.

known as alpha, **α**. It is the threshold for deciding whether a statistical test is significant or not.

Significance—Statistical significance means that the magnitude of a number is large relative to the variability in the measurement of the number such that the number couldn't have occurred by chance. A statistically significant result may or may not be meaningful. Conversely, a meaningful result may or may not be statistically significant. Results that are significant-but-meaningless or insignificant-but-meaningful occur when the statistical test is not designed appropriately, such as there being too few or too many samples.

Significant insignificance—When a statistical test is judged to be statistically significant but the difference is not actually meaningful.

Sill—The maximum spatial variance in a location-dependent (regionalized) variable.

Simple correlation coefficient—The Pearson correlation coefficient. It is a number between -1 and +1 that represents the linear dependence of two variables.

Simple probability—The likelihood of an event occurring, calculated by dividing the number of favorable outcomes by the total number of possible outcomes.

Simple regression model—A linear regression model involving one continuous-scale independent variable and one continuous-scale dependent variable.

Simpson's paradox—An association between two variables emerges, disappears, or reverses when the data are divided into groups.

Simulation—A computer-based experiment that uses a probability model to represent a phenomenon, and then performing the same analysis on many randomly generated datasets and aggregating the results.

Single-choice question—A type of close-ended survey question consisting of a vertical or horizontal list of unrelated responses.

Six Ps of establishing causality—Six reasons why it's useful to establish causality: promote; prevent; prepare; prosecute; pontificate; and probe.

Glossary.

Six sigma—Six sigma refers to a broad range of activities and tools used to address quality in a process or organization. Six sigma techniques are used to address the objectives of corporate leaders and project managers, such as customer satisfaction, worker productivity, fewer defects, reduced waste, increased profits, more market share, and so on. The tools of six sigma include traditional management tools, like people skills, planning, and critical-path scheduling but then also incorporate data and analysis. Six sigma relies heavily on hypothesis testing, process control charts, correlation and regression, ANOVA, cause-and-effect diagrams, pareto charts, decision matrices, flowcharts, failure modes and effects analysis, check sheets, histograms, box-and-whisker diagrams, surveys, affinity diagrams, benchmarking, brainstorming, relations diagrams, tree diagrams, matrix diagrams, process decision program charts (PDPC), and many other analytical tools.

Skewness—The symmetry of a frequency distribution along the axis of the variable's values.

Slices—Segments of a pie chart.

Slippery slope—An argument that claims an initial event or action will trigger a series of other events leading to an extreme or undesirable outcome, usually without any evidence.

Sloganeering—Spreading misinformation by presenting overly simple answers, ideas, sayings, or memes in response to complex questions. Also called reductionism and oversimplification.

Smoothing models—Models that minimize variability to expose patterns in data, especially time-series data. Moving-average models are one type of smoothing model.

Snapshot—A single point in time or a condition.

Snowball sampling—An approach to identifying individuals who are hard to find and hard to extract information from by using information provided by existing subjects. Also referred to as network sampling and respondent-driven sampling.

Solicitation polls—Polls conducted as a ruse in an effort to collect money donations, email addresses, or some other way of allowing the pollster to engage the participant further.

SOP—Standard operating procedure. A written document detailing steps to take in a complex process.

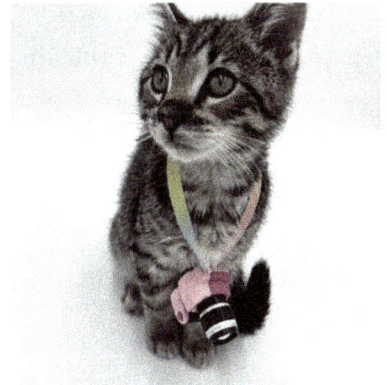

Glossary.

Sparklines—A type of data icon created by Edward Tufte. Sparklines are small lines representing data without any axes for reference. They are used to show general changes in several variables.

Spatial correlation—Similarities in a regionalized variable attributable to the locations of the measurements.

Spatial dependence—The relatedness of location-based data. Spatially-independent data have no correlation with data from locations around it. Examples include population demographics from bordering territories. Spatially-dependent data are correlated with nearby measurements of the same variable based on how far apart the locations are. Spatially dependent metrics are called regionalized variables. Topography, for example, is a regionalized variable.

Spatial probability sampling—A form of systematic sampling of a geographic area in which the size of the grid spacing is designed to identify objects of a specified size, shape, and probability of detection. The scheme is also called search sampling. It is used in environmental studies to locate hot spots of contamination.

SPC—Statistical process control. Statistical techniques involving control charts and other techniques for detecting patterns of defects in manufacturing or business processes.

Spearman rho correlation coefficient—Measures the strength of a data relationship where one variable is measured on an ordinal scale and the other variable is measured on an interval or ordinal scale.

Special cause variability—The not previously observed and unpredictable deviations from expected norms that are often attributable to unique events both within and outside of a system.

Specificity—One of Hill's criteria for causality. A relationship is more likely to be causal if there is no other likely explanation.

Spectral analysis—A mathematical technique for separating patterns in time series data into sine and cosine functions of different frequencies and amplitudes. It is used to try to identify important patterns in data series. Also referred to as frequency domain analysis.

Spherical mean—The average location of data points located on a sphere.

Glossary.

Spider diagrams—Extensions of Venn or Euler diagrams that show linkages between common elements. The term spider diagram may also refer to variants of radar charts and mind maps.

Spline—Splines are mathematical algorithms for interpolation that produce smooth surfaces while minimizing residuals at data points. Sometimes referred to as piecewise polynomials.

Sponsor—The individual or organization who initiated, funded, or promoted a statistical study. This is important to consider as a potential source of bias.

Sports statistics—The application of probability, descriptive statistics, and exploratory data analysis to sports. It was pioneered by Earnshaw Cook in the 1960 in baseball with his questioning of traditional beliefs and strategies using statistics over anecdotes. His work evolved into sabermetrics in the 1980s, which expanded rapidly to other sports as computer technologies became readily available. Also called sports analytics.

Spread plots—Any of a diverse set of statistical graphics that display information on dataset ranges and frequencies.

Spurious relationship—Relationships in which **A** appears to cause or influence **B**, but really does not. Often the reason is that the relationship is based on anecdotal evidence that is not valid more generally. Sometimes spurious relationships may be some kind of relationship that isn't understood or data that are misinterpreted. Spurious relationships may also be based on urban legends, biased assertions, or coincidences.

Stacked-bar charts—A bar chart that shows groups of data as a single bar, either as frequencies or percentages.

Standard deviation—The amount of variation in a set of data values expressed in the same units as the data.

Standard error of estimate—A statistical measure of the precision of predictions made by a regression model. It represents the average distance between predicted values and observed values of the dependent variable. Also known as the standard error of the regression, root mean square error, and SEE.

Glossary.

Standard error—The variability of a sample statistic (e.g., mean) compared to the entire population. It is calculated as the sample standard deviation divided by the square root of the sample size.

Standard operating procedures—A set of step-by-step instructions for performing a complex or repeated activity.

Standardized regression coefficients—Regression coefficients that are adjusted mathematically so that they are comparable despite the original units of the variables. Also called beta coefficients.

Standardize—The use of data transformations to normalize, index, or otherwise adjust data to have more statistically convenient scales. These scale changes do not affect patterns of relationships between variables even though the data values change.

Star plots—A plot depicting multivariate data for a single observation, similar to a *radar plot*.

Static models—Models that do not have a time component.

Statistic—A number, usually carrying information about some sample, population, or phenomenon.

Statistical assumptions—Fundamental assumptions made in conducting statistical tests and other analyses, the most common of which are representativeness, additivity, linearity, Normality, and equal variances.

Statistical comparisons—Same as statistical hypothesis tests.

Statistical graphics—Any of hundreds of methods to provide visual displays of information about statistical properties, proportions or amounts of data groupings, frequencies and shapes of data distributions, relationships between data points, and non-numerical concepts.

Statistical intervals—Statistical intervals are just another way to express the results of a statistical calculation. The interval provides a new perspective for looking at the estimated value by essentially substituting the interval range for the standard deviation.

Glossary.

Statistical literacy—Statistical literacy is understanding the process of statistical thinking.

Statistical models—Mathematical models developed using statistical techniques, like regression, that include a probabilistic error term.

Statistical significance—Rejecting the null hypothesis of no difference in a statistical test.

Statistical skills—The whats and hows of statistics involving calculations, like probabilities, descriptive statistics, and simple tests of hypotheses. Skills are learned by repetition. After Stats 101, skills like designing data matrices for a particular analysis are much more important than the calculations themselves, which are usually carried out by software.

Statistical surveys—A formal, elaborate, and comprehensive data-gathering effort that is conducted with as much statistical rigor as possible. In contrast to a census, survey participants are sampled from a population rather than having the entire population participate. The terms poll and survey are often used synonymously, although historically, polls were shorter and less rigorous. Surveys are sometimes referred to as statistical surveys to distinguish them from land surveys.

Statistical tests—Same as statistical comparisons and hypothesis tests.

Statistical thinking—The process of how statisticians explore a phenomenon in a population. It's inductive reasoning with numbers.

Statistical window—A type of matrix plot used to organize statistical information. Examples of statistical windows include definition windows, variance windows, performance windows, pick charts, and windows on scatter plots.

Statistically significant—In statistics, statistically significant means that a difference is unlikely to have occurred by chance. It indicates that the observed difference or effect is likely to be real and not just attributable to random variation in the data. Statistical significance is determined by a low p-value, typically below 0.05.

Glossary.

Statistician—A highly trained and focused scientist whose mission in life is to convert data from incomplete and messy datasets into information and then knowledge.

Statistics—The science of collecting and analyzing numerical data in large quantities for the purpose of inferring information about a population from a representative sample of the population. May also refer to more than one statistic for describing data.

Stats 101—An introductory college course in statistics.

Stats with Cats—A primer on how to apply statistics at home and work, aimed at individuals who have already completed Stats 101.

Stats with Kittens—A primer on what statistics is all about, aimed at individuals who are currently taking, plan to take, or haven't yet decided on taking Stats 101.

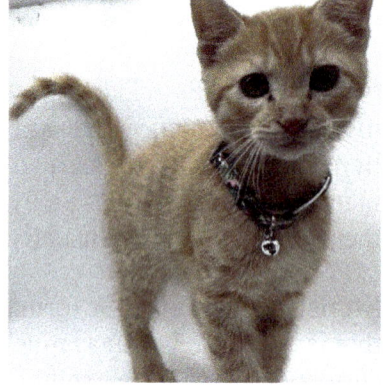

Status quo fallacy—When information is believed to be legitimate because it is well-accepted in society. Also called argument-to-tradition and procrustean fallacy.

STEM—Acronym for Science, Technology, Engineering, and Mathematics.

Stem-leaf diagram—A way to arrange data to show the frequencies of individual data values, in which the first few digits (stem) and the last digit (leaf) replace the bars of a horizontal histogram.

Stepwise approach—The term stepwise can refer to two ideas. Stepwise regression involves a semi-automatic process of selecting independent variables for a model. Stepwise storytelling involves presenting a series of simple graphics, each of which illustrate a single finding. The creator of the graphics steps the audience through each graph to lead them to the conclusion.

Stereotyping—Where a broad assertion is applied to specific cases without support. Also called sweeping generalization.

Stimulated data relationships—Event or condition **A** causes or influences event or condition **B** but only in the presence of event or condition **C**.

Stochastic-empirical models—Models based on experimental observations that include an error term to provide solutions that incorporate uncertainty into an analysis. Statistical models are examples of stochastic empirical models.

Glossary.

Stochastic—Random.

Stratified sampling—Subdividing a population into groups, called strata, based on inherent characteristics and then randomly selecting samples from each stratum in proportion to the number of individuals in the stratum. Also called proportional sampling.

Straw man—A logical fallacy that involves misrepresenting an argument (i.e., the straw man) to make it easier to attack.

Straw polls—Informal, single-question polls for discerning opinions of groups.

Strength—One of Hill's criteria for causality. A relationship is more likely to be causal if the correlation coefficient is large and statistically significant.

Structural equation modeling—A type of multivariate analysis with a wide variety of applications including causal modeling, constrained regression modeling, and confirmatory and higher-order factor analysis.

Structured data—Data that are easily formatted into a matrix.

Study design—The perspective of a study in terms of the role of the experimenter and the function of time. In observational studies, the experimenter observes naturally occurring conditions and events. In experimental studies, the experimenter controls the conditions and events. In cross-sectional studies, data are collected at the same time (or at least over a finite duration). In longitudinal studies, data are collected over a long period of time.

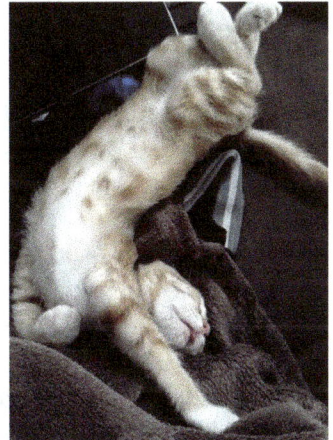

Stylostatistics—The statistical analysis of style, such as in literature and art.

Subject—A word used to refer to an individual sample (e.g., observation, record, case, individual, or what the entity is, such as patient, respondent, or student).

Subpopulation— A portion of a population, usually extracted to analyze characteristics of that portion of the population.

Subsurface maps—Spatial diagrams that represent subsurface areas either parallel to the surface (subsurface maps), or perpendicular to the surface (cross-sections), or both (block diagrams). These diagrams are mostly used to show information about geology or natural resources.

Summary analysis—A simple introduction to a dataset before conducting a specific, routine test or analysis with a narrowly defined hypothesis. SAs are primarily computational. The terms SA, IDA, and EDA mean different things to different statisticians.

Sunburst diagram—Used to display a data hierarchy using concentric rings in which each ring corresponds to a level in the hierarchy and is proportional in size to the data hierarchy. Also known as a ring chart or radial treemap.

Sun-ray plots—A plot depicting multivariate data for a single observation, similar to a radar plot.

Suppressed-data relationships—Event or condition **A** causes or influences event or condition **B** but not in the presence of event or condition **C**.

Surface plot—A two or three-dimensional depiction of data triplets (x, y, z).

Surrogate variable—Metrics used in place of conventional metrics that are too time-consuming or expensive to collect. The surrogate must be highly correlated with the metric it is replacing.

Survey drop-out—Survey dropout rate refers to the percentage of respondents who begin a survey but fail to complete it. This can be due to various factors like survey length, lack of interest, technical issues, or poor design.

Survey fatigue—When participants in a survey become tired, bored, or disinterested, leading to lower response rates, rushed or inaccurate answers, or even complete disengagement from the survey process.

Survey panels—Databases of candidate survey invitees, including their demographic characteristics, that are used as frames in future surveys. People who want to express their opinions through surveys can join panels online.

Survey—A measurement device consisting of questions, which is used for collecting data from human subjects about their opinions and perceptions.

Survivorship bias—Survivorship bias involves using data that remains after some unaccounted for filtering process.

Glossary.

Sweeping generalization—When a broad assertion is applied to specific cases without support.

Symmetry—The property of a frequency distribution describing the similarity of the tails (ends) of the distribution.

Taboo science—A type of bad science in which areas of research are limited or even prohibited either by governments or funding organizations.

Tag clouds—The first example of word frequency graphics that appeared in the mid-2000s.

Tail—The extreme ends of a distribution where the distribution approaches the x-axis.

Take-it-or-leave-it—A choice of accepting an offer or not, sometimes called a Hobson's Choice.

t-Distribution—A family of continuous probability distributions similar to a Normal binary distribution except that there is more area in the tails of the distribution to allow for analysis of smaller sample sizes.

Temporality—One of Hill's criteria for causality. A relationship is more likely to be causal if the effect always occurs after the cause.

Ternary diagram—A two-dimensional graph for three variables arrayed in the form of a triangle or a three-dimensional graph for four variables arrayed in the form of a triangular prism (trigonal prism) which has a triangle as a top and bottom. Soil scientists use a ternary diagram in soil texture classification.

TerraByte—Charlie Kufs.

Tertiary source—A tertiary source compiles primary data and secondary data so it is more readily available. A tertiary source may not be comprehensive or consistent.

Test statistics—Test statistics are values calculated for hypothesis tests based on the concept of the size of differences divided be their uncertainty. The value of the test statistic is then compared to a standard. If the test statistic is greater than the standard, it means that the difference is larger than what might have been expected

by chance, and is said to be statistically significant. If the test statistic is not greater than the standard, it means that the null hypothesis cannot be rejected so the difference is said to be non-significant. Today, statistical software reports exact probabilities, p-values, so comparing test statistics to a tabled value is no longer necessary.

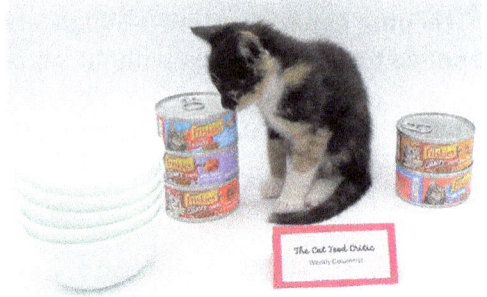

Testimony evidence—Statements from individuals who are witnesses or experts. The evidence is intangible and may be subjective. In some situations, however, it is the only evidence that is available to support a claim.

Testing dataset—A portion of a dataset set aside to test a statistical model that was developed independently from a separate portion of the full dataset called the training dataset.

Tests—Statistical comparisons of differences as a function of uncertainty.

Tetrachoric correlation coefficient— Measures the strength of a data relationship where both variables are measured on binary scales.

TGTBT—Too good to be true. When something is thought to be unlikely because it is especially fortuitous.

Theoretical distribution—A probability distribution derived from mathematical formulas or logical reasoning rather than from observed data. Also known as a parametric distribution.

Theoretical models—Models based on scientific laws and mathematical derivations rather than data. To calibrate a theoretical model, the form of the model (i.e., the equation) is fixed and the inputs are adjusted so that the calculated results adequately represent actual observations.

Theoretical statisticians—Individuals who use advanced mathematics to develop and evaluate statistical tests and other procedures for the analysis of data.

Theory—In science, theories provide explanations for why events or results happen.

Three Fs of statistical graphics—Foundation, Framework, and Facade.

Three monkeys' fallacy—When people decide not to pay attention to a presentation or even any information related to it. Also called deliberate ignorance and closed-mindedness.

Glossary.

Three Rs of variance control—Reference, Replication, and Randomization.

Three-way ANOVA—An analysis-of-variance model that includes three main effects.

Threshold data relationships—Event or condition **A** causes or influences event or condition **B** only when **A** is above a certain level.

Tick marks—Small lines used on chart axes to divide scale intervals.

Timeline—A visual representation of events in chronological order.

Time-series analysis—Analysis of time-dependent data. There are many methods for analyzing time-series data including: repeated measures ANOVA, intervention analysis, ARIMA, seasonal decompositions, time-series regression, and smoothing algorithms.

Time-series charts—A line chart or scatter plot showing time on the horizontal axis and the magnitude of the variable on the vertical axis.

Time-series regression models—Linear regression can be used to develop time-series models using five types of functions: linear trends; sinusoidal cycles; lags and leads; differences; and non-temporal predictors. The advantage of this method is that the software is readily available and more people are familiar with the concepts of regression than with other types of time-series modeling. The disadvantage is that they can be arduous to fit properly.

TNR—The ability of a medical test to detect the absence of a condition. Also called the test's true negative rate, specificity, and probability of nondetection.

Tolerance interval—A statistical interval that provides a specified confidence that specified proportions of a population will be included within its limits.

Too good to be true—Something that seems so favorable that it's hard to believe it's real. The phrase is commonly used to express doubt about the authenticity or reliability of something that seems exceptionally positive. It often implies that there might be a hidden drawback or negative aspect that isn't immediately apparent.

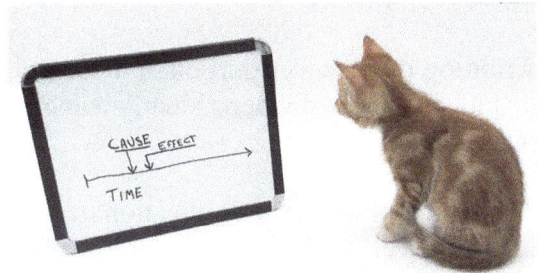

Glossary.

Tooth-fairy science—A type of bad science in which data are reproducible and statistically significant but there is no understanding of why or how the phenomenon exists.

Top-down—An overall view of a situation, also called the big picture. In statistics, a top-down approach involves overall descriptive statistics and dashboards, tests of model effectiveness, and report summaries and conclusions. It is the opposite of a bottom-up view that focuses on details.

Total enumerative sampling—The strategy of identifying characteristics or conditions that are important to the study and sampling entities until a predetermined number having those characteristics or conditions is reached. Also called quota sampling and consecutive sampling.

TPR—The ability of a medical test to detect a condition. Also called the test's sensitivity, probability of detection, and true positive rate.

Track-record—Documentation of past performance taken as evidence of likely future performance.

Training dataset—A portion of a dataset set aside to develop a statistical model so that it can be tested independently using a separate portion of the full dataset called the testing dataset.

Transformation—The application of a mathematical function to data for a variable to give the variable more desirable properties (e.g., linearity, Normality, equal variances).

Treatments—The conditions or actions tested in an experimental study

Tree diagram—A depiction of hierarchies, process steps, sequential decisions, probability calculations, or other segmented items. It may be scaled or not.

Treemap—A chart for hierarchical or matrix data that displays amounts or relationships in cells of different sizes and colors arrayed in rows and columns.

Trend—A pattern in graphed data, appearing as a line or curve, progressing either upward or downward.

Trend-surface interpolation—A spatial interpolation method that estimates grid-node values from a multiple regression model of location coordinates.

Trial—A single repetition of a probability experiment.

Glossary.

Trimean—A measure of central tendency calculated by adding the 25th percentile plus twice the 50th percentile (i.e., the median) plus the 75th percentile, all divided by four.

Trimmed mean—A measure of central tendency, calculated by removing the same number of observations from both ends of a data distribution and averaging the remaining values. Used to address extreme values and censored data.

Triplicates—Three samples collected at the same time, location, and condition.

True Negative Rate—The ability of a medical test to detect the absence of a condition. Also called the test's specificity, probability of nondetection, or TNR.

True Positive Rate—The ability of a medical test to detect a condition. Also called the test's sensitivity, probability of detection, or TPR.

t-Test—The same as a z-test except that the probability of a Type I error is calculated from the t-distribution instead of the Normal distribution. The t-distribution has more of its area in the tails than a Normal distribution so it is often used to compensate for a small number of samples.

Tufte—Edward Tufte (1942-present) is an American statistician noted for his ideas on information design. He coined the terms data-ink ratio, chart junk, and sparklines. He has written **The Visual Display of Quantitative Information** (1983, 2001), **Envisioning Information** (1990), **Visual Explanations** (1997), and **Beautiful Evidence** (2006).

Two-population test—Statistical tests that involve comparing means (or other population parameters) from two independent populations.

Two-sided test, two-tailed tests—Statistical tests that involve both ends of a data distribution. Also called non-directional tests.

Two-step clustering—A statistical method for identifying data clusters in two steps, centroid clustering followed by hierarchical clustering.

Two-way ANOVA—An analysis-of-variance model that includes two main effects.

Glossary.

Two-way plots—Bivariate plots, also called x-y plots.

Type I error—False positive, the probability of rejecting a null hypothesis when it is true.

Type II error—False negative, the probability of not rejecting a null hypothesis when it is false.

Uncertainty—Variability. Inexactness. Error.

Uniform distribution—A theoretical frequency distribution in which the probability of occurrence is equal for all values.

Unimodal—Refers to frequency distributions that have only one "high" point. The Normal distribution is a unimodal distribution. Frequency distributions with two peaks are called bimodal. Frequency distributions with more than two peaks are called multi-modal.

Unintended consequences—Unfavorable events that occur as the results of an action. Famous examples include the Hawthorne effect (people changing their behavior because they are being observed) and the cobra effect (a solution to a problem that makes the problem worse).

Units-of-measurement—The benchmark that describes how the measurements relate to the phenomenon being measured. This is different from scale, which is how the numbers relate to each other.

Univariate tests—Statistical tests involving only one dependent variable.

Unstandardized regression coefficients—Coefficients in a regression model that remain in their original units of measurement (i.e., have not been standardized). Also called raw coefficients.

Unstructured data—Data that need extensive preprocessing to format as a matrix before analysis.

Upper confidence limit—The higher value defining the interval around a statistic that is estimated from a sample and has a certain confidence of including the true value for the population.

Vagueness—A prominent cause of unreliability in survey questions. Vague questions may not bias results in a certain direction but they do add unnecessary variation. Some respondents answer the question as it was meant and others will answer the question as they think it was meant. The results is a mishmash of opinions unsuitable for analysis.

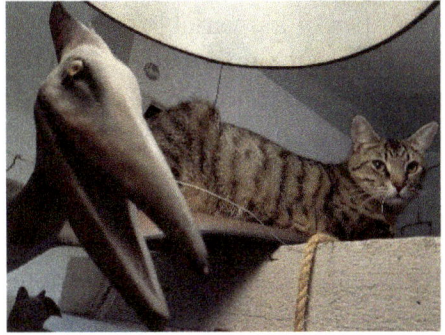

Validation—The process of ensuring that the data were generated in accordance with quality assurance specifications. Also used to refer to the process of making sure a model is representative of real-world conditions.

Validity, big data—A characteristic of big data that refers to how appropriate the data is for its purpose.

Value, big data—A fundamental characteristic of big data that refers to the importance of the data and the benefits they can provide.

Variability, big data—A characteristic of big data that refers to its inconsistency.

Variability, common cause—In six-sigma, predictable variations that can be observed in a process.

Variability, special cause—In six-sigma, variations that are significant departures from the norm, often having extreme values.

Variability—Uncertainty. Inexactness. Error.

Variable triage—Deciding which independent variables to include in a regression model when there are too many for the model to be efficient. Approaches to variable triage include stepwise regression, variable reduction techniques, and regression diagnostics (coefficient of determination, standard error of estimate, F-test, AIC, BIC, Mallows' Cp criterion, standardized regression coefficients, t-tests, variance inflation factors, and partial regression leverage plots).

Variable-reduction techniques—Statistical methods used to reduce the number of independent variables to include in a regression model. Some techniques, like cluster analysis, group variables based on the information they share so some can be eliminated without losing much information. This approach preserves the original identities of the variables. Another approach is to recombine the information the variables contain into new variables and then using only the new variables that contain the most information. This is what factor analysis and principal-components analysis do.

Variables—Variables are the columns of a data matrix that contain the pieces of information from or about each of the samples. In a regression model, the variable that characterizes the phenomenon to be modeled is called the criterion variable, response variable, outcome variable, y-variable, or most commonly, the dependent variable. Variables that are used to test, predict, or explain the dependent variable in the model are called predictor variables, explanatory variables, grouping variables, x-variables or most commonly, independent variables.

Variance control methods—Reference, replication, and randomization.

Variance Inflation Factor—A diagnostic statistic in regression used to measure the severity of multicollinearity. It is the variance of a parameter estimate in the full model compared to the variance when it is the only term in the model.

Variance windows—Like a Rumsfeld window in that they explore knowns and unknowns, but with a focus on controlling extraneous variance in the process of collecting data. Using a variance window involves categorizing sources of variability in the panes of the window based on how well each source is understood and how it could be controlled.

Variance—A measure of dispersion. Also, a difference from a benchmark in a data report.

Variety, big data—A fundamental characteristic of big data that refers to diversity of data types.

Variogram—Graph of the spatial variance on the y-axis versus the distance between samples on the x-axis, used in geostatistics.

Variogramming—Analyzing the relationship between spatial variance and the distance between samples. Same as semivariogramming.

Velocity, big data—A fundamental characteristic of big data that refers to the speed at which data is generated, gathered, and analyzed.

Venn diagrams—Diagrams that use overlapping shapes, usually but not always circles, to show common traits in related ideas or groups.

Veracity, big data—A fundamental characteristic of big data that refers to the accuracy, quality, and reliability of data.

Verification—The process of ensuring that each value in the dataset is identical to the value that was originally generated. Also used to refer to the process of guaranteeing that a model is producing the correct results.

Violations of assumptions—When basic assumptions in statistical modeling—linearity, additivity, independence of errors, Normality of errors, and equality of variances (also called homogeneity)—are not met sufficiently so that results are inaccurate.

Violin plot—A graph of the distributions of numeric data for one or more groups using density curves. The width of each curve corresponds with the approximate frequency of data points in each region.

Visualization, big data—A characteristic of big data that refers to its ability to be graphed or analyzed in other ways.

Visualization—Plots, charts, graphs, and diagrams created to make sense of data visually and to explore data interactively. The process of visualization is often semi-automatic in which different options are tested for interpretability through the use of software.

Volatility, big data—A characteristic of big data that refers to how quickly data elements age to the point that they are no longer useful and have to be removed from the dataset.

Volume, big data—A fundamental characteristic of big data that refers to the amount of data

Glossary.

Volunteer sampling—When individuals without invitations participate in open surveys without any review by the pollster. Volunteer sampling is mostly limited to surveys conducted on social media for entertainment. Also called self sampling.

Vulnerability, big data—A characteristic of big data that refers to how susceptible the data are to data breaches and other security concerns.

Waterfall chart—Used to show how data values increase or decrease from an initial point. The horizontal axis is usually an ordinal variable, most often time, but could also represent nominal or ordinal-scale variables. The vertical axis is not fixed, or even shown on the chart for that matter. Changes in data values are represented by bars that successively change position in either a positive or negative direction. Each bar is essentially its own floating axis. The bars in a waterfall chart are usually color-coded to represent either positive or negative changes.

Web chart—A radar chart, also called a spider chart, polar chart, and Kiviat chart.

Weibull distribution—A continuous theoretical probability distribution used to model times to failure and times between events.

Weighted mean—A measure of central tendency calculated by assigning more important data values weights greater than 1, multiplying the data values by their weights, and averaging the products divided by the sum of the weights.

Weighting—The process of applying a multiplier to data values to give the values either more or less influence in a subsequent calculation.

Weight—The number (constant or mathematical expression) that a data value for a sample is multiplied by to give the value either more or less influence in a subsequent calculation.

Where there's smoke there's fire—Where-there's-smoke-there's-fire (also called hasty conclusion or jumping to a conclusion) is where a conclusion is reached without considering enough evidence.

Whiskers—The lines extending above and below the box in a vertical box-and-whisker plot.

Glossary.

Window diagrams—Analytical windows are a type of matrix plot that can contain data, tables, graphs, or text arranged in ways that facilitate understanding.

Winsorized mean—A measure of central tendency used to address extreme values and censored data.

Winsorizing—The process of adjusting for extreme values by replacing the same number of extreme values on each side of the distribution with the next "non-extreme" value.

Wisdom—Wisdom is the understanding of the fundamental principles governing knowledge. Knowledge is information created through the analysis of data.

Within-groups variance—Variability in an ANOVA design caused by differences within individual treatment groups not attributable to independent variables. Also called error variance.

Within-subjects ANOVA—An ANOVA design in which repeated measurements of the independent variables are taken on each subject. Also called a repeated-measures ANOVA.

Word clouds—Graphical representations of how many times specific words appear in a passage of text. The relative frequency of the words is shown using text size (and sometimes colors and fonts).

Wordles—The second generation of word frequency graphics that appeared in the 2010s. They were preceded by tag clouds in the mid-2000s, It evolved into word clouds a decade later as internet capabilities improved. Not related to the game Wordle that first appeared in 2021.

Worst case fallacy—When an argument is critiqued by focusing on highly unlikely events or outcomes rather than the most probable outcomes. This fallacy uses fear and speculation about an improbable outcome to dismiss a reasonable course of action. It is also used often in deterministic data modeling to compensate for not considering variance. Also known as a just-in-case fallacy.

X-variables—Independent variables.

X-Y plots—Graphs having two perpendicular axes. Usually scatter plots or line plots.

Y-variable—Dependent variable.

z-Test—A statistical comparison between a difference in a variable, usually either between two means or between a mean and a constant, relative to the variability of the variable. The null hypothesis is usually that the difference is zero. Large differences (relative to the variability) are considered evidence that the null hypotheses should be rejected. The probability that a difference of a similar magnitude could have occurred by chance, a Type I error, is calculated from the Normal distribution.

That's a lot of definitions.

Place
TRUE
Predict
Diagram
Graph
Individual
Source
Population
More
Differ
Regression
Axis
Area
Study
Deviate
Box
Analytics
Correlate
Range
Sample
Size
Line
Part
Collect
Question
Standard
Chart
Need
Number
Scale
Model
Frequency
Measure
Test
Variable
Control
Take
Inform
Topic
Statistician
Experiment
Dataset
Comparison
Research
Representative
Trend
Time
Chapter
Bias
Value
Term
Vote
Data
Change
Design
Pattern
Level
Relate
t-test
Equal
Group
Event
Effect
Random
Probability
Limit
Person
Result
Error
Process
Mean
Average
Rate
Point
Statistics
Response
MoE
Coefficient
Distribution
Participate
Science
Hypothesis
Method
Plot
Pollster
Bar
Variance
Step
Normal
Poll
Causation
Analysis
Independent
Survey
Year
Example
Descriptive
Relationship
Concept

www.ingramcontent.com/pod-product-compliance
Lightning Source LLC
Chambersburg PA
CBHW061929190326
41458CB00009B/2696